AF664659

Atlantis Thinking Machines

Volume 7

Series editor

Kai-Uwe Kühnberger, Osnabrück, Germany

Aims and Scope of the Series

This series publishes books resulting from theoretical research on and reproductions of general Artificial Intelligence (AI). The book series focuses on the establishment of new theories and paradigms in AI. At the same time, the series aims at exploring multiple scientific angles and methodologies, including results from research in cognitive science, neuroscience, theoretical and experimental AI, biology and from innovative interdisciplinary methodologies.

For more information on this series and our other book series, please visit our website at: www.atlantis-press.com/publications/books

AMSTERDAM—PARIS—BEIJING
ATLANTIS PRESS
Atlantis Press
29, avenue Laumière
75019 Paris, France

More information about this series at http://www.springer.com/series/10077

Tarek R. Besold · Marco Schorlemmer
Alan Smaill
Editors

Computational Creativity Research: Towards Creative Machines

Editors
Tarek R. Besold
Institute of Cognitive Science
University of Osnabrück
Osnabrück
Germany

Alan Smaill
CISA, School of Informatics
University of Edinburgh
Edinburgh
UK

Marco Schorlemmer
IIIA-CSIC
Bellaterra, Barcelona
Spain

ISSN 1877-3273
Atlantis Thinking Machines
ISBN 978-94-6239-084-3 ISBN 978-94-6239-085-0 (eBook)
DOI 10.2991/978-94-6239-085-0

Library of Congress Control Number: 2014957325

Printed on acid-free paper

Foreword

How Computational Creativity Began

Computational creativity (CC, for short) is the use of computers to generate results that would be regarded as creative if produced by humans alone. Strictly speaking, this includes not only art, but also innovative scientific theories, mathematical concepts, and engineering designs. But the term is often used—as I shall do, here—to apply mainly to results having *artistic* interest.

CC was glimpsed on the horizon over 170 years ago, when Ada Lovelace said of Charles Babbage's Analytical Engine that it "might compose elaborate and scientific pieces of music of any degree of complexity or extent" [41, p. 270]. Indeed, the general principle had been glimpsed in the early eleventh century, when Guido d'Arezzo—who also invented the basis of tonic solfa and of today's musical notation—wrote a system of formal rules for composing hymns. But whereas d'Arezzo had not had computing machines in mind, Lovelace did.

A century later, Alan Turing was producing (as a joke) programmed love-letters on Manchester's MADM computer [37]; and haikus would soon be generated on Cambridge's EDSAC machine (see below). Even more to the point (or so it might seem), "creativity" was identified as one of the chief goals in the document planning the Dartmouth Summer School of 1956 [42]. That meeting was where artificial intelligence was officially named, and where hopes for computational modeling first reached beyond a tiny coterie.

By the 1950s, then, there was no shortage of heralds.

Heralds Unheard

But their trumpeting did not announce a triumphant procession entering the halls of fame. The forefronting of creativity in the Dartmouth document did not establish CC as a research project. For several decades, creativity was largely ignored by the very people who had declared it a prime goal for AI.

That is not quite fair: problem-solving in general involves creative thinking, and AI soon abounded with intriguing attempts to model that. The pioneering Logic Theorist was first run in the final week at Dartmouth, followed by its cousin GPS shortly afterwards [54–56]. Arthur Samuel's checkers-player was even earlier: it had been implemented in 1949, was learning how to beat its human opponents by 1955, and had appeared on TV in February 1956—a full six months before the Dartmouth meeting [64]. Moreover, when Simon [65], 40 years later, sought to convince sceptics that AI is respectable *as a science*, he took creative problem-solving as his core topic—defiantly subtitling his paper "AI on the Topics of Intuition, Insight, and Inspiration".

But if creative thinking was arguably a key research theme of professional AI right from the start, there was more emphasis on the "thinking" than on the "creative". Relatively few of the early AI workers focussed on what is normally thought of as creativity. Amarel [1] wrote a fascinating paper showing that creating new representations of a given problem-space could make problem-solving increasingly efficient—but he could not say how those novel representations themselves were generated. Evans [21] wrote an interesting model of analogical thinking, applied to problems typically included in IQ tests. Gelernter's [24] geometry-program apparently came up with a highly insightful proof about isosceles triangles (only *apparently*: the achievement here was much less striking than it seemed—[5, pp. 116–123]). In the 1970s, scientific discovery was modelled by a team led by Simon [35, 36]. And story-writing was attempted too, albeit with painfully crude results [33, 51, 52].

The most convincing early computer models of creativity were built by people outside mainstream AI. For instance, Lejaren Hiller, a professional chemist who was also a Master of Music, initiated the Illiac suite in the 1950s [27]. The Illiac music was generated from rules defining various styles (including sixteenth-century counterpoint and 12-tone music, and a variety of dynamics and rhythms), sometimes combined with tone-pairs chosen by chance. In the early 1960s, the avant-garde artists of IRCAM, in Paris, started experimenting with computer music—though focussing more on instrumentation than composition. And in 1968, the IFIP meeting in Edinburgh included a competition for computer-composed music—which led to the founding of the Computer Arts Society (CAS).

AI workers in general did not identify with these ventures. Nor—despite the mention of "creativity", "invention", and "discovery" in the Dartmouth proposal, plus occasional asides about "creative" problem-solving—did they see their own work as at all relevant. That is why in the early 1970s, when I was writing a book on the field [2], my AI colleagues were bewildered on being told that I was devoting one whole chapter to creativity. Several complained: "But there isn't any work on creativity!" (My answer was twofold: (1) If AI is relevant to intelligence in general, then there *should* be!, and (2) much of the existing AI work would in fact turn out to be relevant, once researchers started thinking seriously about creativity.)

The difference in attitude between the AI professionals and the artists (whether professional or amateur) is partly explained by differences in motivation. Like AI itself, CC can be driven by two distinct motivations. Individual workers may be excited by both, but very often are interested only in one. Within AI in general, the goal may be scientific or technological: understanding the mind or producing useful tools. Analogously, CC practitioners may aim either to illuminate the psychological processes underlying creativity or to produce aesthetically interesting results. Most professional computer artists have scant interest in the first of those goals. Rather, they want to produce artworks. The pioneers of AI ignored art-focused CC because they were interested in modelling mental processes. (Even when they hoped to build useful tools, they typically started from some psychological theory or introspection about how people think.)

Early Efforts in CC Art

A few examples of very early CC art were individually exhibited in the mid-1960s. The first of these—in 1963—were Joan Shogren's computer-generated drawings at San Jose, California [53, p. 66]. Others, which received more attention, were by Georg Nees, Michael Noll, and Frieder Nake, and shown in Stuttgart and New York in 1965. They were highly abstract, quasi-geometrical, designs, generated by programs written not by professional artists but by mathematicians.

A wide range of early CC work was featured soon afterwards at London's seminal *Cybernetic Serendipity* exhibition in 1968 [59]. This presented a variety of experiments with the new computer technologies, both analogue and digital. They included kinetic sculptures by Gordon Pask and Edward Ihnatowicz, algorithmic graphics by the people mentioned above, and haikus generated by Margaret Masterman's group in Cambridge [49, 50]. The show aroused a good deal of excitement—some of which was largely unmerited. (Consider this haiku, for instance: All green in the leaves/I smell dark pools in the trees/Crash the moon has fled; its "success" is deceptive, being due to the reader's well-practised habit of projecting meaning into obscure poetry.)

By the following year, CC was being initiated not only by scientists and mathematicians, who happened to have access to computers (for example, Manfred Mohr at the Paris Institute of Meteorology), but also by aspiring—and some already-established (see below)—artists. Several of these were directly influenced by the *Cybernetic Serendipity* exhibition. For instance, this event inspired two young painters who were already committed to modernism—Ernest Edmonds and Paul Brown—to turn to computers to produce their artworks. (For modernists, the very *impersonality* of computers was an advantage.) Edmonds soon complemented his actual artwork by discussing the nature of computer art in general terms [20].

Some art-inspired CC efforts came from architects. The pioneer here was John Frazer (influenced by Pask), whose work—as a student—on CAD and "intelligent" environments won an Architectural Association prize as early as 1969. With his wife Julia Frazer, he went on to provide more elaborate computer-generated (and eventually interactively evolved) designs for buildings and urban centres [23].

Other architects wrote "shape grammars" that could generate designs for Palladian villas or Prairie Houses [26, 34, 67].

Two highly visible, and highly influential, CC artists of the early-1980s were Harold Cohen in the graphic arts and David Cope in music. Both men had embarked on a decades-long—and still continuing—project in the late 1960s, wherein their CC-oeuvre grew increasingly impressive [11–14, 16–19, 44]. And both had to learn to do their own coding, using equipment that was state of the art at the time, but seems hugely primitive today. Cope even had to invent an early equivalent of the MIDI code for transferring musical notation to the computer.

Cohen was remarkable not least for the fact that he abandoned a hugely successful career as a pots-and-brushes painter for CC art. He felt that this would enable him to understand more clearly just why he was making certain aesthetic decisions rather than others. Encouraged by the AI scientist Edward Feigenbaum, he moved to Stanford—and then to San Diego—to use the computing facilities there.

His early introduction to the then young AI community was partly responsible for his visibility. But above all was the aesthetic merit of his artwork. He produced (and is still producing) increasingly striking versions of his program AARON, which gradually became more able to represent 3D scenes. At first, AARON's images hinted at three-dimensionality, containing objects interpretable as arbitrarily-shaped rocks. Later, they represented foliage, and increasingly realistic human figures and faces. From the mid-1990s, his main focus has been on colour. The later versions of his program can generate colours with extraordinary success ("I am a first-rate colourist. AARON is a world-class colourist"—Cohen p.c.). But even when AARON was able to produce only black-and-white drawings of shapeless rocks, Cohen's work was being shown in major galleries around the world.

Cope, also, was a recognized practising artist: namely, a composer (and also a professor of musicology at UC Santa Cruz). His program, "Emmy", composed new pieces in many different styles—from Western to Thai, baroque fugues to modern jazz, Bach to Joplin; and it could produce new styles by mixing existing styles together. The program explored generative structures (formally defined by Cope *qua* musicologist), as well as recombining specific musical motifs.

In building the data-base, Cope first selected brief "signatures" as being characteristic of the human composer concerned. These exemplified melody, harmony, metre, and ornament. Next, Emmy used statistical techniques to identify the core features of the signatures. Finally, the program—guided by general musicological principles defined by Cope—used these features to build new structures. The early versions of Emmy were implemented using the traditional methods of symbolic AI; from 1990, Cope added a connectionist network, giving increased flexibility.

Whereas Cohen's AARON, at least after the early days, never produced an artwork that was aesthetically unacceptable, Cope's Emmy sometimes did (for examples, see [18, pp. 182–3, 385–390]). Nevertheless, it more often produced highly convincing compositions (the 1,000 "best" are listed in [19, pp. 385–9]). These impressed even sceptical critics who heard them (one reviewer invited to

Emmy's first public concert in 1989 actually *refused* to hear them, so upset was he by the very idea of computer-generated "pastiches": [19, p. 345])—including Douglas Hofstadter, who had long complained that symbolic AI work on creativity was shallow and trivial [29, pp. 55–193]. (In fact, Hofstadter was *over*-impressed. He reported playing one of Emmy's "Chopin" mazurkas and finding it "unmistakably *Chopin-like* in spirit, and... *not emotionally empty*" [28, p. 39]. Here, he was making the same mistake as those over-enthused by the haikus at *Cybernetic Serendipity*: the emotion was being projected onto the score by Hofstadter himself.)

In the late 1980s, new types of computing became available. These included PDP connectionism [43, 63] and evolutionary programming. Evolutionary computing had been outlined over twenty years earlier by John Holland, who formally defined the first genetic algorithm in the early-1970s [31, 32]. But his algorithm had not been implemented, because of lack of computer power. By the 1990s, the necessary machines existed. Both these new AI methodologies attracted interest from CC artists.

For example, Richard Brown's *Starfish*, exhibited at the turn of the century, used connectionism to build what appeared to be a huge multi-coloured starfish trapped inside a translucent marble table. Whenever someone approached it, or made a movement or a sound nearby, the trapped starfish moved—and in amazingly lifelike ways. In fact, this was an illusion: the image of the starfish was projected downwards onto the (empty) table from the ceiling. It was represented as a neural network, which responded to sounds and movements in the vicinity by altering the shape of the starfish's limbs.

As for evolutionary CC art, early examples were produced by Sims [66] and William Latham [68]—and others soon followed [45, 46, 70]. Even Cohen eventually experimented with evolutionary techniques, to produce a constantly varying range of coloured imagery. One way in which the individual examples varied was the extent to which they allowed relatively radical mutations to happen. Sims, for instance, allowed one image-generating mini-program to be nested into, or concatenated with, another. In consequence, the visual transformations observed were sometimes so great as to destroy the family resemblance even between neighbouring generations. Latham, by contrast, allowed only relatively superficial mutations—such as changing one numeral into another. His CC artworks were, therefore, instantly recognizable as "Lathams", and even as falling within a particular Latham style.

Even those CC artists who kept faith with the more traditional (symbolic) AI methods could take advantage of the recent increases in computer power. Edmonds, for example, produced increasingly ambitious interactive works integrating colour and music. And Paul Brown, who had been working with cellular automata for many years, was able to define increasingly complex CA systems.

Meanwhile, Back at the Ranch…

All those professional artists had relied on mainstream AI to provide their CC methodologies. Meanwhile, AI itself—when it considered creativity at all—was still mostly concentrating on creative *reasoning*.

For example, Douglas Lenat modelled heuristic thinking in mathematics, and wrote a program that originated an engineering design (for a 3D computer chip) that was awarded a US patent [38, 39]. Simon's studies of scientific theorizing continued, and integrated the planning of experiments into the reasoning concerned [36].

Some AI researchers, however, were now aiming at art-relevant questions. Work on "case-based" analogical reasoning was integrated into tasks such as story-telling [69]—and roundly critiqued by some [8]. Literary applications were central, too, to two book-length issues of the *Stanford Humanities Review* devoted to AI and creativity, wherein Simon's views were discussed in some detail [22, 25]. Hofstadter's long-standing critique of AI work on creativity continued [29, pp. 55–193]; and the ideas first embodied in his "Copycat" model of analogy were part-implemented in "Letter Spirit", for the design of alphabetic fonts [30, 48, 57, 58].

In short, professional AI was at last coming closer to CC. Besides the AI exercises modelling specific cases of artistic interest, creativity in general was more often discussed. For instance, AI-concepts and models were used to provide an analysis of three types of creativity—based on unfamiliar combinations, and on the exploration and transformation of familiar thinking-styles [4, 5]. With the increase of interest among the AI community, creativity was chosen as one of the keynote topics for the 1997 IJCAI conference in Japan [3]. And, as remarked above, Simon [65] took creativity as the test-case in his retrospective defence of AI as a whole.

Conclusion

Today, CC is thriving. It has given rise to various genres of computer art, besides fostering countless individual styles [7]. The ever-widening range of CC art is evident in several continuing conferences, such as the "Creativity and Cognition" series instituted by Edmonds, and in a number of descriptive books (e.g. [10, 47, 70]) and histories (e.g. Brown et al. [9]). And one recent CC model is interesting not only as a study of creativity, but also as an impressive integration of a range of AI techniques—including complex computer vision [15].

Ada Lovelace should be happy. For her "scientific pieces of music" are now many and various. Some of these throw light on how human beings create music. Some are generated by "stand-alone" computer programs, untouched by human hand. Others are partly interactive. Some are evolutionary. Yet others, being live-coded, are brought into existence before one's very eyes (and ears) by the composer's writing the program line by line to create the music on the hoof. Even *expressiveness* in music can be now generated by computer (early attempts were made by Riecken [60–62], but the most sophisticated is a remarkable program due to Longuet-Higgins [40]).

The readers of this book should be happy, too. For many current CC examples are discussed in the following pages. They are often intriguing, sometimes exciting, and usually provocative.—*Enjoy!*

East Sussex, UK Margaret A. Boden

References

1. Amarel, S.: On representations of problems of reasoning about actions. In: Michie, D. (ed.) Machine Intelligence, vol. 3, pp. 131–172. Edinburgh University Press, Edinburgh (1968)
2. Boden, M.A.: Artificial Intelligence and Natural Man. Basic Books, New York (1977)
3. Boden, M.A.: Creativity and artificial intelligence. Artif. Intell. **103**, 347–356 (1998)
4. Boden, M.A.: The Creative Mind: Myths and Mechanisms. Weidenfeld & Nicolson, London (1990)
5. Boden, M.A.: The Creative Mind: Myths and Mechanisms, 2nd edn. (revised/expanded). Routledge, London (2004)
6. Boden, M.A.: Mind as Machine: A History of Cognitive Science. Oxford University Press, Oxford (2006)
7. Boden, M.A., Edmonds, E.A.: What is generative art? In: Boden, M.A. (ed.) Creativity and Art: Three Roads to Surprise, pp. 125–163. Oxford University Press, Oxford (2010)
8. Bringsjord, S., Ferrucci, D.A.: Artificial Intelligence and Literary Creativity: Inside the Mind of BRUTUS, a Storytelling Machine. Lawrence Erlbaum, Mahwah (2000)
9. Brown, P.H., Gere, C., Lambert, N., Wilson, C.: White Heat Cold Logic: British Computer Art 1960–1980. MIT Press/Leonardo, Cambridge (2009)
10. Candy, L., Edmonds, E.A. (eds.): Explorations in Art and Technology. Springer, London (2002)
11. Cohen, H.: Harold Cohen: Drawing, Exhibition Catalogue. San Francisco Museum of Modern Art, San Francisco (1979)
12. Cohen, H.: On the Modelling of Creative Behavior, RAND Paper P-6681. RAND Corporation, Santa Monica (1981)
13. Cohen, H.: The further exploits of AARON painter. In: Franchi, Guzeldere (eds.), pp. 141–160 (1995)
14. Cohen, H.: A million millennial medicis. In: Candy, Edmonds (eds.), pp. 91–104 (2002)
15. Colton, S.: The painting fool: stories from building an automated painter. In: McCormack, d'Inverno (eds.), pp. 3–38 (2012)
16. Cope, D.: Computers and Musical Style. Oxford University Press, Oxford (1991)
17. Cope, D.: The Algorithmic Composer. A-R Editions, Madison (2000)
18. Cope, D.: Virtual Music: Computer Synthesis of Musical Style. MIT Press, Cambridge (2001)
19. Cope, D.: Computer Models of Musical Creativity. MIT Press, Cambridge (2006)
20. Cornock, S., Edmonds, E.A.: The creative process where the artist is amplified or superceded by the computer. Leonardo **6**, 11–16 (1973)
21. Evans, T.G.: A program for the solution of geometric-analogy intelligence test questions. In: Minsky, M.L. (ed.) Semantic Information Processing, pp. 271–353. MIT Press, Cambridge (1968)
22. Franchi, S., Guzeldere, G. (eds.): Constructions of the mind: artificial intelligence and the humanities. Stanford Humanit. Rev. **4**(2), 1–345 (1994) (Spec. Suppl.)

23. Frazer, J.H.: An Evolutionary Architecture. Architectural Association Publications, London (1995)
24. Gelernter, H.L.: Realization of a geometry-proving machine. In: Feigenbaum, E.A., Feldman, J. (eds.) Computers and Thought, pp. 134–152. McGraw-Hill, New York (1959/1963) (First published Paris: UNESCO House, 1959)
25. Guzeldere, G., Franchi, S. (eds.): Bridging the gap: where cognitive science meets literary criticism (Herbert Simon and respondents). Stanford Humanit. Rev. **4**(1), 1–164 (1994) (Spec. Suppl.)
26. Hersey, G., Freedman, R.: Possible Palladian Villas (Plus a Few Instructively Impossible Ones). MIT Press, Cambridge (1992)
27. Hiller, L.A., Isaacson, L.M.: Experimental Music: Composition with an Electronic Computer. McGraw-Hill, New York (1959)
28. Hofstadter, D.: Staring Emmy straight in the eye—and doing my best not to flinch. In: Cope (ed.), pp. 33–82 (2001)
29. Hofstadter, D., FARG (The Fluid Analogies Research Griup).: Fluid Concepts and Creative Analogies: Computer Models of the Fundamental Mechanisms of Thought. Basic Books, New York (1995)
30. Hofstadter, D., McGraw, G.: Letter spirit: esthetic perception and creative play in the rich microcosm of the roman alphabet. In: Hofstadter, FARG (eds.), pp. 407–66 (1995)
31. Holland, J.H.: Outline for a logical theory of adaptive systems. J. Assoc. Comput. Mach. **9**, 297–314 (1962)
32. Holland, J.H.: Adaptation in Natural and Artificial Systems. University of Michigan Press, Ann Arbor (1975)
33. Klein, S., Aeschlimann, J.F., Balsiger, D.F., Converse, S.L., Court, C., Foster, M., Lao, R., Oakley, J.D., Smith, J.: Automatic Novel Writing: A Status Report. Technical Report 186, University of Wisconsin Computer Science Department, Madison (1973)
34. Koning, H., Eizenberg, J.: The language of the prairie: frank lloyd wright's prairie houses. Environ. Plan. B **8**, 295–323 (1981)
35. Langley, P.W.: BACON.1: A general discovery system. In: Proceedings of the Second National Conference of the Canadian Society for Computational Studies in Intelligence, pp. 173–80. Toronto (1978)
36. Langley, P.W., Simon, H.A., Bradshaw, G.L., Zytkow, J.M.: Scientific Discovery: Computational Explorations of the Creative Process. MIT Press, Cambridge (1987)
37. Lavington, S.H.: A History of Manchester Computers. NCC Publications, Manchesteer (1975)
38. Lenat, D.B.: The ubiquity of discovery. Artif. Intell. **9**, 257–86 (1977)
39. Lenat, D.B.: The role of heuristics in learning by discovery: three case-studies. In: Michalski, R.S. Carbonell, J.G., Mitchell, T.M. (eds.) Machine Learning: An Artificial Intelligence Approach, pp. 243–306. Tioga, Palo Alto (1983)
40. Longuet-Higgins, H.C.: Artificial intelligence and musical cognition. Philos. Trans. R. Soc. Lond. Ser. A **349**, 103–13 (1994) (Special Issue on 'Artificial Intelligence and the Mind: New Breakthroughs or Dead Ends?', eds. M. A. Boden, A. Bundy, and R. M. Needham)
41. Lovelace, A.: Notes by the translator. In: Hyman, R.A. (ed.) Science and Reform: Selected Works of Charles Babbage, pp. 267–311. Cambridge University Press, Cambridge, 1989 (1843) (reprinted)
42. McCarthy, J., Minsky, M.L., Rochester, N., Shannon, C.E.: A proposal for the dartmouth summer research project on artificial intelligence. In: Chrisley, R.L. (ed.) Artificial Intelligence: Critical Concepts, vol. 2, pp. 44–53. Routledge, London (1955)
43. McClelland, D.E., Rumelhart, J.L.: Parallel Distributed Processing: Explorations in the Microstructure of Cognition, ii. Psychological and Biological Models, vol. 2. MIT Press, Cambridge (1986)
44. McCorduck, P.: AARON's Code: Meta-Art, Artificial Intelligence, and the Work of Harold Cohen. W.H. Freeman, New York (1991)
45. McCormack, J.: Art and the mirror of nature. Digital Creativity **14**(1), 3–22 (2003)

46. McCormack, J.: Impossible Nature: The Art of John McCormack. Australian Centre for the Moving Image, Melbourne (2004)
47. McCormack, J., d'Inverno, M. (eds.): Computers and Creativity. Springer, Berlin (2012)
48. McGraw, G.: Letter spirit (part one): emergent high-level perception of letters using fluid concepts. Ph.D. thesis, Indiana University (1995)
49. Masterman, M.: Computerized haiku. In: Reichardt, J. (ed.) Cybernetics, Art, and Ideas, pp. 175–83. Studio Vista, London (1971)
50. Masterman, M., McKinnon-Wood, R.: In: Reichardt (ed.), pp. 54–55 (1968)
51. Meehan, J.: Using planning structures to generate stories. Am. J. Comput. Linguist. Microfiche **33**, 77–93 (1975)
52. Meehan, J.: TALE-SPIN and micro TALE-SPIN. In: Schank, R.C., Riesbeck, C.K. (eds.) Inside Computer Understanding: Five Programs plus Miniatures, pp. 197–258. Lawrence Erlbaum, Hillsdale (1981)
53. Nake, F.: Construction and intuition: creativity in early computer art. In: McCormack, d'Inverno (eds.), pp. 61–94 (2012)
54. Newell, A., Shaw, J.C., Simon, H.A.: Empirical explorations with the logic theory machine. In: Proceedings of the Western Joint Computer Conference, vol. 15, pp. 218–39 (1957) (Reprinted with added subtitle, A case study in heuristics. In: Feigenbaum, E.A., Feldman, J. (eds.): Computers and Thought, pp. 109–33. McGraw-Hill, New York 1963)
55. Newell, A., Shaw, J.C., Simon., H.A.: Elements of a theory of human problem solving. Psychol. Rev. **65**, 151–66 (1958)
56. Newell, A., Simon, H.A.: The logic theory machine: a complex information-processing system. IRE Trans. Inf. Theory IT-**2**(3), 61–79 (1956)
57. Rehling, J.A.: Letter spirit (part two): modeling creativity in a visual domain. Ph.D. Thesis, Indiana University (2001)
58. Rehling, J.A.: Results in the letter spirit project. In: Dartnall, T. (ed.) Creativity, Cognition, and Knowledge: An Interaction, pp. 273–82. Praeger, London (2002)
59. Reichardt, J.: Cybernetic Serendipity. Studio International, London (1968)
60. Riecken, R.D.: Wolfgang: musical composition by emotional computation. In: Schorr, H., Rappaport, A. (eds.) Innovative Applications of Artificial Intelligence, pp. 251–69. AAAI Press, Menlo Park (1989)
61. Riecken, R.D.: WOLFGANG—a system using emoting potentials to manage musical design. In: Balaban, M., Ebcioglu, K., Laske, O. (eds.) Understanding Music with AI: Perspectives on Music Cognition, pp. 206–236. MIT Press, Cambridge (1992)
62. Riecken, R.D.: The wolfgang system: a role of "emotions" to bias learning and problem-solving when learning to compose music. In: Trappl, R., Petta, P., Payr, S. (eds.) Emotions in Humans and Artifacts, Chap. 10. MIT Press, Cambridge (2002)
63. Rumelhart, D.E., McClelland, J.L.: Parallel Distributed Processing: Explorations in the Microstructure of Cognition, i. Foundations, vol. 1. MIT Press, Cambridge (1986)
64. Samuel, A.L.: Some studies in machine learning using the game of checkers. IBM J. Res. Dev. **3**, 211–29 (1959) (Reprinted In: Feigenbaum, E.A., Feldman, J. (eds.) Computers and Thought, pp. 71–108. McGraw-Hill, New York (1963))
65. Simon, H.A.: Explaining the ineffable: ai on the topics of intuition, insight, and inspiration. In: Proceedings of the Fourteenth Joint Conference on Artificial Intelligence, vol. i. pp. 939–48 (1995)
66. Sims, K.: Artificial evolution for computer graphics. Comput. Graphics **25**(4), 319–28 (1991)
67. Stiny, G., Mitchell, W.J.: The palladian grammar. Environ. Plan. B **5**, 5–18 (1978)
68. Todd, S.C., Latham, W.: Evolutionary art and computers. Academic, London (1992)
69. Turner, S.R.: The Creative Process: A Computer Model of Storytelling and Creativity. Lawrence Erlbaum, Hillsdale (1994)
70. Whitelaw, M.: Metacreation: Art and Artificial Life. MIT Press, London (2004)

Contents

Part I Theory

Part II Practice

Introduction

Computational Creativity, Concept Invention, and General Intelligence

Computational Creativity, Concept Invention, and General Intelligence in their own right all are flourishing research disciplines producing surprising and captivating results that continuously influence and change our view on where the limits of intelligent machines lie, each day pushing the frontiers a bit further. By 2014, all three fields also have left their marks on everyday life: Machine-composed music has been performed in concert halls, automated theorem provers are accepted tools in enterprises' R&D departments, and cognitive architectures are being integrated in pilot assistance systems for next generation airplanes.

Still, although the corresponding aims and goals are clearly similar (as are the common methods and approaches), the developments within the individual areas have happened mostly individually within the respective communities and without closer relationships to the goings-on in the other two disciplines. In order to overcome this gap and to provide a common platform for interaction and exchange between the different directions, the International Workshops on "Computational Creativity, Concept Invention, and General Intelligence" (C3GI) were initiated.

At ECAI-2012 and IJCAI-2013, the first and second edition of C3GI each gathered researchers from all three fields, presenting recent developments and results from their research, thus bridging the disciplinary boundaries through dialogue and joint debates. Some of the chapters contained in this book are based on expanded versions of accepted contributions to the workshops, giving an impression of current questions under discussion and showing how the original research projects developed since. Additionally, selected contributions by renowned researchers in the relevant fields have been included in this volume, complementing and completing the collection of works from the workshops to give a full representative (although clearly not exhaustive) overview of creativity-related work in the relevant fields.

When looked at individually, the chapters give an account of the state-of-the-art in their respective areas, discussing both, theoretical approaches as well as implemented systems. When taken together and looked at from an integrative perspective, the book in its totality offers a starting point for a (re)integration of Computational Creativity, Concept Invention, and General Intelligence, making visible common lines of work and theoretical underpinnings, and revealing chances and opportunities arising from the interplay of the three fields.

This clearly promotes one certain interpretation of Computational Creativity more than others, putting emphasis on the more "cognitively-oriented" line(s) of work and, thus, favoring systems and theories which aim to produce models or simulations of human creative capacities over projects trying to develop a human-independent, computational flavor of creativity and the corresponding artistic fields. Still, as in our reading Computational Creativity both historically and currently has strong conceptual and methodological ties to Artificial Intelligence and research in cognitive systems in particular, whilst not devaluing differing views by any means, we are willing to accept this bias. We consider it possibly beneficial as it could help to focus efforts and resources, and allow for a clearer definition of and a more homogeneous approach to the overall research program.

Summary of Contributions

The book is divided into two main categories dedicated to theory and practice of Computational Creativity, respectively. These two main parts are framed by an introductory preface, giving an account of the history and origins of the research program, and a postface speculating on possible future developments in the field.

In her opening chapter on How Computational Creativity Began, Margaret Boden outlines her personal view on the first years of research in Computational Creativity, tracing the idea of computing machines performing creative tasks back at least to Ada Lovelace and following the development of the scientific effort from the 1950s until the present.

In the first chapter of the *Theory* part, Simon Colton et al. present an investigation into the nature and status of the different stakeholder groups in computational creativity research and practice. They put forward several points deemed important in determining the way in which society accepts creative software, focusing especially on the role of the general public, Computational Creativity researchers, and fellow creatives in this context, and outline several questions the Computational Creativity research community should address.

In their chapter on *Weak and Strong Computational Creativity*, drawing upon an analogy to Searle's well-known notions of weak and strong artificial intelligence, Mohammad Majid al-Rifaie and Mark Bishop investigate computational creativity in swarm intelligence systems, tying the discussion to the concepts of freedom and constraint in a system and linking up to topics of understanding and embodiment.

A formal proof that a certain notion of high-level general intelligence unavoidably entails creativity is then provided by Selmer Bringsjord, connecting the two fields of Computational Creativity and (strong) Artificial Intelligence even closer. Dan Ventura subsequently investigates three different systems for producing creative artifacts in different domains, aiming to obtain a general-purpose computationally creative agent and a description of the agent's behavior in an archetypical creativity algorithm which might even help to shed new light on issues in computational complexity theory.

The question of how models of creativity and, especially, analogy need to address what has become known as the "tailorability concern": How can models (and corresponding systems) generalize out of the realm of manually constructed toy examples with guaranteed solution into real domain-independent frameworks?, is dealt with by John Licato et al., before Bipin Indurkhya describes his experiences in conceptualizing and building several computer-based creativity-support systems, offering a discussion of general lessons learned and implications for the question of how far creativity can be approached in an algorithmic computational manner.

The last chapter of the *Theory* part, contributed by Geraint Wiggins and Jamie Forth, then deals with "Information Dynamics of Thinking" (IDyOT), their computational theory of creativity as everyday reasoning from learned information, revisiting many of the topics already touched upon in previous chapters and pointing out how many of the corresponding results and insights can be integrated into a cognitive architecture with a view to developing a minimalist theory of cognition encompassing many human behaviors within as simple a functional framework as possible.

The opening chapter of the part of the book dealing with practical matters and systems of computational creativity describes a way of *Accounting for Creativity Within a Psychologically Realistic Cognitive Architecture*: Ron Sun and Sebastian Helie present a psychological process model and simulation of human creative problem solving based on the CLARION cognitive architecture, accompanied by their Explicit–Implicit Interaction theory of creative problem solving as the underlying unified framework.

This is followed by Oliver Kutz et al.'s chapter on the *Formalisation, Use-Cases, and Computational Support for Conceptual Blending*, discussing conceptual blending on all levels, from the theoretical and formal underpinnings up to a system for supporting the computation and evaluation of novel and creative blends.

Tony Veale addresses issues in the creation of meaningful and poetic instances of rhetorical forms, advocating a conceit-driven approach to computational poetry for overcoming current limitations computational systems experience in tasks demanding the creative use of language. The problem of open-ended elaboration in creative metaphor generation with computational means is then discussed by John Barnden who presents the ATT-Meta approach and outlinines how certain broad and important types of creativity in metaphor can be dealt with. In the following chapter, Hugo Goncalo Oliveira and Amílcar Cardoso show how PoeTryMe, a platform for the automatic generation of poetry, can be used for generating poetry in Portuguese, flexibly taking into account user-provided parameters and constraints.

The succession of contributions on language-related computational creativity is concluded by Rafael Pérez y Pérez's chapter on the engagement-reflection computer model of creative writing employed to implement agents capable of generating individual or collaborative plots, exemplifying its use in his MEXICA-impro system where two computer agents collaborate to produce a joint narrative.

Simon Ellis et al. then describe Handle, a computational system in the domain of music usable as conductor, jazz musician, and (soon) as film composer. One of the key characteristics of their approach is a strong commitment to formal computational logic as basis for their framework, leading to the development of a music calculus. In the second music-centered chapter, David Cope discusses definitions of creativity and subsequently introduces computer programs creatively providing output respectively based on a given situation.

Totally different application domains are being addressed by the final three chapters of the second main part of the book: Florian Pinel et al. present their *Culinary Computational Creativity System* which, based on a data-driven approach, uses association principles and an evaluation step for compiling novel and creative recipes and step by step cooking instructions. Following this, GAIA, a meta-reasoning based, CAD-like environment for the conceptual design of game-playing agents intended for the support of design creativity by encompassing a rapid generation, testing and revision approach is described by Ashok Goel and Spencer Rugaber. Concludingly, Petros Stefaneas et al.'s work on collective discovery events on the web using codelets for distributed mathematical problem-solving is outlined. Their efforts aim to bridge from theoretical aspects of collective, web-based discovery to the practical side, aiming at eventually designing cognitively-inspired systems to support collective problem-solving.

The *Postface* and last chapter of the book is *A Personal Perspective into the Future of Computational Creativity Research* by Pablo Gervás, exploring several possible future paths for the field based on his experiences from the last 15 years of Computational Creativity.

Acknowledgments

In conclusion, we want to thank all the authors who contributed to this book for their patience, their dedication, and all the time and effort they put into their respective contributions. We are grateful to all of them for their willingness to join this project, drafting their chapters, undergoing a reviewing process, incorporating the feedback into new versions of their chapters and eventually compiling the present final versions.

August 2014

Tarek R. Besold
Marco Schorlemmer
Alan Smaill

Part I
Theory

Chapter 1
Stakeholder Groups in Computational Creativity Research and Practice

Simon Colton, Alison Pease, Joseph Corneli, Michael Cook, Rose Hepworth and Dan Ventura

Abstract The notion that software could be independently and usefully creative is becoming more commonplace in scientific, cultural, business and public circles. It is not fanciful to imagine creative software embedded in society in the short to medium term, acting as collaborators and autonomous creative agents for much societal benefit. Technologically, there is still some way to go to enable Artificial Intelligence methods to create artefacts and ideas of value, and to get software to do so in interesting and engaging ways. There are also a number of sociological hurdles to overcome in getting society to accept software as being truly creative, and we concentrate on those here. We discuss the various communities that can be considered stakeholders in the perception of computers being creative or not. In particular, we look in detail at three sets of stakeholders, namely the general public, Computational Creativity researchers and fellow creatives. We put forward various philosophical points which we argue will shape the way in which society accepts creative software. We make various claims along the way about how people perceive software as being creative or not, which we believe should be addressed with scientific experimentation, and we call on the Computational Creativity research community to do just that.

S. Colton (✉) · J. Corneli · M. Cook · R. Hepworth
Computational Creativity Group, Department of Computing, Goldsmiths College, University of London, London, UK
e-mail: s.colton@gold.ac.uk
http://ccg.doc.gold.ac.uk

A. Pease
School of Computing, University of Dundee, Dundee, UK

D. Ventura
Computer Science Department, Brigham Young University, Provo, USA

T.R. Besold et al. (eds.), *Computational Creativity Research: Towards Creative Machines*, Atlantis Thinking Machines 7, DOI 10.2991/978-94-6239-085-0_1

1.1 Introduction

It seems uncontroversial to state that one of the long-term goals of research into Computational Creativity is to see creative software embedded in society: Apple's iTunes will one day compose new music for us, rather than just recommending it; Microsoft's PowerPoint will suggest jokes for a speech we're writing; videogames will be constructed on the fly to fit our preferences and mood; software will routinely make scientific discoveries; and household appliances will be endowed with creative abilities, like a refrigerator able to concoct a recipe to fit its contents. It is also uncontroversial to point out that another long-term goal of the field is to further our understanding of human creativity, both individually and in societies, through computer simulation.

Computational Creativity researchers have made steady progress towards software which creates, by employing, advancing and inventing novel Artificial Intelligence, natural language processing, graphics, audio and other techniques for creative purposes. There is, of course, much progress still to be made technically, so that software can be creative and be seen to be creative, in order for consumers to be provided with valuable artefacts and enjoyable creative experiences. In addition to the technological hurdles faced, it is clear that certain sociological issues stand in the way of progress. That is, people naturally tend towards thinking that nuts-and-bolts, bits-and-bytes machines will never have a creative spark, and different sets of people instantiate this tendency in different ways. Through much engagement and outreach, we have come to the conclusion that understanding people's conceptions of software being creative is an important tool to be used towards the long-term goal of understanding human creativity, and that favourably guiding these conceptions will be essential in bringing about the long-term goal of embedding creative software in society.

In largely separate tracks of research, we have examined how creative software is perceived by three different types of *creativity stakeholders*—people who may have something to gain or lose from software which is creative—from a practical and a philosophical perspective. We address the different types of creativity stakeholders in general in Sect. 1.3, and concentrate in the rest of the chapter on three particular types. In particular, in Sect. 1.4, we address members of the general public exposed to creative software. Following this, in Sect. 1.5, we address observer issues within Computational Creativity research itself. Finally, in Sect. 1.6 we address videogame designers, as an exemplar of a focused community of creative individuals within which creative software has begun to make an impact. We posit that, because of the different issues that each stakeholder community raises with creative software, it currently helps to study them independently, and suggest approaches to altering the perception that people have of software in these groups in different ways. However, by bringing together these strands for the first time here, we can begin to discuss more unified approaches to the presentation of software written to be autonomously creative.

Throughout this chapter, we propose hypotheses about how each set of stakeholders perceive software as being creative or not, based on practical experiences, philosophical studies and theoretical advances. We believe that our arguments in favour of these claims are sufficiently strong for them to be taken to the next level and tested scientifically through observer-based experimentation—and that the hypotheses provide an agenda the Computational Creativity research community cannot ignore. To conclude in Sect. 1.7, we suggest some practical ways in which these claims (which are presented as numbered hypotheses) could be investigated. In order to explain and support the claims we make, in the next section, we first present a philosophical perspective on the notion of creativity, which will introduce ideas that underpin the material in the rest of the chapter.

1.2 A Perspective on Creativity

We hold that creativity is a *secondary* and *essentially contested* quality of a person, and that linguistic usage of terms related to creativity can often be declarative illocutionary speech acts. We unpack these assertions below. Firstly, we believe that attributions of creativity are contextualist, having no truth value which is independent of context, perception and interpretation. In this way we see creativity attributions as analogous to the Lockean notion of a secondary quality [1]. Locke distinguished *primary* and *secondary qualities*, where the former are taken to be intrinsic to an object, for example, its mass, and the latter are understood to be perception-dependent, for example, colour. While these Lockean qualities are directly tied to sensory perception, as opposed to the aesthetic and social category of creativity, the distinction is still a useful one here, since it highlights different types of properties. Dennett's intentional stance [2] is also of interest here: we may adopt a "creativity stance" towards a person and interpret their work *as though they were being creative*, in order to better understand (rather than predict) their behaviour. Likewise, we may find that the "creativity stance" provides a new way of understanding the behaviour of a piece of software which goes beyond the physical details of the program.

Gallie introduced *essentially contested concepts* as those for which "the proper use . . . inevitably involves endless disputes about their proper uses on the part of their users" [3, p. 169], to which Gray added that the disputes ". . . cannot be settled by appeal to empirical evidence, linguistic usage, or the canons of logic alone" [4, p.344], and Smith noted that ". . . all argue that the concept is being *used inappropriately* by others" [5, p.332]. In the *Cambridge Handbook of Creativity*, Plucker and Mabel assert that:

> . . . despite the abundance of definitions for creativity and related terms, few are widely used and many researchers simply avoid defining the relevant terms at all. [6, p.48]

Clearly, certain notions such as *art* are essentially contested concepts, looking at the multitude of articles written each year in the popular and cultural press asking: "But is it Art?" Indeed, Gallie points out that the statement: "This picture is painted in oils" can be disputed whilst the disputants nevertheless agree on the proper usage of the terms involved, whereas the assertion "This picture is a work of art" is likely to be contested

> ... because of an evident disagreement as to—and the consequent need for philosophical elucidation—of the proper general use of the term "work of art" [3, pp.167].

As a recent example, the question of whether videogames should be classed as art was raised by a Guardian art critic [7], to which the Guardian games editor responded:

> Here is a good way to tell if a critic is having a moment of madness: they will attempt to define art. The greatest philosophers in history have floundered on the question, many simply avoided it altogether, preferring to grapple with more straightforward questions—like ... the existence of God. Art is ethereal, boundless, its meaning as transient as the seasons. When you think you have grasped it, it slips through your fingers [8].

While this is only one example, it serves as an exemplar of the kinds of debates that occur daily about the nature of art. While it is true that the preoccupation with expressing creativity is a relatively modern aspect of the visual arts, if the notion of *art* is indeed essentially contested within our culture, then the notion of the *creativity* that went into producing a given artwork should be seen accordingly. In particular, a selection of criteria for what counts as creativity is required in any coherent scheme for understanding and evaluating creativity in art. This is the perspective advanced by Jordanous [9], with which we agree—although we also agree with her point that there is unlikely to be broad and lasting agreement about just what the precise criteria of creativity actually are. We can further justify the idea that proper usage of the term *creativity* involves endless debate about its proper usage by reference to the multitude of volumes written about improving, managing and assessing creativity in people, organisations and society. Indeed, as a society, we are better off if we do not agree about what creativity means—in the sense that the disputes we have about this are an engine for change and progress, and it would surely be stultifying if we all suddenly agreed on this most important of concepts. While it is problematic for various areas of study—not least Computational Creativity—that creativity is an essentially contested quality of any person, it is something we need to embrace and even celebrate. For more in-depth discussion of these issues, see Jordanous [9, Chap.3]. We may ask, in practice, what does it mean to *say* someone or something is "creative"? Austin informally introduced the notion of an *illocutionary act* as a locution that also serves to perform another action [10]. Searle further categorised such speech acts into: *assertives*, *directives*, *commissives*, *expressives* and *declarations* [11]. *Declarations* in particular are understood to change reality in accordance with the proposition stated. An example of such a speech act is: "I pronounce you husband and wife." We believe that—in certain circumstances—people can bestow the reality of a person being creative simply by stating it. To see this, we recall the contested nature of creativity, and the assumption that there is no general consensus about what makes someone creative. It follows that people who are not particularly invested in the creativity (or lack thereof) of someone else may be swayed by the declarative speech act of a third party in a position of authority. When Nicholas Serota, long time director of the Tate art museums and galleries, says that a piece is a great work of art, that work becomes (at least temporarily) a great work. When he states that a particular artist is unusually creative, who are we to argue? Given that the sentence 'X is creative' is often shorthand for: 'Most people agree that they perceive X to be creative', such authorities can essentially bring into being the creativity of X, regardless of whether X perceives him/herself as creative or not.

1.3 Communities of Creativity Stakeholders

In order to understand the different groups of creativity stakeholders, the relationships between them, and the ways in which meaning is continually being created, negotiated and re-created, we can look to sociology. In Latour's *Actor Network Theory* [12], he describes such stakeholders and diverse social groups as actors in a network. Meaning is created socially via actors who cluster into diverse stakeholder groups. These groups are in constant flux, as relationships, actors and ideas within the groups change and come into conflict with each other. Latour holds that understanding such dynamics in the network is essential to understanding processes of innovation and knowledge-creation in science and technology. The process by which a network is formed and comes to be represented as a single entity is called *translation*, and is a key concept in the Actor Network Theory. Translation consists of various phases: the initial formation of a programme and identification of actors in a new network with a novel, shared goal (*problematisation*); the strengthening of the network via formal and informal means (*interessement*); ways of evolving the network and providing structures for new members to join (*enrolment*); and acquiring the resources and power to build an effective institution which can achieve its goal (*mobilisation*).

In the case of Computational Creativity, relevant creativity stakeholders include researchers, the wider AI community, funding bodies, experts in the psychology of human creativity, neuroscientists, artists, art critics, journalists, philosophers, educators, the public, and so on. Each group has accompanying visions, beliefs and goals, in which they have, to a varying degree, invested (and which, to a varying degree, define them as a group). We hold that understanding such different perspectives and their interactions is essential if software is ever to be deemed creative by mainstream consumers of cultural artefacts. In this section, we consider these stakeholder groups and in particular use Latour's notion of translation to look at how Computational Creativity researchers have evolved into a community. We also look at some of the relationships between the groups, both in the context of Computational Creativity and the wider scientific arena.

1.3.1 The Computational Creativity Stakeholders

Members of the Computational Creativity community are largely people with a background in Artificial Intelligence or computer science and an interest in creativity.

They are usually professional academics with the infrastructure of a university supporting them. AI is itself a young field—originating in the 1950s—and, since initial attempts to build general intelligence machines, has fragmented into many different specialisations and subdisciplines: once established, these then form the *internal environment* for any new area, in terms of providing ideas, methods and concepts, and at times, competition. Academic measures of the health of such subdisciplines include the amount of funding awarded, the number of lectureships or professorships in the field, the existence of a journal and an international conference series, and other scientifically respectable incentivisation schemes and recognition.

It was against this backdrop that AI researchers with an interest in creativity found themselves in the late 1990s. Given their background, they were not only accustomed to the idea that machines can be intelligent, but their very livelihood depended on that premise. So it was not, perhaps, such a huge leap to the idea that machines can be creative. However, since there was no infrastructure supporting research into Computational Creativity, early researchers largely had to establish their reputation in different (possibly related) areas of AI and build up the Computational Creativity community almost on their own time, sometimes taking considerable career risks to do so.

Latour's notion of *translation* can help us to understand how the community formed. *Problematisation* occurred when a few core people identified the goal of building creative software as a subdiscipline of AI. Between them, they had the influence and organisational power to make *Creativity in AI and Cognitive Science* the theme of the AISB'99 Convention (co-chaired by Geraint Wiggins, Helen Pain and Andrew Patrizio). This featured a keynote address by Margaret Boden, a cognitive scientist known for her popular writing on creativity in people and in machines [13, 14]. The initial symposium was followed up by four further events[1] held at AISB'00 - AISB'03, and a series of workshops on creative systems at major AI conferences. We present an extract from the editors' introduction to the *Proceedings of the Symposium on Creative and Cultural Aspects of AI and Cognitive Science*, held at AISB in 2000 in Fig. 1.1. This was the *interessement* phase. These were further consolidated with the International Joint Workshops on Computational Creativity (IJWCC), held 2004–08, during which time the community grew from twenty, or so, to double that (*enrolment*). Finally, the community was considered healthy enough, strong enough and large enough to launch the first International Conference on Computational Creativity in 2010. For a history of the field up to this stage, see [15] in a special issue of the AI magazine on Computational Creativity.

The community continues to evolve and grow, with the series having recently held its Fifth Annual International Conference (2014), with around 90 delegates. In order to organise and guide the international series, a Steering Committee was set up consisting of anyone who had chaired an IJWCC event, and they formed the Association for Computational Creativity (ACC) in 2010 and set out rules which enabled new members to join and old members to leave the Association (*mobilisation*). Landmark

[1] *Creative and Cultural Aspects of AI and Cognitive Science* (2000) and then simply *AI and Creativity in Arts and Science* (2001–2003).

Following on from the successful AISB'99 Convention, whose theme was the study of creativity in AI and Cognitive Science, the purpose of this symposium is to bring together researchers interested in all AI and cognitive aspects of creativity and cultural enterprise. The aim of holding one unified meeting, instead of several simultaneous smaller ones, is to promote communication between those studying different aspects of creativity, and this has been mostly achieved: the papers (and the corresponding presentations) are for the most part grouped by their relation to creativity in general, rather than to a particular domain. The exceptions to this are two papers on important low-level aspects of modelling musical creativity.

The four sections which have naturally arisen, then, are headed Exploratory Creativity, Modelling & Supporting Creative Processes, Techniques for Modelling Musical Creativity, and Methodology & Evaluation.

Fig. 1.1 An excerpt from the preface of the Proceedings of the Symposium on Creative and Cultural Aspects of AI and Cognitive Science, held at AISB in 2000, written by Geraint Wiggins. Note the 'natural' emergence of themes within the field, although of course these are very much subject to the Call for Papers, the communities who received the call, the instructions given to the reviewers, the reviewers themselves and the editor's vision

events during this time included the first ever award of a Chair in Computational Creativity (to Geraint Wiggins, in 2004, by Goldsmiths, University of London) (only one of two—the other being held by Simon Colton also at Goldsmiths, University of London, awarded in 2013); the first PhD with Computational Creativity in its title (Anna Jordanous, University of Sussex, 2012 [9]) and the first NSF and EU calls for proposals in Computational Creativity (*CreativeIT, NSF Program Solicitation 09–572* [16] and *Objective ICT-2013.8.1, Technologies and scientific foundations in the field of creativity* [17, p.81]). The process has been carefully managed throughout, with an eye on political as well as intellectual developments. Social factors have also played a key role, being inextricably linked to internal development of scientific knowledge [18].

1.3.2 Other Creativity Stakeholders

Each of the other stakeholder groups will have a similarly fascinating history. Some, such as the EU funding body, are tightly bound and have a formal definition of themselves and their goals. Others, such as the general public—for whom the concept of *translation* is meaningless—are much more loosely defined. Of note is who the decision makers are in each of these groups. In the Computational Creativity community, it is clear that a few people have had a huge influence, and it is likely that this is also the case for other groups of stakeholders. It may be worth considering these in detail, especially from a point of view of motivation and power. For instance, Boden's way of seeing creativity dominated the first decade of the community growth. Likewise, a few core individuals working for the EU had the influence to prioritise research into Computational Creativity, and to fund around €10 m worth of projects.

- **The general public**
When describing what they do, to a layman, most researchers into Computational Creativity will probably have experienced reactions such as: "A computer that is creative might be dangerous—it might kill us"; "Creativity is a celebration of humanity, and the very idea of Computational Creativity cheapens that"; "I read a poem or listen to music to communicate with another human being. I don't want to communicate with a computer, I want a live human connection", and so on. It is important to determine where these ideas come from, whether they are grounded in anything, whether we should try to counter them, and if so, how? While such emotional responses are not necessarily negative, it might be the case that they hinder reasoned debate. Public perception of Computational Creativity derives from multiple sources, including journalistic coverage (or lack of it), science fiction narratives, opportunities to consume computationally created artefacts and so on. We look further at observer issues in the general public in Sect. 1.4.
- **Fellow creatives**
Creative people sometimes voice the worry that "Computers are going to put us out of a job". This group is similar to the general public in terms of influences and attitudes. It seems that artists might be being encouraged to worry about software replacing them, because such sensationalist stories sell newspapers. We study a particular community of creative people, namely videogame designers in Sect. 1.6.

1.3.3 Relationships Between the Different Stakeholder Groups

There have been several interactions between the Computational Creativity community and members of the public and fellow creatives. For instance, Colton and Ventura hosted a festival of Computational Creativity in 2013, *You Can't Know my Mind* [19], and other events have followed on from this. Historical relationships between scientists and the public can also elucidate current interactions. In other fields, there have been some explicit campaigns to manufacture doubt, by parties who are threatened by specific scientific advances. For instance, the tobacco industry tried to discredit and discourage the notion that smoking is bad for our health; likewise the fossil fuel industry did the same in the case of global warming. Here we see that a few powerful actors can sometimes bring an entire body of established scientific knowledge into question.

Ravetz argues that scientific ignorance may in some ways be as prone to social construction as scientific knowledge [20, 21], cited in [22, p.37]. Stocking and Holstein [22] explore different perceptions that journalists have of their roles, concluding tentatively that journalists construct scientific ignorance consistent with their own interests. Even without such dark agendas, there are other examples from the history of science in which public perceptions conflict with scientific thinking and have been managed, or controlled, in order to bring them into line with current scientific results. Famous examples in which scientific advances have challenged our image of ourselves and our universe include Copernicus's heliocentric model, which challenged our view that the earth is the centre of the universe; Darwin's theory of evolution, which challenged concepts of what it means to be human, to be distinct from other animals, and the notion that our existence has a higher purpose; and Lemaître et al.'s Big Bang theory, which challenged the view that the universe is a stable, stationary entity. In all of these cases the scientists faced their own challenges of reconciling their findings with their religious or world views, and then a process of outreach was necessary in order to gain wider social acceptance. Thus, we see Thomas Huxley—"Darwin's bulldog"—promoting Darwin's theory in the face of many varied and negative responses to it (some of which are recorded in [23]) and helping it to gain wider acceptance, transitioning from scientific to social fact. Today, people in the fields of genetically modified food and stem cell research endeavour to gain wider social acceptance in the form of media coverage and well-funded outreach programmes aimed at educating both school children and the wider community.

Computational Creativity is in a particularly difficult position, since its main research question concerns an essentially contested concept. On certain understandings, the question "can machines be creative?" may be answered negatively, without further elaboration or debate. Thus, we see part of the job of the Computational Creativity community consisting in the delivery of outreach programmes, in which creative software is demonstrated and explained, and the artefacts it has produced exhibited in a setting in which consumers of creative artefacts might begin to appreciate them. In [24], Franzen et al. explore the impact that such dissemination activities can have on scientific progress, and argue that the right name, image or metaphor has the power to make or break relations between a scientific discipline and the public. For instance, consider Dolly the sheep from the Roslin Institute in Edinburgh and Ida the primate fossil from the Messel Pit in Germany. These names make it easier for the discoveries to be visualised and discussed. Arbib and Hesse go further, stating that "scientific revolutions are, in fact, metaphoric revolutions" [25, p.156], cited in [26, p.5].

In addition, then, to sociological narratives, it is important to consider language use by each stakeholder group. The role of spin doctors is well-known in the political arena, in which those who bestow power are influenced in their thinking by vocabulary, metaphors and frames. In our case, the public have the power to bestow or withhold the word "creative" when describing software. Thus, we need to consider the language that we use. Lakoff [27] argues that we fit new information into pre-existing frames, which are built up slowly over time, and if we don't have appropriate frames, then we might misunderstand the information. Using the wrong frame, which is triggered by specific vocabulary, even to deny a message, only reinforces the frame. Thus, rather than trying to argue that "creative software is not scary", we should build up our own vocabulary, frames and metaphors for thinking about it.

Hypothesis 1 Different stakeholder groups (including Computational Creativity researchers, the general public, domain creatives, psychologists, philosophers, educators, critics, journalists, bureaucrats, etc.) assess creativity in software differently, and there is no one-size-fits-all approach to presenting what software does and what it produces in the best way to increase perception of creativity.

Given this, we believe it is currently appropriate to study stakeholder groups separately, as we do in the following sections.

1.4 Observer Issues with the General Public

We introduce here three notions, namely *essential behaviours, the humanity gap* and *software accounting for its actions*. We believe these are important in understanding how people generally react to the idea of software being creative, and thus are important in managing and shaping those reactions. To end the section, we present a case study in handling public perception of creativity in software, and we introduce another notion, namely that of *accountable unpredictability*.

A working definition of the field of Computational Creativity research as a subfield of Artificial Intelligence research given in [28] is as follows:

> The philosophy, science and engineering of computational systems which, by taking on particular responsibilities, exhibit behaviours that unbiased observers would deem to be creative.

While this definition is not universally accepted (with a challenge to focus on system-level creativity rather than individual responsibilities given in [29]), variations of it have been used to describe the field for many years.

The usage of the word 'unbiased' in the above definition hints at a problem encountered in evaluating projects where generative software produces *artefacts* (poems, paintings, sonatas, recipes, theorems, etc.) for human consumption. In particular, people generally have natural biases against, but also occasionally in favour of, artefacts produced by computers over those produced by people. In particular, negative, so called 'silicon', biases have been observed under experimental conditions [30, 31]. Hence, in stipulating that observers must be unbiased, the definition above emphasises a scientific approach to evaluating progress in the building of creative systems, whereby experimental conditions are imposed to rule out, or otherwise cater for, such biases. One such experimental setup is the *Turing-style comparison test*, where computer-generated and human-produced artefacts are mixed and audience members make choices between them with zero context given about the processes involved in their production. It is seen as a milestone moment if audiences cannot tell the difference between the artefacts produced by people and those produced by a computer. We believe there are many problems in the application of such tests in the general context of presenting the processing and products of creative software, as expanded in the subsections below.

1.4.1 Essential Behaviours

We suggest not asking people if they believe software is behaving creatively, but rather concentrating on whether they perceive the software to be acting *uncreatively*. Using our standpoint above that the notion of creativity is essentially contested [3], we expect that no matter how sophisticated our software gets, we will not see consensus on such matters. However, we have found that people agree much more on notions of uncreativity: if a program doesn't exhibit certain behaviours onto which certain words can be projected, then it is easy to condemn it as being uncreative. Building on the foundational arguments given in [32], we propose that audience members can too easily label software as uncreative if they are unable to project any of the following words onto the behaviours they perceive software to be exhibiting:

> skill, appreciation, imagination, learning, intentionality, accountability, innovation, subjectivity and reflection

We have found that assessing the level of projection of these words onto the behaviours of software can help us to gauge people's opinions about (the lack of) important higher-level aspects of software behaviour, such as autonomy, adaptability and self-awareness. Note that we make no claim about the above behaviours being sufficient for a perception of creativity: a necessary set of behaviour types for avoiding the uncreativity label is not the same as a sufficient set of behaviour types for gaining the creativity label. This mis-interpretation of our aims for highlighting the above essential behaviours has propagated somewhat, for instance in [33].

Hypothesis 2 Creativity in people and software is essentially contested and secondary, and hence it might be advantageous to work on people's perception of *uncreativity* in software, as this is easier to predict/manage. Software exhibiting the essential behaviour types highlighted above is necessary for it to avoid being labelled as uncreative. Eventually, when there are no good reasons to label software as uncreative, people may choose to label it as creative.

1.4.2 The Humanity Gap

One could argue that, given the particularly human-centric nature of creativity, and that a human connection is paramount in much of the arts, it is simply inappropriate to use the term 'creative' to describe software. The status quo is that we currently haphazardly apply human terminology related to creativity to software, which often requires the projection of other human qualities onto software, such as it being juvenile, which is inherently error prone, given that computers are patently not people. Another option is to ignore the non-human nature of software and concentrate on what it produces, rather than on what it is, or what it does. To begin to address the kind of silicon biases described above, researchers often compare the interpretation of computer-generated and human-produced artefacts in a rather extreme "blind experiment" situation in which knowledge about the personality of the artist and their practice is entirely missing. The philosophical grounding of such an approach [34, 35] matches the motivation of several art movements [36, 37] and many individual artists who have expressed a desire for their work to be taken at face value (see [38] for examples and further discussion).

We argue that in modern culture, a curious thing can happen when artists attempt to remove all reference to themselves and their process from discussions about the artistic (and commercial) value of their work. That is, in the absence of such information, people may tend to fill in the gaps about personality and process, and may do so in ways which bolster the credibility of an artist and increase the perceived value of his/her works. Indeed, one could argue that—in the same way that artists invite people to interpret the imagery in artworks in their own way by not prescribing what people should see/read/hear, in refusing to provide meta-level details about personality and process, artists, writers and musicians are actually (purposefully or not) inviting art lovers to invent interesting and engaging back-stories about who they are and what they do.

In such a context of non-disclosure, the comparison of the situation for computer-generated artefacts with the situation for human-produced artefacts is not particularly favourable. The vast majority of people have little or no idea about programming or programs, and may even harbour a desire not to find out about these things. Thus, when invited to assess a computer generated painting or poem, say, without background knowledge, they are denied any opportunity to invent a back-story, as they cannot project personality traits or romantic situations onto the computer, and cannot enter into any dialogues. More importantly, this situation can lead to people realising how much they value the human connection, whether actual or imagined, in such situations. We posit that there is a *humanity gap* that must be faced by Computational Creativity researchers who want their software to enhance society by being creative for artistic and utilitarian purposes.

Turing-style experiments, which epitomise the practice of non-disclosure, are intended to reduce variables so that a scientific study of the value of computer generated artefacts can be undertaken. One could argue that these contexts are intended to help people realise how much they value the aesthetic appeal of art, literature and music, regardless of other factors. This may be true, but we believe that such tests can actually help people realise how little they can relate to the computational origin of artefacts. In [39], we raise other issues with Turing-style comparison studies: in particular, we suggest that they encourage naïvety and pastiche generation in creative software. As a final point, it is clear that such experimental conditions are not sustainable if we are to enhance society with creative software. In the long term, biases about machine creation need to be embraced and managed, rather than factored out through experimental setups.

Hypothesis 3 Turing-style comparison tests serve to highlight the humanity gap, and while they might serve short-term scientific gain, they are damaging to the long-term goal of embedding creative software in society.

1.4.3 Software Accounting for Its Actions

We argue in [32, 40] that people take into account how a person or software operates when they assess the value of the output it produces. To address this issue, we advocate a development path to follow when building creative software: (i) the software is given the ability to provide additional, meta-level, information about its process and output, e.g., giving a painting or poem a title; (ii) the software is given the ability to write commentaries about its process and its products; (iii) the software is given the ability to write stories—which may involve fictions—about its processes and products; (iv) the software is given the ability to engage in dialogues with people about what it has produced; how and why. Indeed, giving software the ability to discuss its creative works would mirror Turing's original proposal for an intelligence test [34] to a greater extent than tests focusing only on consumer perception of artefacts. As a preliminary example, in [41], we demonstrated a poetry generation system which is able to provide commentaries about its poetry, and how and why it produced a particular poem.

As we discuss in [42], in a computational setting, there are advantages to software being immersed in environments where serendipity might occur. However, accounting for lucky events that trigger creative acts may actually lessen the celebration and hence the impact that the acts have. It is important to note that people tend not to describe their processes and products in the explicit way we advocate for software, preferring to maintain some level of mystery. Nevertheless, we believe that, at this stage in the development of computationally creative systems, it is important to address the humanity gap—without aspiring to eliminate it. *Framing* [40] serves to highlight that intelligent processing was used to produce artefacts, which is an important first step. Given that audience members will typically not be able to come up with an interesting backstory without some scaffolding, positive acts of framing are likely to have more fruitful impact than an overall air of mystery.

Another possible way to address the humanity gap is to manage people's expectations about the level of humanity they will encounter through a computationally produced artefact. In the same way that when people buy an *e-book* they know they are not going to get a physical object, we advocate telling audiences that they are reading a *c-poem*, and hence—in the knowledge that it was produced computationally—they will get a reduced human connection. We can go further in *re-imagining* traditional artefacts, for instance in suggesting that a c-poem is actually a doublet of texts, one which resembles a traditional poem and another which provides a commentary about the motivations, actions and results of the software's processing. We believe this will highlight the humanity gap, but that it will do so in such a way as to help people to engage with and appreciate the creative process, and better enjoy the artefacts produced by software.

Hypothesis 4 The humanity gap can be addressed by re-imagining the nature of creative artefacts, to manage expectations of humanity. In particular, it is advantageous for software to account for its processes and products through additional material such as a commentary.

1.4.4 A Case Study in Automated Portraiture

As part of an exhibition with The Painting Fool[2] system [43] in 2013, we enabled the software to produce portraits for people live in a gallery, as described in [19]. Managing the expectations and perceptions of the observers was a key aspect of this project. To this end, we hung posters describing the behaviour of the software as exhibiting aspects of intentionality, imagination, skill, appreciation, reflection and learning (six of the essential behaviours described above). Moreover, the software's actions and output were tailored to support the perception of these behaviours and an impression of creativity in the software by observers present in the exhibition, especially those sitting for a portrait.

Portraits were painted with people sitting in front of a laptop. It was immediately made clear that (i) the software was modelling a 'mood' to direct its painting, and (ii) the sitter was very much a tool for the software, not the other way around. This was achieved by opening remarks from the software such as: "Thank you for being my model. I'm in a negative mood right now, so I would like you to express a sad emotion." This was followed by The Painting Fool explicitly directing the sitter, while video recording them. A still image was then extracted where the sitter was expressing an emotion. Machine vision techniques were applied to remove the background, into which was substituted one of 1,000 abstract art images, to which one of 1,000 image filters was applied. The filter was chosen to increase the chances that the resulting image might reflect a changing simulated mood gained through reading newspaper articles, as described in [19]. The same filter was applied to the face of the sitter placed in the foreground, producing in a few seconds an image conception, or sketch for the portrait, such as the first image of Fig. 1.2.

Following this, a canvas appeared on screen, and a hand holding either a pencil, paint brush or pastel stick made virtual marks on the canvas leading to a non-photorealistic rendering of the background and foreground of the portrait, taking between 2 and 10 minutes, depending on the style. An example portrait is given at the bottom of Fig. 1.2, which was printed and given to the sitter, along with the commentary (the whole of Fig. 1.2). The most important aspect of the commentary is the expression of intention, by first showing a conception of the type of portrait the software aimed to produced, then showing what it produced and finally analysing and criticising—using machine vision techniques described in [44]—its results with respect to its aims.

The purpose of the exhibition was cultural, not scientific, and no experimentation was undertaken. From our experience, however, we contend that the behaviours exhibited by the software and explained in poster form enabled people to be surprised by the resulting portrait (and many of the 100 or so sitters in the exhibition were very surprised), while still projecting creativity onto the software. This upheld the aim of the *You Can't Know my Mind* exhibition: as it used some intelligence, and could explain its actions, it was somewhat appropriate to employ the word 'mind' with reference to The Painting Fool. However, as the process was unpredictable due to

[2] Online presence: www.thepaintingfool.com.

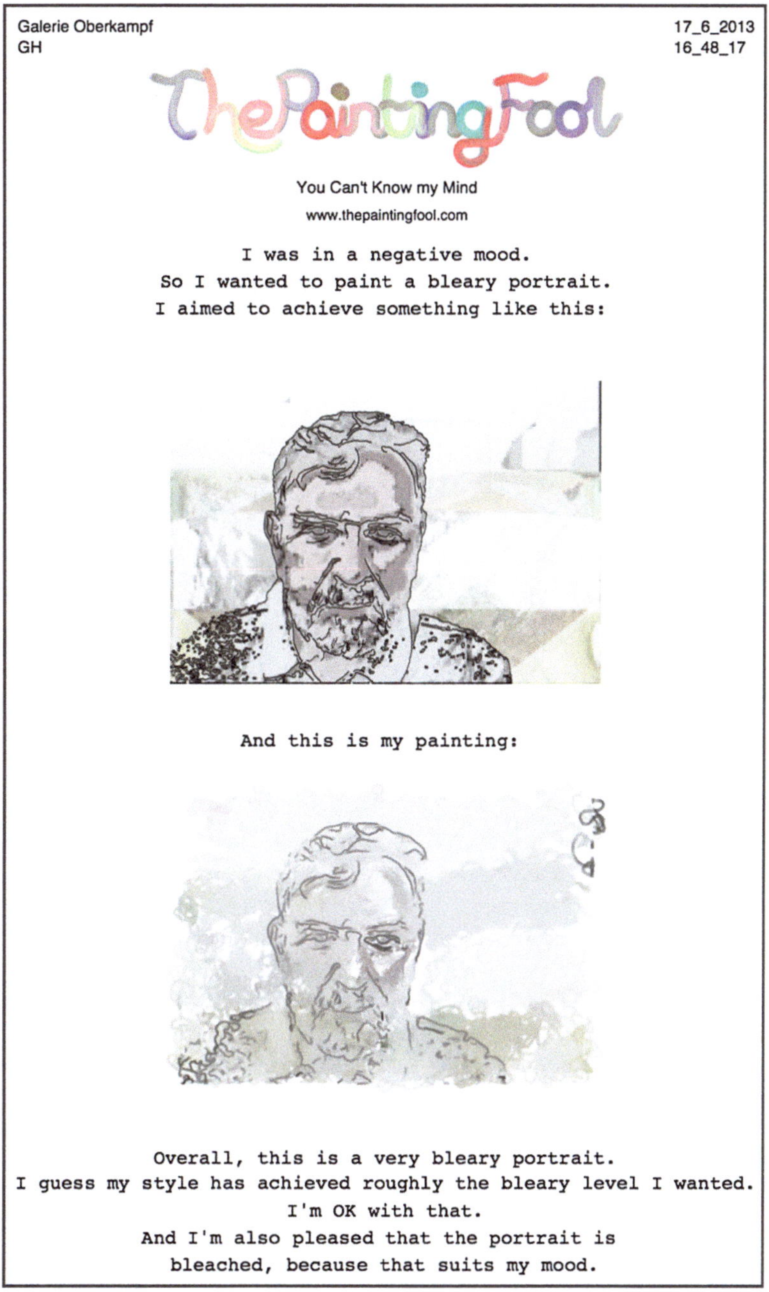

Galerie Oberkampf 17_6_2013
GH 16_48_17

ThePaintingFool

You Can't Know my Mind

www.thepaintingfool.com

I was in a negative mood.
So I wanted to paint a bleary portrait.
I aimed to achieve something like this:

And this is my painting:

Overall, this is a very bleary portrait.
I guess my style has achieved roughly the bleary level I wanted.
I'm OK with that.
And I'm also pleased that the portrait is
bleached, because that suits my mood.

Fig. 1.2 Example commentary by The Painting Fool, from the *You Can't Know my Mind* exhibition, Paris, June 2013

the dynamic nature of the software's changing mood, it was impossible to know this mind, and people realised that some software is written not to be a tool, but to be a creative individual. In fact, when in the most negative of moods, The Painting Fool refused to paint a portrait and sent the (often shocked) sitter away, citing a particularly depressing keyphrase in a particularly distressing newspaper article that it had recently read.

In these cases, The Painting Fool pointed out explicitly: "No random numbers were used in coming to this decision". This is because we feel that *accountable unpredictability* is important for creative systems. That is, we have found that when people realise that a certain important event has happened or an important artefact has been produced because of a random act, any dialogue (perceived or real) comes to an abrupt halt, and detracts from the creative experience. In contrast, unpredictability through accountable actions such as reading newspaper articles can add a great deal to a creative experience, at the very least by providing additional talking points.

Hypothesis 5 Accountable unpredictability enhances the experience people have when told about software creating an artefact, whereas random number based unpredictability detracts from the experience.

1.5 Formally Capturing Progress in Creative Systems

Naturally, another major set of stakeholders in the notion of software being creative are the Computational Creativity researchers who aim to write such systems, and use them to study creativity in people and machines. As they are familiar with the issues of simplistic arguments for and against creativity in software, these stakeholders require more formalism in any argumentation put forward to support the hypothesis of increased creativity in software.

We have focused on formalising the general notion of *progress* in Computational Creativity research. To do this, we first introduced the FACE and IDEA descriptive models in [45, 46]. The FACE model categorises generative acts by software into those at (g)round level, during which base objects are produced, and (p)rocess level, during which methods for generating base objects are produced. These levels are sub-divided by the types of objects/processes they produce: F_g denotes a generative act producing some framing information, A_g denotes an act producing an aesthetic measure, C_g denotes an act producing a concept and E_g denotes an act producing an example of a concept. Generative acts producing new processes are defined accordingly as F_p, A_p, C_p and E_p. Tuples of generative acts are collated as *creative acts*, and various calculations and recommendations are suggested in the model with which to compare creative systems. We developed the IDEA model so that creative acts and any impact they might have could be properly separated. We defined various stages of software development and used an ideal audience notion, where people are able to quantify changes in well-being and the cognitive work required to appreciate a creative act and the resulting artefact and/or process.

The majority of researchers develop software using only themselves as an evaluator, because observer-based models are too time-consuming to use on a day-to-day basis. These informal in-house evaluation techniques generally do not capture the global aims of the research project, or of the field (e.g., producing culturally important artefacts and/or convincing people that software is acting in a creative fashion). This can lead to situations where systems are presented as feats of engineering, with little or no evaluation at all [9]. In [47], we argue that assessing *progress* is inherently a process-based problem, and hence we concentrate our formalism on processes, tempered with aspects of artefact evaluation. In the subsections below, we present this formalism with worked examples, followed by a case study describing the development of an evolutionary art system.

1.5.1 Formal Assessment of Progress

We combine the most useful aspects of the IDEA and FACE models, the list of essential behaviours described in Sect. 1.4.1, and certain aspects of assessing artefact value in a diagrammatic formalism for evaluating progress in the building of creative systems. We focus on the creative acts that software performs, the artefacts it produces and the way in which audiences perceive it and consume its output. We simplify by assuming a development model where a single person or team develops the software, with various major points where the program is sufficiently different for comparisons with previous versions. We aim for the formalism to be used on a daily basis without audience evaluations, to determine short term progress, but for it also to enable fuller audience-level evaluations at the major development points. We also aim for the formalism to help determine progress in projects where there are both *weak* and *strong* objectives, focused, respectively, on the production of increasingly higher valued artefacts, and on increasing the perception of creativity people have of the system. We found that the original FACE model didn't enable us to properly express the process of building and executing generative software. Hence another consideration for our formalism is that it can capture various timelines both in the development and the running of software in such a way that it is fairly obvious where the programmer contributed creatively and where the software did likewise.

1.5.2 Diagrammatic Capture of Timelines

Taking a realistic but abstracted view of generative software development and deployment, we identify four types of timeline. Firstly, generative programs are developed in **system epochs**, with new versions being regularly signed off. Secondly, each process a program undertakes will have been implemented during a **development period** where creative acts by programmer and program have interplayed. Thirdly, at run-time, data will be passed from process to process in a series of creative and

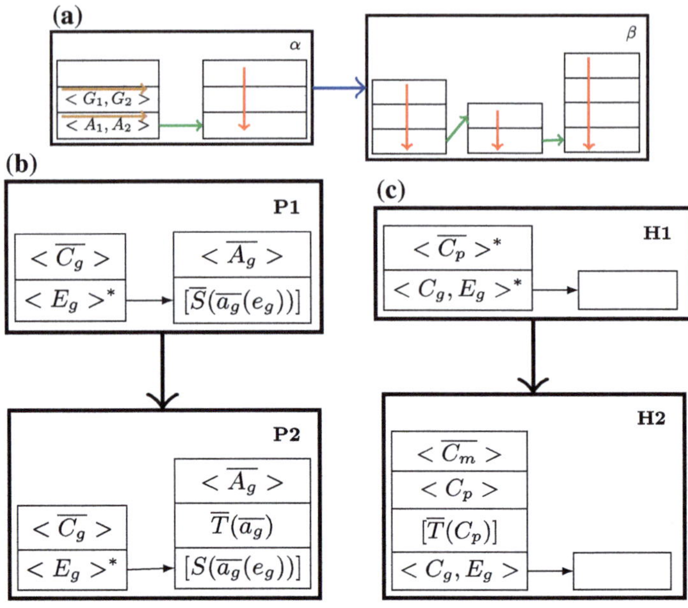

Fig. 1.3 **a** Key showing four types of timelines; **b** progression of a poetry system; **c** progression of the HR system

administrative **subprocesses** performed by software and programmer. Finally, each subprocess will comprise a sequence of generative or administrative **acts**.

We capture these timelines diagrammatically: the four different kinds of transitions are highlighted with coloured arrows in Fig. 1.3a. The blue arrow from box α to β represents a change in epoch at system level. The red arrows overlapping a *process stack* represent causal development periods. The green arrows represent data being passed from one subprocess to another at run-time. The brown arrows represent a series of generative/administrative acts which occur within a subprocess. Inside each subprocess box is either a < creative act > from the FACE model (i.e., a sequence of generative acts), or an [administrative act] which doesn't introduce any new concept, example, aesthetic or framing information/method. Administrative acts were not originally described in the FACE model, but we needed them to describe certain progressions during software development. For our purposes here, we use only T to describe a translation administrative act often involving programming, and S to describe when an aesthetic measure is used to select the best from a set of artefacts. We employ the FACE model usage of lower-case letters to denote the output from the corresponding upper-case generative acts. Furthermore, we extend the FACE notion of (g)round and (p)rocess level generative acts with (m)eta level acts during which process generation methods are invented. As in the original description of the FACE model, we use a bar notation to indicate that a particular act was undertaken by the programmer. We use a superscripted asterisk (*) to point out repetition.

As a simple example diagram, Fig. 1.3b shows the progression from poetry generator version P1 to P2. In the first version, there are two process stacks, hence the system works in two stages. In the first, the software produces some example poems, and in the second the user chooses one of the poems (to print out, say). The first stack represents two timesteps in development, namely that (a) the programmer had a creative act $< \overline{C_g} >$ whereby he/she came up with a concept in the form of some code to generate poems, and (b) the software was run to produce poems in creative acts of the form $< E_g >^*$. The second stack represents the user coming up with an idea for an aesthetic, e.g., preferring lots of rhyming, in creative act $< \overline{A_g} >$, and then applying that aesthetic $\overline{a_g}$ him/herself to the examples produced by the software, in the *selection* administrative act $[\overline{S}(\overline{a_g}(e_g))]$, which maps the aesthetic $\overline{a_g} : \{e_g\} \rightarrow [0, 1]$ over the generated examples, and picks the best one. In the P2 version of the software, the programmer undertakes the *translation* act $[\overline{T}(\overline{a_g})]$, writing code that allows the program to apply the rhyming aesthetic itself, which it does at the bottom of the second stack in box P2.

Figure 1.3c shows a progression in the HR automated theory formation system [48] which took the software to a meta-level, as described in [49]. HR operates by applying production rules which invent concepts that categorise and describe input data. Each production rule was invented by the programmer during creative acts of the type $< \overline{C_p} >$, then at run-time, HR uses the production rules to invent concepts and examples of them in $< C_g, E_g >^*$ acts. In the meta-HR version, during the $< \overline{C_m} >$ creative act, the programmer had the idea of getting HR to form theories about theories, and in doing so, generate concept-invention processes (production rules) in acts of the form $< C_p >$. The programmer took meta-HR's output and translated it via $[\overline{T}(C_p)]$ into an implemented production rule that HR could use, which it does at the bottom of the stack in box H2.

1.5.3 Comparing Diagrams and Output

Examining the transition from one epoch-level diagram to another should provide some shortcuts to estimate audience reactions, especially when these are linked to strong objectives. As with the original FACE model, the diagrams make it obvious where creative or administrative responsibility has been handed over to software, namely where an act which used to be barred has become unbarred, i.e., the same type of generative act still occurs, but it is now performed by software rather than programmer. For instance, this happened when the $\overline{S}$ became an S in Fig. 1.3b and when the $\overline{C_p}$ became a C_p in Fig. 1.3c. At the very least, in these cases, an unbiased observer would be expected to project more autonomy onto the software, and so progress in the strong sense has likely happened. In addition, the diagrams make it obvious when software is doing more processing in the sense of having more stacks, bigger stacks or larger tuples of acts in the stack entries. Moreover, the diagrams make it clear when more varied or higher-level creative acts are being performed by the software. All of these features have the potential to convince audience members that software is being more sophisticated, and can be taken as a preliminary indicator of progress.

When dealing with actual external evaluation, where people don't know what the software does, we suggest that the diagrams above (or verbalisations/simplifications of them) can be used to describe the software to audience members, to explain what the software does, and what the programmer has done in the project. Audience members can then be asked whether they would project any of the essential behaviours from Sect. 1.4.1 onto any of the creative acts undertaken by the programmer or by the system. Thus, one method for estimating progress from version v1 of a creative system to version v2 that takes into account features of both processing features and artefact quality would be:

- show audience members the diagrams for v1 and v2 as above, and explain the acts undertaken by the software, then
- show audience members the output from v1 and v2, and,
- ask each person to compare the pair of product and process for v1 with that of v2.

A statistical analysis could then be used to see whether the audience as a whole evaluates the output as being better, worse or the same, and whether they think that the processing is better, worse or the same in terms of the software seeming less uncreative.

1.5.4 A Case Study in Evolutionary Art

Evolutionary art—where software is evolved which can generate abstract art—has been much studied within Computational Creativity circles [50]. Based on actual projects which we reference, we hypothesise here the various timelines of progress that could lead from a system with barely any autonomy to one with nearly full autonomy. Figure 1.4 uses our diagrammatic approach to capture three major lines of development, with the (hypothetical) system in box 8 representing finality, in the strong sense that the software can do very little more creatively in generating abstract art. Since features from earlier system epochs are often present in later ones, we have colour-coded individual creative acts as they are introduced, so the reader can follow their usage through the systems. If an element repeats with a slight variation (such as the removal of a bar), this is highlighted. Table 1.1 is a key to the figure, which describes the most important creative and administrative acts in the systems. Elements in the key are indexed with a dot notation: *system.process-stack.subprocess* (by number, from left to right, and top to bottom, respectively). System diagrams have repetitive elements, so that the timelines leading to its construction and what it does at run-time can be read in a stand-alone fashion.

Following the first line of development, system 1 of Fig. 1.4 represents an entry point for many evolutionary art systems: the programmer invents ($\overline{C_p}$) (or borrows) the concept formation process of crossing over sets of mathematical functions to

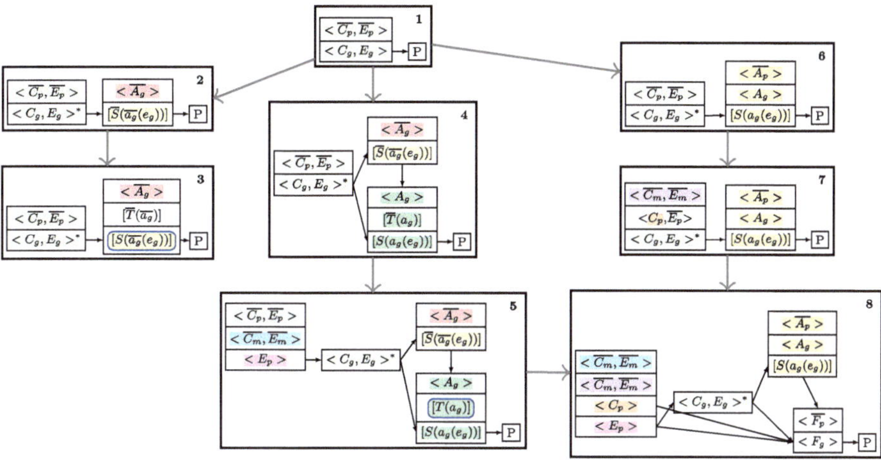

Fig. 1.4 The progression of an evolutionary art program through eight system epochs, taken from [47]

produce offspring sets. He/she also has an idea ($\overline{E_p}$) for a *wrapper* routine which can use such a set of functions to produce images. He/she then uses the program to generate (C_g) a set of functions and employ the wrapper to produce (E_g) an image which is sent to the (P)rinter. The crossover and subsequent image generation is repeated multiple times in system 2, and then the programmer—who has invented ($\overline{A_g}$) their own aesthetic—chooses a single image to print. In system 3, as in the poetry example above, the programmer translates their aesthetic into code so the program can select images. This is a development similar to that for the NEvAr system [51].

Following the second line of development, in system 4, the programmer selects multiple images using his/her own aesthetic preferences, and these become the positives for a machine learning exercise as in [52]. This enables the automatic invention (A_g) of an aesthetic function, which the programmer translates by hand $\overline{T}(a_g)$ from the machine learning system into the software, as in [53], so the program can employ the aesthetic without user intervention. In system 5, more automation is added, with the programmer implementing their idea ($\overline{C_m}$) of getting the software to search for wrappers, then implementing this ($\overline{E_m}$), so that the software can invent (E_p) new example generation processes for the system.

Following the final line of development, in system 6, we return to aesthetic generation. Here the programmer has the idea ($\overline{A_p}$) of getting software to mathematically invent fitness functions, as we did in [54] for scene generation, using the HR system [48] together with The Painting Fool [43]. In system 7, the programmer realises ($\overline{C_m}$) that crossover is just one way to combine sets of functions, and gives ($\overline{E_m}$) the software the ability to search a space of combination methods (C_p). The software does this, and uses the existing wrapper to turn the functions into images. System 8 is the end of the line for the development of the software,

Table 1.1 Key to Fig. 1.4

ID	Event	Explanation
1.1.1	$\overline{C_p}$	The programmer invents the idea of crossing over two sets of mathematical functions to produce a new set of mathematical functions.
1.1.1	$\overline{E_p}$	The programmer implements a wrapper method that takes a set of mathematical functions and applies them to each (x, y) co-ordinate in an image to produce an RGB colour.
1.1.2	C_g	The software generates a new set of functions by crossing over two pairs of functions.
1.1.2	E_g	The software applies these functions to the (x, y) co-ordinates of an image, to produce a piece of abstract art.
2.2.1	$\overline{A_g}$	The programmer had in mind a particular aesthetic (symmetry) for the images.
2.2.2	$\overline{S}(\overline{a_g}(e_g))$	The programmer uses his/her aesthetic to select an image for printing.
3.2.2	$\overline{T}(\overline{a_g})$	The programmer took their aesthetic and turned it into code that can calculate a value for images.
3.2.3	$S(\overline{a_g}(e_g))$	The software applies the aesthetic to select one of a set of images produced by the software.
4.3.1	A_g	The software uses machine learning techniques to approximate the programmer's aesthetic.
4.3.2	$\overline{T}(a_g)$	The programmer hand-translates the learned aesthetic into code.
4.3.3	$S(a_g(e_g))$	The software applies the new aesthetic to choosing the best image from those produced.
5.1.2	$\overline{C_m}$	The programmer has the idea of getting the software to search through a space of wrapper routines.
5.1.2	$\overline{E_m}$	The programmer implements this idea.
5.1.3	E_p	The software invents a new wrapper.
5.4.2	$T(a_g)$	The software translates the machine-learned aesthetic itself into code.
6.2.1	$\overline{A_p}$	The programmer has the idea of getting the software to invent a mathematical fitness function.
6.2.2	A_g	The software invents a novel aesthetic function.
6.2.3	$S(a_g(e_g))$	The software selects the best image according this aesthetic.
7.1.1	$\overline{C_m}$	The programmer has the idea of getting the software to invent and utilise new function combination techniques, generalising crossover.
7.1.1	$\overline{E_m}$	The programmer implements this idea so that the software can invent new combination techniques.
7.1.2	C_p	The software invents a novel combination technique.
8.4.1	$\overline{F_p}$	The programmer has the idea of getting the software to produce a commentary on its processes and the images it produces.
8.4.2	F_g	The software produces a commentary about its process and product.

as it brings together all the innovations of previous systems. The software invents aesthetic functions, innovates with new concept formation methods that combine mathematical functions, and generates new wrappers which turn the functions into images. Finally, the programmer has the idea ($\overline{F_p}$) of getting the software to write commentaries, as in [41], about its processing and its results, which it does in generative act F_g.

Tracking how the system diagrams change can be used to estimate how audiences might evaluate the change in processing of the software, in terms of the extended creativity tripod described above. Intuitively, each system represents progress from the one preceding it, justified as follows:

1 → 2: $< C_g, E_g > \rightarrow < C_g, E_g >^*$

Simple repetition means that the software has more *skill*, and the introduction of independent user selection shouldn't change perceptions about *autonomy*.

2 → 3: $\overline{S} \rightarrow S$

By reducing user intervention in choosing images, the software should appear to have more *skill* and *autonomy*.

1 → 4: Introduction of A_g and $S(a_g(e_g))$ acts

Machine learning enables the generation of novel aesthetics (albeit derived from human choices), which should increase perception of *innovation*, *appreciation* and *learning*, involving more varied creative acts.

4 → 5: Introduction of an E_p act, $\overline{T} \rightarrow T$

Wrapper generation increases the variety of creative acts, and may increase perception of *skill* and *imagination*.

1 → 6: Introduction of A_g and $S(a_g(e_g))$ acts

The software has more variety of creative acts, and the invention and deployment of its own aesthetic—this time, without any programmer intervention—should increase perception of *intentionality* in the software.

6 → 7: Introduction of a C_p act

Changes in the evolutionary processes should increase perceptions of *innovation* and *autonomy*.

5, 7 → 8: Introduction of an F_g act

Framing its work should increase perceptions of *accountability* and *reflection*.

With all strands brought together, the programmer does nothing at run-time and can contribute little more at design time. The software exhibits behaviours onto which we can meaningfully project words like skill, appreciation, innovation, intentionality, reflection, accountability and learning, which should raise impressions of autonomy, and make it difficult to project uncreativity onto the software.

Hypothesis 6 The diagrammatic formalism given above—or some extension of it—is sufficient to capture the creative acts performed in building and running any kind of generative software. Moreover, when this is used alongside audience evaluation of the artefacts produced, a formal assessment of progress in creative software development can be achieved.

1.6 Software as Part of a Creative Community

For each domain in which creative software operates, there is a community of people who have a stake in the notion of whether software working in that domain is perceived as creative. As described in this section, we have recently started to embed our software in such a community, for various reasons, including the study of how people react to it and to the work that it produces. These experiments will form part of a larger study of how people accept (or not) creative technologies that undertake activities which used to be the purview of people only.

1.6.1 Accountable Subjectivity

Applying aesthetic judgements and expressing preferences contribute to the perception of a person or piece of software as being creative. Aesthetics and preferences allow a creative entity, be it a person or software, to express founded judgement (even if we regard the judgement as worthless, or subjectively 'wrong') on creative artefacts, both those created by the entity itself and those created by others. It can also serve as a driving force behind future creation, allowing someone to work towards goals that they have set themselves and strengthening claims of intentionality.

Despite this, little work has been done to build systems which can generate aesthetic preferences of their own and apply them intelligently. One reason for this may be the uncomfortable clash between the subjective and the objective that so often affects research in Computational Creativity. The notion of 'optimality' in many creative domains, particularly those associated with the arts, is a contentious one and leads to much criticism of systems which attempt to quantify the quality of an artefact. The idea of having a system quantify the quality of an *opinion* on creative artefacts is equally controversial, if not more so. Similarly, in the past, the question of how to quantify the degree to which a system is creative was also a subjective and controversial task. In this case, researchers such as Ritchie found it useful to use metrics which dealt with abstract notions of creativity without directly laying out objective measures of quality for any particular artefact or medium. Ritchie's criteria are described in [55], and have been used in many evaluations of creative systems in a variety of different fields and media.

We propose here a similar set of criteria which apply to aesthetics or preferences rather than creative systems. By using abstract metrics, we can avoid talking about aesthetic measures in objective ways, while retaining a meaningful vocabulary with which to describe different kinds of aesthetics. These metrics can be used to evaluate *aesthetic comparator functions*, namely binary functions which take two examples of a type of object, and then return -1, 0 or 1 depending on whether the first object is preferred less, the same as, or more than the second object. Assuming we have an aesthetic function f, and a set of objects the function expresses a preference over, O, we define the following criteria which can be used to differentiate aesthetic functions from one another. Note that these metrics do not necessarily represent a linear gradient of quality—different types of aesthetic function may be desirable in different scenarios.

The first metric is *specificity*. Specificity captures the degree to which the aesthetic represents a *total order* over the set of objects O. If an aesthetic can offer a definite preference (that is, a nonzero result) for many of the objects, it will have a high specificity, and vice versa. High-specificity aesthetics might suggest the aesthetic is experienced or well-developed in some way, if it is able to make clear distinctions between many different artefacts.

The second metric is *transitive consistency*. This captures how self-contradictory the aesthetic function is. Suppose we have three artefacts: A, B and C, and our function f. We can write $A < B$ to indicate that B is preferred to A. We might expect that if $A < B$ and $B < C$ then $A < C$. Transitive consistency measures what proportion of O this holds for. In some scenarios, we might want a high transitive consistency, as this indicates a lack of contradictions in the preferences being expressed. However, in some scenarios, preferences can be complex and multi-objective, and it might be the case that transitivity does not hold for highly subjective opinions about artefacts produced by creative acts.

The third metric is *agreement*. Instead of being expressed in terms of a single aesthetic function, agreement is expressed about two different aesthetics, which we can call f and g. Agreement measures the proportion of the object set O that f and g agree on. This can be *strict*, in which case f and g must return exactly the same value for two objects to be said to agree. Alternatively, agreement can be *non-strict*, in which case f and g can either return the same value, or one of the functions can return zero (no preference) to be said to agree. Informally, agreement lets us assess how closely two aesthetic functions are aligned with each other. Of course, they may be in close agreement for very different reasons—this metric simply establishes similarity in the result of the subjective judgements.

Hypothesis 7 The perception of creativity in software which produces artefacts within a creative community will be increased if the software can exhibit subjective judgements about its own work and that of others, and defend those judgements in an accountable way. This can be seen as part of a bigger picture of software exhibiting a personality, in order to be accepted into a creative community.

1.6.2 A Case Study in Automatically Designed Videogames

A *game jam* is a contest where entrants attempt to make a videogame from scratch in a short period of time, normally with the added restriction of a theme which developers must incorporate into their game somehow. Ludum Dare is one of the largest regularly occurring game jams in the game development community, taking place three times a year and garnering over 2,000 entries in December 2013, where developers were given the theme 'You Only Get One'. The ANGELINA system is an automated videogame designer developed to investigate issues surrounding Computational Creativity in a ludic and interactive context [56]. Many different versions of ANGELINA have been developed, working with various different kinds of game, technologies and user guidance. The most recent iteration, ANGELINA-5, was designed to enter game jams, by allowing it to be given just a theme in plain text as a starting point. This theme is then interpreted by ANGELINA-5 and used to influence the design of the game.

ANGELINA-5 entered Ludum Dare for the first time in December 2013, the 28th edition of the event. One of the objectives was to investigate the reactions of various groups of people to a piece of creative software entering such a contest. To gain more insight into these groups, we entered two games designed by ANGELINA-5 to Ludum Dare 28. In the first submission, *To That Sect*,[3] we included a commentary generated by ANGELINA-5 to illustrate the actions of the system, as well as multiple paragraphs describing the research behind ANGELINA-5 and identifying the game as the creation of a piece of software. In the second submission,[4] we anonymised ANGELINA-5's commentary to remove references to it being software-based, edited it for grammar, and added no supplementary explanation about the software, the

[3] *To That Sect* game: www.tinyurl.com/tothatsect.

[4] *Stretch Bouquet Point* game: www.tinyurl.com/stretchpoint.

Table 1.2 Percentile rankings for ANGELINA-5's two games entered into Ludum Dare 28, and its single entry to Ludum Dare 29 *(Jet Force Gemini)*

	To That Sect	Stretch Bouquet Point	Jet Force Gemini
Overall	36	29	23
Fun	34	30	26
Audio	73	43	74
Graphics	43	33	36
Mood	77	39	80
Innovation	64	33	59
Theme	32	30	26
Humour	48	59	51

Note that higher percentile rankings indicate higher achievements. There were 780 submissions in the LD28 track, and 1,004 entries in the LD29 track

origin of the game, or anything to connect the game with a digital author. The ratings process for Ludum Dare takes place in the 22 days following the contest, and is conducted as a peer review system, where each entrant is asked to rate and review games by other entrants. Ratings are given as marks out of five for eight categories: Audio, Graphics, Mood, Theme, Humour, Fun, Innovation and Overall.

The results for the two entries by ANGELINA-5 can be seen in Table 1.2. While we were unable to get specific vote data, we do know that 70 people rated *To That Sect*, the non-anonymised submission, while 26 people rated *Stretch Bouquet Point*.[5] While it is impossible to calculate confidence intervals for these ratings without the vote data, we can see that they differ by hundreds of positions for some categories such as Mood and Audio. We can also see a noticeable difference in the comments left by some of the reviewers underneath both submissions, in terms of their tone and attitude when dealing with each game. Many commentators indirectly criticise the anonymised game, with comments such as "You made me feel something there. Don't make me put it into words though". Other commentators made more obvious statements of criticism or praise, such as "This was a rather annoying experience" or "This game feels dreamy. The audio is intense." Only one comment included both praise and criticism. We attribute the indirect or sarcastic comments to an unwillingness to potentially criticise a person for performing poorly, even though other reviewers were less tactful. Ludum Dare is often used as a learning experience for amateur developers, and many children enter using simple game creation tools. We believe many reviewers felt uncomfortable with direct criticism for this reason.

By contrast, comments on *To That Sect* were more balanced in nature, often offering both praise and criticism in equal amounts, e.g., "Angelina seems really good at creating an atmosphere with both sound and visuals. But the game part of it seems a bit lacking still". In the description of the game, we asked people to rate it as they would any other Ludum Dare entry, hoping to dissuade people from reviewing the

[5] This is due in part to ANGELINA-5's small following on the internet, which promoted the non-anonymised submission more than normal.

concept of ANGELINA-5 rather than the game itself. Nevertheless, many reviewers suggest that their scores were influenced by their appraisal of ANGELINA-5 as a novel system, rather than what it was capable of creating, e.g., "creating a program to create your game ... [is] certainly not something you see every day. On that front alone, this gets a lot of points for innovation". These results suggest that reviewers were unable to separate the creator from the artefact, and were incapable of reviewing the game as if created by a person. For instance, *To That Sect* rated 282nd of 780 for Innovation. These ratings are subjective, and it is hard for us to objectively assess them. However, we do not believe there is anything particularly innovative about *To That Sect*. As such, we must attribute this high ranking to reviewers assessing the game as a product of ANGELINA-5. It seems that reviewers projected (human) innovation in the ANGELINA project onto the game it produced.

We can compare the results of ANGELINA-5's debut in Ludum Dare with the results garnered from a second entry to the game jam in April 2014, Ludum Dare 29. This time ANGELINA-5 was only entered into the game jam once, with the game *Jet Force Gemini*, created in response to the theme "Beneath The Surface". As with the non-anonymised entry in Ludum Dare 28, *Jet Force Gemini* was entered with a commentary describing some of the decisions contributing to the design process. Table 1.2's rightmost column shows the results for *Jet Force Gemini* in contrast to the entries in Ludum Dare 28. The number of entries in Ludum Dare 29 was nearly 30 % larger than Ludum Dare 28; ANGELINA's percentile scores drop for four of seven specialised categories, and fall dramatically in the *Overall* rating.

We believe this is evidence of the relationship between the observers and ANGELINA shifting over time. While some of the comments underneath *Jet Force Gemini* indicate that the reviewer is encountering ANGELINA-5 for the first time (which is unsurprising, since the number of reviewers account for less than 1 % of total Ludum Dare entrants) others explicitly note that they are reviewing ANGELINA-5's games for a second time. One states that 'I'm sorry to say that I can't really see improvements from last time', indicating that there is either an expectation of growth on the part of the software, or an expectation that the software's author will grow the software over time. Despite many of the other comments being generally positive, the drop in ratings suggests that people perhaps feel less compelled to rate ANGELINA-5 highly for novelty value alone. Given that Ludum Dare is a community built on the idea of improving creative skills through regular practice, it is interesting to note the expectation of growth shown by some reviewers. We hypothesise that this may be a factor which is particularly important for creative individuals in assessing creativity, as opposed to other types of observer.

We can also examine reactions to particular elements of ANGELINA-5's work and compare it to critiques of similar games. One comment on *To That Sect* states "If it [had] added shooting at the statues that you must avoid and a goal how much ships you have to collect, it would have been better. It felt like playing [an] 'art-message' type of game". *LITH*[6] is a game entered into the competition by a human designer, where the player navigates a maze and collects bags of gold coins, while avoiding

---

[6] *LITH* game: www.tinyurl.com/lith-ludum.

patrolling robots. They can escape to an exit at any stage, with their score being the amount of gold collected. While not an exact duplicate, the rules of *LITH* are very similar to those of *To That Sect*, i.e., search for as many objects of a certain type as possible, while avoiding another object, then exit. *LITH* was entered in the same track as ANGELINA-5's games, and ranked 95th Overall, 125th for Fun, and 274th for Theme. None of the comments on *LITH* reference the game's rulesets in a critical way. Notably, *LITH* ranks 259 places above *To That Sect* for Theme. This is significant, as the *LITH* designer justifies its theme in a fairly thin way, by saying simply that the player only has one opportunity to save their score (which they do by ending the game, as in *To That Sect*). The games are by no means identical: *LITH*'s level is more closed in to accentuate a feeling of claustrophobia, but the similarities are many. This analysis suggests a fundamental difference in how people evaluate a game when they have knowledge and when they have no knowledge of its designer and design process.

Hypothesis 8 There can be both positive and negative biases at work when people consume artefacts in the knowledge that computers created them. By managing both cases in a creative community context, we can increase perception of software as being creative and enjoyment of the artefacts produced. This increase will be further fuelled if the software shows clear growth in sophistication in the field, and expresses this through its processes and products.

1.7 Conclusions and Future Work

Simply stated, one of the main aims of research into Computational Creativity is to one day see creative software properly embedded into society. To achieve this aim, larger sectors of society need to join the effort, including creative communities within the arts and sciences, the creative industries, technology firms, and the next generation of Artificial Intelligence researchers. Hence, we need to convince certain sets of stakeholders that creative software is no fantasy, but a potential reality that will bring benefits to society. As described above, we have studied three sets of stakeholders, namely the general public, fellow Computational Creativity researchers, and a specific community of creative people, namely videogame designers. These studies have enabled us to make concrete hypotheses related to how stakeholder communities perceive creativity in software, and how best to manage that perception in the future. Based on our immersion in the stakeholder communities mentioned, we have argued above in favour of the truth of the hypotheses, with extended discourse and argumentation given in [38, 47] amongst other papers. We believe it is now time to turn the hypotheses into experiments designed to see whether the ways in which sets of stakeholders perceive and react to creative software fit our beliefs.

Our first hypothesis is pitched somewhat at a meta-level, in that it proposes that different stakeholder groups see creative systems differently and their perception of software behaviour could and should be managed in a bespoke manner. We can therefore imagine an experiment where we present the processes and products from creative software to different stakeholder groups and assess their reaction to see if there is indeed a difference in how different groups react, learning from analyses of the results. Hypothesis 2 encompasses much of our philosophical position on the notion of creativity being essentially contested and secondary in nature. One can imagine restricting participants in an experiment to fairly constrained groups, and testing whether there is general (healthy) disagreement about the nature of creativity in people and software or not, and further testing whether there is more consensus about software being uncreative. To properly test Hypothesis 2, we would need to ask participants about the essential behaviours—such as intentionality, learning and reflection—they perceive to be taking place in software and see how it affects their perception of uncreativity in the system.

Our third hypothesis makes a bold statement: that blind comparison tests damage the long-term goal of embedding creative software in society, by emphasising the evident humanity gap. If this effect is true, it would be borne out by a Turing-style test where, when people are told that it was software that produced an artefact that they particularly liked, they were also asked about whether their perception of the creative act and/or the artefact had changed in light of the new knowledge. More pointed questions about the nature of any change in perception could lead to insights about how to manage the humanity gap in future projects. This would lead into an experiment to address Hypothesis 4, where computer generated artefacts were presented as re-imagined pieces with specific management of the relative lack of humanity in the generation of the artefacts. The re-imagining would specifically include commentaries and other framing information produced by the creative system. If Hypothesis 4 is correct, people would appreciate the re-imagined versions of artefacts more than those presented merely as computer-generated versions from the human oeuvre.

By proposing that random number generation detracts from an experience of a creative act, whereas more accountable unpredictability can benefit the experience, Hypothesis 5 is more specific than those preceding it. We can imagine an experiment where one set of participants are told that a particularly impressive creative act (in terms of the processing performed and/or the resulting artefacts) was because of a random event, and another set are given interesting framing information about what led—in a non-random way—to the same unpredictably good creative act. If the latter group appreciated the creative act and its results more than the former group, the truth of the hypothesis would be upheld.

We have already started work on testing Hypothesis 6, i.e., that the formalism presented in [47], can capture notions of progress when building creative systems. That is, we have used the formalism to capture abstracted timelines leading to the building of certain creative systems, and timelines where that software operates and produces artefacts of value. However, to convince the Computational Creativity researcher stakeholders of the value of the formalism, we need to work with them to capture the essence of their approaches to implementing and operating creative software. Moreover, our audience evaluation model is far from complete. We plan to employ the criteria specified in [55], for more fine-grained evaluations of the quality, novelty and typicality of artefacts. We will also import audience reflection evaluation schemes from the IDEA descriptive model, e.g., change in well-being, cognitive effort and emotional responses such as surprise and amusement.

The final two hypotheses we present above relate to communities of creative people into which creative software is implanted. To address Hypothesis 7, we will need to implement software behaviours which can meaningfully be described as subjective, and we plan to do so with the ANGELINA videogame generation system, and others such as The Painting Fool automated artist. With such systems, we can experiment to see whether members of the creative community are more impressed by subjective software or not. Such an experiment could be simultaneously used to address the final hypothesis, with knowledge of the computational origins of artefacts systematically withheld in order to see whether positive or negative biases hold in different creative communities. Similarly, experiments where participants are told about the intellectual growth of a system could be carried out, to see if this influences their impression of the software. An analysis of the findings from such experiments could help pave the way for software to be full members of these kinds of communities.

Looking at the three stakeholder groups studied here, we see some emerging generalities. In particular, looking at behaviours where systems exhibit subjectivity and intentionality, it seems clear that in all three groups, *personality modelling* in software has the potential to increase the impression that people have of what software does and, in turn, what it produces. This is part of a new understanding of creative acts as being potentially interesting, even dramatic, episodes of activity which can amuse and engage people, rather than a means to the end of producing an artefact of value. This is in contrast with the traditional idea that the value of the output from software can increase people's appreciation of the creativity it exhibits. While the traditional view is often correct, it is not the only model of managing perceptions of creativity in software.

The hypotheses presented here are only a subset of those which should be proposed and addressed in the future of Computational Creativity research. Not addressing such issues would be a mistake, as stakeholder perception of creativity in software will in part dictate the number of researchers and businesses coming into the field. Done badly, handling of stakeholder perceptions could stall the forward progress achieved towards embedding creative software in society. As a recent controversial example, online retailer Amazon briefly sold T-shirts with slogans such as *"Keep Calm and Rape a Lot"* [57]. The T-shirt company responsible posted an apology on its website, and insisted that the offending articles were "automatically generated using a scripted computer process running against hundreds of thousands of dictionary words." This may be the first example of computer generated artefacts causing such offence and a company—while taking responsibility—blaming generative software for poor quality artefacts, while tacitly acknowledging that the software had taken on unsupervised creative responsibilities in their workplace.

Situations where software is employed independently for creative purposes in commerce and elsewhere are likely to become more commonplace in the future. As a more positive example, IBM researchers have recently undertaken research to explore the commercial potential of Computational Creativity [58], with particular emphasis on culinary creativity [59, 60]. Creative software will make great inventions and make terrible mistakes in the future, and this will lead to a re-evaluation of humanity as being the centre of the creativity universe. Managing stakeholder perceptions of creativity in software will be paramount in making this transition as smooth and as fruitful for society as possible.

Acknowledgments Some of the work presented here was originally explored in [38, 47], and we are very grateful to the organisers of the AISB 2014 symposium on Computing and Philosophy, and the organisers of the 2014 International Conference on Computational Creativity. We wish to thank the many researchers with whom we have discussed the views presented in this chapter, especially our colleagues in the Computational Creativity group at Goldsmiths College. This research has been funded by EPSRC grants EP/L00206X and EP/J004049, and with the financial support of the Future and Emerging Technologies (FET) programme within the Seventh Framework Programme for Research of the European Commission, under FET-Open Grant numbers: 611553 (COINVENT) and 611560 (WHIM).

References

1. Locke, J.: An Essay Concerning Human Understanding. Oxford University Press, Oxford (1975)
2. Dennett, D.: Three kinds of intentional psychology. In: Perspectives in the Philosophy of Language: A Concise Anthology, pp. 163–186. Broadview Press, Peterborough (2000)
3. Gallie, W.: Essentially Contested Concepts. Proc. Aristot. Soc. **56**, 167–198 (1956)
4. Gray, J.N.: On the contestability of social and political concepts. Polit. Theory **5**(3), 331 (1977)
5. Smith, K.: Mutually contested concepts and their standard general use. J. Class. Sociol. **2**(3), 329–343 (2002)
6. Plucker, J.A., Makel, M.C.: Assessment of creativity, The Cambridge Handbook of Creativity, pp. 48–73 (2010)
7. Jones, J.: Santa bought me a Playstation. But it's still not art. The Guardian. 7th January (2014)
8. Stuart, K.: Video games and art: why does the media get it so wrong? The Guardian. 8th January (2014)
9. Jordanous, A.K.: Evaluating Computational Creativity: A Standardised Procedure for Evaluating Creative Systems and its Application. PhD thesis, Department of Informatics, University of Sussex (2012)
10. Austin, J.L.: How to do Things with Words. Oxford University Press, Oxford (1975). The William James Lectures delivered at Harvard University in 1955
11. Searle, J.: A taxonomy of illocutionary acts. In: Günderson, K. (ed.) Language, Mind, and Knowledge, vol. 7. University of Minnesota Press (1975)
12. Latour, B.: Science in Action: How to Follow Scientists and Engineers Through Society. Open University Press, Milton Keynes (1987)
13. Boden, M.A.: What is creativity? In: Boden, M.A. (ed.) Dimensions of Creativity, pp. 75–117. MIT Press, Cambridge (1996)
14. Boden, M.A.: The Creative Mind: Myths and Mechanisms. Weidenfield and Nicholson, London (1990)
15. Cardoso, A., Veale, T., Wiggins, G.A.: Converging on the divergent: the history (and future) of the international joint workshops in computational creativity. AI Mag. **30**(3), 15 (2009)
16. National Science Foundation, CreativeIT, Program Solicitation 09-572 http://www.nsf.gov/pubs/2009/nsf09572/nsf09572.htm (2009)
17. The Seventh Framework Programme (2007–2013) of the European Union: Information and Communication Technologies. http://cordis.europa.eu/fp7/ict/docs/ict-wp2013-10-7-2013-with-cover-issn.pdf (2013)

18. Lemaine, G., Macleod, R., Mulkay, M., Weingar, P.: Problems in the emergence of new disciplines. In: Lemaine, G., Macleod, R., Mulkay, M., Weingar, P. (eds.) Perspectives on the Emergence of Scientific Disciplines, pp. 1–26. Maison des Sciences de l'Homme (1976)
19. Colton, S., Ventura, D.: You can't know my mind: a festival of computational creativity. In: Proceedings of the Fifth International Conference on Computational Creativity (2014)
20. Ravetz, J.R.: Usable knowledge, usable ignorance: Incomplete science with policy implications. Knowl. Creat. Diffus. Util. **9**(1), 86–116 (1987)
21. Ravetz, J.R.: Scientific Knowledge and its Social Problems. Transaction Publishers, New Brunswick (1996)
22. Stocking, S.H., Holstein, L.W.: Manufacturing doubt: journalists' roles and the construction of ignorance in a scientific controversy. Public Underst. Sci. **18**, 23–42 (2009)
23. Ellegard, A.: Darwin and the General Reader: The Reception of Darwin's Theory of Evolution in the British Periodical Press, 1859–1872. University Of Chicago Press, Chicago (1990)
24. Franzen, M., Weingart, P., Rödder, S.: Exploring the impact of science communication on scientific knowledge production: an introduction. In: Rödder, S., Franzen, M., Weingart, P. (eds.) The Sciences' Media Connection—Public Communication and its Repercussions, pp. 3–16. Springer, Dordrecht (2012)
25. Arbib, M.A., Hesse, M.B.: The Construction of Reality. Cambridge University Press, Cambridge (1986)
26. Fahnestock, J.: Rhetorical Figures in Science. Oxford University Press, New York (1999)
27. Lakoff, G.: Why it matters how we frame the environment. Environ. Commun. **4**(1), 70–81 (2010)
28. Colton, S., Wiggins, G.A.: Computational creativity: the final frontier? In: Proceedings of the European Conference on AI (2012)
29. Johnson, C.: Is it time for computational creativity to grow up and be irresponsible? In: Proceedings of the Fifth International Conference on Computational Creativity (2014)
30. Eigenfeldt, A., Burnett, A., Pasquier, P.: Evaluating musical metacreation in a live performance context. In: Proceedings of the Third International Conference on Computational Creativity (2012)
31. Moffat, D., Kelly, M.: An investigation into people's bias against computational creativity in music composition. In: Proceedings of the Third Joint Workshop on Computational Creativity (2006)
32. Colton, S.: Creativity versus the perception of creativity in computational systems. In: Proceedings of the AAAI Spring Symposium on Creative Intelligent Systems (2008)
33. Bown, O.: Empirically grounding the evaluation of creative systems: incorporating interaction design. In: Proceedings of the Fifth International Conference on Computational Creativity (2014)
34. Turing, A.M.: Computing machinery and intelligence. Mind **59**, 433–460 (1950)
35. Wimsatt, W.K.: The Verbal Icon: Studies in the Meaning of Poetry. Number 123. University Press of Kentucky, Kentucky (1954)
36. Evans, K.: The Stuckists: The First Modernist Art Group. Victoria Press (2000)
37. Lambourne, L.: The Aesthetic Movement. Phaidon, London (1996)
38. Colton, S., Cook, M., Hepworth, R., Pease, A.: On acid drops and teardrops: observer issues in computational creativity. In: Proceedings of the 7th AISB Symposium on Computing and Philosophy (2014)
39. Pease, A., Colton, S.: On impact and evaluation in computational creativity: a discussion of the turing test and an alternative proposal. In: Proceedings of the AISB Symposium on Computing and Philosophy (2011)
40. Charnley, J., Pease, A., Colton, S.: On the notion of framing in computational creativity. In: Proceedings of the Third International Conference on Computational Creativity (2012)
41. Colton, S., Goodwin, J., Veale, T.: Full FACE poetry generation. In: Proceedings of the Third International Conference on Computational Creativity (2012)
42. Pease, A., Colton, S., Ramezani, R., Charnley, J., Reed, K.: A discussion on serendipity in creative systems. In: Proceedings of the Fourth International Conference on Computational Creativity (2012)

43. Colton, S.: The painting fool: stories from building an automated painter. In: McCormack, J., d'Inverno, M. (eds.) Computers and Creativity, pp. 3–38. Springer, Berlin (2012)
44. Norton, D., Heath, D., Ventura, D.: Finding creativity in an artificial artist. J. Creat. Behav. **47**(2), 106–124 (2013)
45. Colton, S., Pease, A., Charnley, J.: Computational creativity theory: the FACE and IDEA descriptive models. In: Proceedings of the Second International Conference on Computational Creativity (2011)
46. Pease, A., Colton, S.: Computational creativity theory: inspirations behind the FACE and the IDEA models. In: Proceedings of the Second International Conference on Computational Creativity (2011)
47. Colton, S., Pease, A., Corneli, J., Cook, M., Llano, T.: Assessing progress in building autonomously creative systems. In: Proceedings of the Fifth International Conference on Computational Creativity (2014)
48. Colton, S.: Automated Theory Formation in Pure Mathematics. Springer, London (2002)
49. Colton, S.: Experiments in meta-theory formation. In: Proceedings of the AISB'01 Symposium on Artificial Intelligence and Creativity in Arts and Science (2001)
50. Romero, J., Machado, P.: The Art of Artificial Evolution: A Handbook on Evolutionary Art and Music. Springer, Berlin (2008)
51. Machado, P., Cardoso, A.: All the truth about NEvAr. Appl. Intell. **16**(2), 101–118 (2002)
52. Li, Y., Hu, C., Minku, L., Zuo, H.: Learning aesthetic judgements in evolutionary art systems. Genet. Program. Evol. Mach. **14**(3), 315–337 (2013)
53. Colton, S.: Evolving a library of artistic scene descriptors. In: Proceedings of the EvoMusArt Conference (2012)
54. Colton, S.: Automatic invention of fitness functions with application to scene generation. In: Proceedings of the EvoMusArt Workshop (2008)
55. Ritchie, G.: Some empirical criteria for attributing creativity to a computer program. Minds Mach. **17**(1), 67–99 (2007)
56. Cook, M., Colton, S., Gow, J.: Automating game design in three dimensions. In: Proceedings of the AISB Symposium on AI and Games (2014)
57. McVeigh, T.: Amazon acts to halt sales of 'Keep Calm and Rape' T-shirts. The Guardian. 2nd March (2013)
58. Jagmohan, A., Li, Y., Shao, N., Sheopuri, A., Wang, D., Varshney, L., Huang, P.: Exploring application domains for computational creativity. In: Proceedings of the Fifth International Conference on Computational Creativity (2014)
59. Pinel, F., Varshney, L., Bhattacharjya, D.: A culinary computational creativity system. In: This Edition (2014)
60. Shao, N., Murali, P., Sheopuri, A.: New developments in culinary computational creativity. In: Proceedings of the Fifth International Conference on Computational Creativity (2014)

Chapter 2
Weak and Strong Computational Creativity

Mohammad Majid al-Rifaie and Mark Bishop

Abstract In the spirit of Searle's definition of weak and strong artificial intelligence, this paper presents a discussion on weak computational creativity in swarm intelligence systems. It addresses the concepts of *freedom* and *constraint* and their impact on the creativity of the underlying systems. An analogy is drawn on mapping these two 'prerequisites' of creativity onto the two well-known phases of exploration and exploitation in swarm intelligence algorithms, followed by the visualisation of the behaviour of the swarms whose performance are evaluated in the context of arguments presented. The paper also discusses that the strong computational creativity is presented in ways emphasising that genuine creativity implies 'genuine understanding' and other cognitive states, along with autonomy—asserting that without 'Strong Embodiment', computational systems are not genuinely autonomous.

2.1 Introduction

In recent years, studies of the behaviour of social insects (e.g. ants and bees) and social animals (e.g. birds and fish) have proposed several new metaheuristics for use in collective intelligence resulting from social interaction.

Among the many works in the fields are research on swarm painting (e.g. [8, 26, 37, 38]), ant colony paintings (e.g. [21, 25, 34]) and other multi-agent systems (e.g. RenderBots [32] and the particle-based non-evolutionary approach of Loose and Sketchy Animation [15]).

In most of the swarm-based work mentioned above (e.g. [8, 21, 26, 37, 38]), the painting process does not re-work an initial drawing, but rather focuses on presenting "random artistic patterns", somewhere between order and chaos [38]. Other classes of research (e.g. by Schlechtweg et al. [32] and Curtis [15]) are based on reworking an initial drawing. There is a significant number of related papers in the area of

M.M. al-Rifaie (✉) · M. Bishop
Goldsmiths, University of London, London SE14 6NW, UK
e-mail: m.majid@gold.ac.uk

M. Bishop
e-mail: m.bishop@gold.ac.uk

T.R. Besold et al. (eds.), *Computational Creativity Research: Towards Creative Machines*, Atlantis Thinking Machines 7, DOI 10.2991/978-94-6239-085-0_2

non-photorealistic rendering; particularly, many papers approach drawing and painting using the optimisation framework (where optimisation and generative techniques are utilised an artistic context). Furthermore, particles have been used for stippling and other aesthetic styles in numerous papers. Turk and Bank's work [36] is an early example of optimising particle positions to control a stroke-based rendering. Hertzmann [22] optimised a global function over all strokes using a relaxation approach. In one of his works, Collomosse [14] used a global genetic algorithm to define a rendering algorithm. More recently, Zhao et al. [41] deployed an optimisation-based approach to study the stroke placement problem in painterly rendering, and presented a solution named stroke processes, which enables intuitive and interactive customisation of painting styles.

This work is an extension of ideas first presented at the Computing and Philosophy symposium at AISB 2011 [2] and subsequently published in the Cognitive Computation journal [6]. In the work discussed herein the impact of freedom and constraint on the concept of 'creativity' is discussed, followed by a discussion on the creativity of swarm intelligence systems. This paper also addresses the thorny issue of 'Weak' verses 'Strong' computational creativity.

2.2 On Art, Freedom and Creativity

For many years there has been discussions on the relationship between art, creativity and freedom; a debate elegantly encapsulated in the famous German prose by Ludwig Hevesi at the entrance of the Secession Building in Vienna:

> Der Zeit ihre Kunst
> Der Kunst ihre Freiheit

That is: "To Time its Art; To Art its Freedom".

which, centuries after, resonates an earlier observation from Aristotle (384–322 BCE) [18] emphasising the importance of freedom, giving rise to further exploration of areas otherwise left untouched (here, having "a tincture of madness") in presenting a creative act.

> There was never a genius without a tincture of madness.

On the other hand Margaret Boden, in [9], more recently argues that creativity has an ambiguous relationship with freedom:

> A style is a (culturally favoured) space of structural possibilities: not a painting, but a way of painting. Or a way of sculpting, or of composing fugues .. [] .. It's partly because of these [thinking] styles that creativity has an ambiguous relationship with freedom.

Considering the many factors constituting the evaluation of what is deemed 'creative', raises core issues regarding how humans evaluate creativity; their aesthetic capacity and potentially that of other animals (e.g. as exhibited in, say, mate-selection). Galanter [19] suggests that perhaps the 'computational equivalent' of a bird or an insect (e.g. in evaluating mate selection) is all that is required for [computational] aesthetic evaluation:

> This provides some hope for those who would follow a psychological path to computational aesthetic evaluation, because creatures with simpler brains than man practice mate selection.

In this context, as suggested in [16], the tastes of the individual in male bowerbirds are made visible when they gather collections of bones, glass, pebbles, shells, fruit, plastic and metal scraps from their environment, and arrange them to attract females [10]:

> They perform a mating dance within a specially prepared display court. The characteristics of an individual's dance or artefact display are specific to the species, but also to the capabilities and, apparently, the tastes of the individual.

However the question of whether *'mate selection behaviour in animals implies making a judgement analogous to aesthetic judgement in humans'* is perhaps (pace Nagel's famous discussion 'What is it like to be a bat? '[27]) a fundamentally unanswerable question.

In contrast, the role of education (or training) in recognising 'good' and 'bad', 'creative' and 'non-creative' has been experimentally probed. A suggestive study investigating this topic by Watanabe [39] gathers a set of children's paintings, and then adult humans are asked to label the "good" from the "bad". Pigeons are then trained through operant conditioning to only peck at good paintings. After the training, when pigeons are exposed to a novel set of already judged children's paintings, they show their ability in the correct classification of the paintings.

This emphasises the role of learning training and raises the question on whether humans are fundamentally trained (or "biased") to distinguish good and/or creative work.

Another tightly related topic to swarm intelligence in this context is the creativity of social systems. Bown in [11] indicates that our creative capabilities are contingent on the objects and infrastructure available to us, which help us achieve individual goals, in two ways:

> One way to look at this is, as Clark does [13], in terms of the mind being extended to a distributed system with an embodied brain at the centre, and surrounded by various other tools, from digits to digital computers. Another way is to step away from the centrality of human brains altogether and consider social complexes as distributed systems involving more or less cognitive elements.

Discussion on creativity and the conditions which make a particular work creative, have generated heated debate amongst scientists and philosophers for many years [31]; for a theoretical review on 'conditions of creativity'; the 'systems' view of creativity; cognitive approaches, etc. see also [35]. Although this article does not aim to resolve any of these issues (or even suggest that the presented work strongly fits and endorses the category of the 'computationally creative realm'), we investigate the performance of a swarm intelligence sketching system which, we suggest, highlights core issues inherent in exploring conceptual/artistic space(s).

2.3 Creativity in Swarms

2.3.1 Freedom versus Constraint

Freedom and constraint have been at the core of several definitions for creativity. Philip Johnson-Laird in his work on freedom and constraint in creativity [23] states:

> ... for to be creative is to be free to choose among alternatives .. [] .. for which is not constrained is not creative.

In swarm intelligence systems, the two phases of exploration and exploitation introduce the freedom and control the level of constraint. Pushing the swarms towards exploration, freedom is boosted; and by encouraging exploitation, constraint is more emphasised. Finding a balance between exploration and exploitation has been an important theoretical challenge in swarm intelligence research and over the years many hundreds of different approaches have been deployed by researchers in this field. In the presented work, two swarm intelligence algorithms are deployed: the algorithm which is responsible for the "intelligent" tracking of the line drawing is Particle Swarm Optimisation (PSO) [17, 24]. This well-known algorithm, which mimics the behaviour of birds flocking, has an internal mechanism of balancing off the exploitation and exploration phases. However due to the weakness of the exploration in this algorithm, our system also deploys another nature inspired algorithm to overcome this weakness, Stochastic Diffusion Search (SDS) [1], which mimics the behaviour of one species of ants (*Leptothorax acervorum*) foraging. Therefore, exploration is promoted by utilising the SDS algorithm, whose impact on different swarm intelligence algorithms has been scientifically reported using various measures and statistical analysis in several publications (e.g. [3–5, 7]).

In the visualisation, the swarms are presented with a set of points (which constitute a line drawing—see Fig. 2.1) and are set to consider these points (one at a time) as their global optimum. In other words, the global optimum is dynamic, moving from one position to another and the swarms aim to converge over this dynamic optimum (Fig. 2.2).

As stated in the introduction, there have been several relevant attempts to create creative computer generated artwork using Artificial Intelligence, Artificial Life and Swarm Intelligence. Irrespective of whether the swarms are considered genuinely creative or not, their similar individualistic approach is not totally dissimilar to those of the "elephant artists" [40]:

> After I have handed the loaded paintbrush to [the elephants], they proceed to paint in their own distinctive style, with delicate strokes or broad ones, gently dabbing the bristles on the paper or with a sweeping flourish, vertical lines or arcs and loops, ponderously or rapidly and so on. No two artists have the same style.

Similarly if the same line drawing (see Fig. 2.1) is repeatedly given to the swarms, the output sketches (e.g. Fig. 2.2) made by the swarms, are never the same (see Fig. 2.4

Fig. 2.1 This figure shows a series of points that make a line drawing; sample line drawing after one of Picasso's sketches

Fig. 2.2 A sketch produced by the swarms, using both SDS and PSO algorithms

to compare different sketches). In other words, even if the swarms process the same input several times, they will not make two identical sketches; furthermore, the outputs they produce are not merely randomised variants of the input. In order to demonstrate this claim qualitatively in an experiment, the output of the swarm-based system is compared against a simple randomised tracing algorithm, where

Fig. 2.3 A sketch produced by the swarms without SDS exploration

each point in the line drawing could be surrounded with lines at a random distance and direction.

In Fig. 2.3, only PSO algorithm is used to produce the sketch. This experiment is run in order to highlight the impact of removing exploration (i.e. 'freedom') which is otherwise induced by the SDS algorithm (Fig. 2.4).

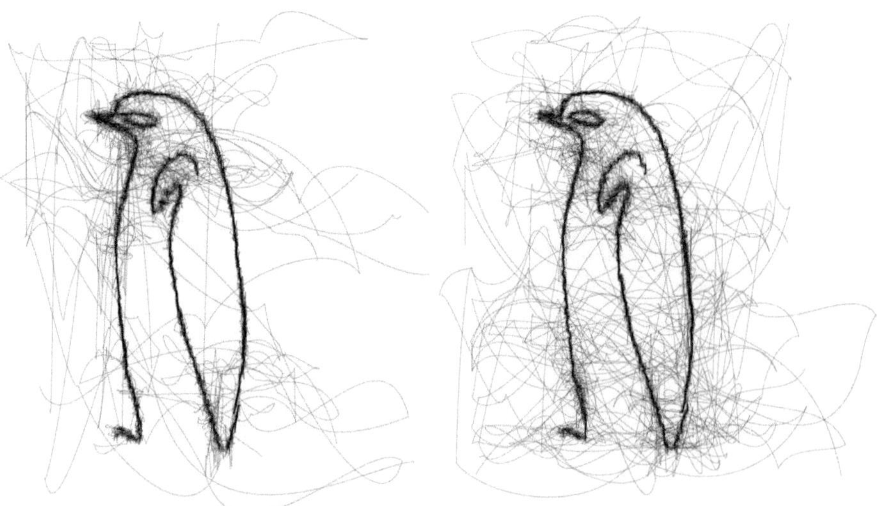

Fig. 2.4 Different sketches of the swarms off a single line drawing

2.3.2 Swarmic Freedom Versus Random Freedom

This part presents an experiment with the goal of contrasting the behaviour of the swarms to that of a group of random agents. In this experiment, the freedom of the swarm (i.e. *Swarmic Freedom*) is maintained by the swarm intelligence algorithms used in the system, whereas the freedom of the agents in the randomised algorithm is controlled by what we call the *Random Freedom*. These definitions are utilised here to highlight the potential of the swarms in exhibiting computational creativity.

The sketches in Fig. 2.5 (top and middle) show two outputs from a simple randomised algorithm when configured to exhibit limited 'random' variations in its

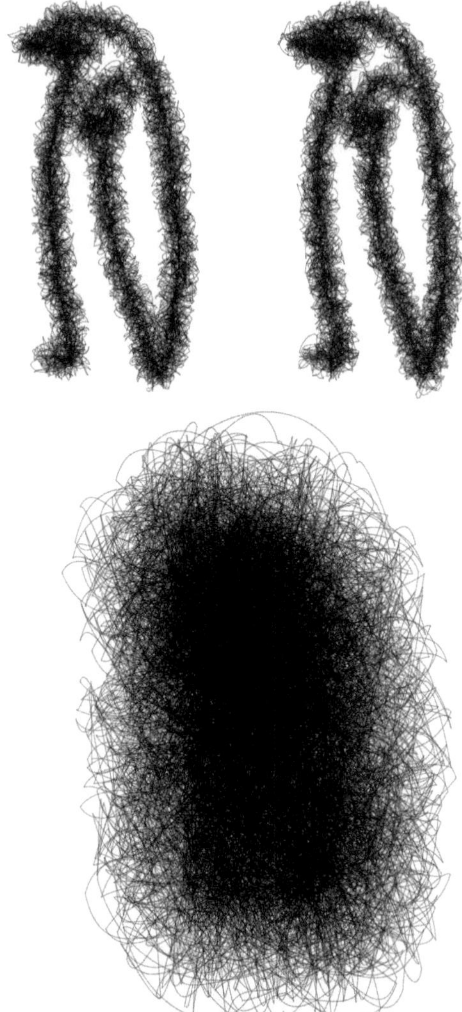

Fig. 2.5 The sketches of the swarms with random behaviour: This figure shows the sketches made with a simple randomised tracing algorithm, using random distance and direction from the lines of the original line drawing. The first two sketches (*top* and *middle*) use the same random distance (e.g. d) and the *bottom* sketch uses the random distance of $d \times 6$

behaviour (i.e. there is only small random distance and direction from the points of the original line drawing); comparing the two sketches, we note a lack of any significant difference between them. Furthermore, when more 'freedom' is granted to the randomised algorithm (by increasing the range in the underlying random number generator, which allows the technique to explore broader areas of the canvas), the algorithm soon begins to deviate excessively from the original line drawing. For this reason such randomisation results in a very poor—low fidelity—interpretation of the original line drawing (Fig. 2.5-bottom). In contrast, although the agents in the swarms are free to access any part of the canvas, the swarm-control mechanism (i.e. Swarm Freedom) naturally enables the system to maintain recognisable fidelity to the original input. In the randomised algorithm, contra the swarms system, it can be seen that simply by giving the agents more randomised behaviour (Random Freedom), they fail to produce more 'creative sketches'.

The Swarmic Freedom or 'controlled freedom' (or the 'tincture of madness') exhibited by the swarm algorithms (induced by the stochastic side of the algorithms) is crucial to the resultant work and is the reason why having the same line drawing does not result in the system producing identical sketches. This freedom emerges, among other influencing factors, from the stochasticity of the SDS algorithm in picking agents for communication, as well as choosing agents to diffuse information; the tincture of madness in the PSO algorithm is induced via its strategy of spreading the particles throughout the search space as well as the stochastic elements in deciding the next move of each particle.

In other words, the reason why the swarm sketches are different from the simple randomised sketches, is that the underlying PSO flocking component-algorithm constantly endeavours to accurately trace the input image whilst the SDS foraging component constantly endeavours to explore the wider canvas (i.e. together the two swarm mechanisms ensure high-level fidelity to the input without making an exact low-level copy of the original line drawing). Although the algorithms (PSO and SDS) are nature-inspired, we do not claim that the presented work is an accurate model of natural systems. Furthermore, whilst designing the algorithm there was no explicit 'Hundertwasser-like' attempt [30] by which we mean the stress on using curves instead of straight lines, as Hundertwasser considered straight lines not nature-like and tried not to use straight lines in his works to bias the style of the system's sketches.

2.4 Weak Versus Strong Computational Creativity

Before approaching the topic of weak or strong computational creativity, the difference between weak and strong AI is highlighted. In strong AI, the claim is that machines can think and have genuine understanding and other cognitive states (e.g. "suitably programmed machines will be capable of conscious thought" [12]); weak AI, in contrast, does not usually go beyond expecting the simulation of human intelligence. I.e. instantiating genuine "understanding" is not the primary concern in weak AI research.

An analogy could be drawn to computational creativity, extending the notion of weak AI to 'weak computational creativity', which does not go beyond exploring the simulation of human creativity; emphasising that genuine autonomy and genuine understanding are not the main issues in conceptualising weak computationally creative systems. Conversely in 'strong computational creativity', the expectation is that the machine should be autonomous, creative, have 'genuine understanding' and other cognitive states.

The concept of 'Strong AI'—instantiating a computational machine with genuine understanding—has provoked many critics, among whom John Searle made perhaps the most famous attack with his Chinese Room Argument (CRA) [33] (for comprehensive discussion see [29]). The central claim of the CRA is that computations alone cannot in principle give rise to understanding, and that therefore computational theories of mind cannot fully explain human cognition. More formally, Searle stated that the CRA was an attempt to prove that syntax (rules for the correct formation of sentences:programs) is not sufficient for semantics (understanding). Combining this claim with those that programs are formal (syntactical), whereas minds have semantics, led Searle to conclude that programs are not minds.

But then it is equally clear that Searle accepts that there is no boundary on a fundamental level to the idea that a machine can think as in Minds, Brains and Programs (MBP) Searle [33], in reply to the inquiry 'Can a machine think?", expressly states that the response is clearly 'yes', as we are precisely such machines.

Clearly Searle did not intend the CRA to target machine intelligence per se, but rather any form of artificial intelligence according to which a machine could have genuine mental states (e.g. understanding Chinese) purely in virtue of executing an appropriate series of computations: what Searle coined 'Strong AI'.

Searle contends that understanding, of say a Chinese story, can never emerge purely as a result of following the procedures prescribed by any machine program, for in the CRA Searle offers a first-individual story delineating precisely how he could instantiate such a system, functioning as the Central Processing Unit of a computer, produce right internal and external state transitions, and even pass a 'Turing test' for comprehension of Chinese, and yet still not comprehend a single word of Chinese.

In Mind, Brains and Programs, Searle [33] states that in Strong AI "the computer is not merely a tool in the study of the mind; rather the appropriately programmed computer really is a mind, in the sense that, computers given the right programs can be literally said to *understand* and have other cognitive states".

Accordingly, in Strong Computational Creativity, we argue that computer is not merely a tool in the study of the creativity; rather the appropriately programmed computer really is creative, in the sense that, computers given the right programs can be literally said to *understand its creation* and have other cognitive states (e.g. teleological and emotive) associated with human creative processes.

In other words, the creative human fundamentally *knows* that s/he is making a mark on paper (or a computer screen) or moulding clay or making sound etc. Whereas merely by following the rules of the program, the computer cannot know that it is doing anything other than manipulating uninterpreted symbols (or squiggles).

We fundamentally suggest that Searle's famous thought experiment also targets the analogous notion of 'strong computational creativity'. I.e. Searle using a similar "room" could get so good at following the rules that the strings of symbols he outputs from the room successfully control a 'Strong' computationally controlled creative art-system, producing works judged to have artistic merit by people outside the room, even though Searle-in-the-room remains ignorant of the produced art and the externally labelled 'art practise'. To paraphrase Dennett's deployment of 'the intentional stance', the computational system is merely instantiating a form of 'as-if creativity' without any real cognitive states, meaning or intentionality. I.e. Any 'creativity' ascribed to the computational system is merely a reflection of the engineer that designed it and the user who operated it in a given social nexus.

2.5 The Body in Question

In our opinion Searle's Chinese room argument suggests that to take the notion of 'Strong creativity' seriously we need to move away from purely computational explanations of creativity and look at how human meaning and creative processes are fundamentally grounded in the human body and society; taking the body, issues of embodiment and our social embedding, much more seriously. And this 'strong' notion of embodiment cannot simply be realised by opting a putative computationally creative system onto a conventional 'tin can robot'.[1]

As Nasuto, Bishop et al. [28] a fortiori argue in their discussion of *Biologically controlled animats*[2] and the so-called *'Zombie' animals*[3] (two examples carefully chosen to lie at polar ends of the spectrum of possible engineered robotic/cyborg systems), merely instantiating appropriate sensorimotor coupling is not sufficient to instantiate meaningful intentional states, as in both cases the induced behavioural couplings are not the effect of the intrinsic nervous system's constraints (metabolic or otherwise) at any level. On the contrary, they are actually the *cause* of extrinsic metabolic demands (made via the experimenter's externally directed manipulations). Since the experimenter drives the sensorimotor couplings in an arbitrary way (from the perspective of the intrinsic metabolic needs of animal or its cellular constituents), the causal relationship between the bodily milieu and the motor actions and sensory readings can never be genuinely and appropriately coupled. Thus Nasuto and Bishop [28] assert that only the 'right type' and 'directionality' of sensorimotor couplings can ultimately lead to genuine understanding and intentionality.

---

[1] Whereby a robot body is imply bolted on to an appropriate AI and the material of that 'embodiment' is effectively unimportant.

[2] Robots controlled by a cultured-array of real biological neurons.

[3] E.g. An animal whose behaviour is 'remotely-controlled', by an external experimenter, say by optogenetics; see also Gradinaru et al. [20], who used optogenetic techniques to stimulate neurons selectively, inducing motor behaviour without requiring conditioning.

For example, considering the case of the optogentically remotely controlled rat, even though the animal still possesses a fully functional biological body (and, arguably, a functioning brain), the fact that it receives external [optogenetic] commands cannot give rise to a genuine understanding of what the animal is experiencing; the rat's brain receives alien input that, at best, may resemble drug-induced decontextualised hallucinations—mere uninterpreted symbols/squiggles and squoggles—which will remain meaningless despite how accustomed the animal may become to this new mode of remotely induced behaviour. The situation is analogous to the alien hand syndrome—where in patients, for example, see their arm perform actions not of their volition or under their control. Such patients never ascribe meaning as to why their arm acted in this way, albeit they can see (and hence comprehend) the actions in the same way as any other observer; in this sense they are 'external observers' of their own limb(s) 'zombie movements'.

In contrast we suggest that fully intentional creative acts, as engaged in the process of genuinely understanding the world, require both a fully functional brain and a fully functional body; so translating the question of the requirements for 'genuine creative thinking' to the defining features of the processes whereby the brain and body interact with the world and society.

In the light of these concerns, until the challenge of the CRA has been *fully* met, we urge caution in ascribing 'Strong' notions of creativity to any fundamentally computational system.

2.6 Conclusion

In this paper, we have discussed the potential of the swarms in exhibiting 'weak computational creativity'. This specific work described herein uses swarm intelligence techniques to explore the difference between using Random Freedom and Swarmic Freedom in the visualisation of the swarms 'tracing' line drawings; this work highlights the features of swarm-regulated difference versus simple-random difference in the production of such 'sketches' by computer. We stressed on the significant impact of both freedom and constraint on the emergent creativity, and presented a discussion on how these two concepts are mapped onto exploration and exploitation, two of the most infamous phases in the swarm intelligence world.

The 'weak computationally creative' artist described herein is the result of merging two swarm intelligence algorithms, preserving freedom (exploration) and constraint (exploitation).

The Strong computational creativity is presented in ways emphasising:

- that genuine creativity implies 'genuine understanding' and other cognitive states.
- autonomy, asserting that without 'Strong Embodiment', computational systems are not genuinely autonomous.

Therefore, we urge caution in ascribing "Strong creativity" to any merely computational system.

References

1. al-Rifaie, M.M., Bishop, M.: Stochastic diffusion search review. Paladyn, J. Behav. Robot. **4**, 155–173 (2013). doi:10.2478/pjbr-2013-0021
2. al-Rifaie, M.M., Bishop, M., Aber, A.: Creative or not? Birds and ants draw with muscles. In: AISB 2011: Computing and Philosophy, pp. 23–30. University of York, York (2011). ISBN: 978-1-908187-03-1
3. al-Rifaie, M.M., Bishop, M., Blackwell, T.: An investigation into the use of swarm intelligence for an evolutionary algorithm optimisation. In: International Conference on Evolutionary Computation Theory and Application (ECTA 2011). IJCCI (2011)
4. al-Rifaie, M.M., Bishop, M., Blackwell, T.: Information sharing impact of stochastic diffusion search on differential evolution algorithm. J. Memet. Comput. **4**(4), 327–338 (2012). doi:10.1007/s12293-012-0094-y
5. al-Rifaie, M.M., Bishop, M., Blackwell, T.: Resource allocation and dispensation impact of stochastic diffusion search on differential evolution algorithm. In: Pelta, D., Krasnogor, N., Dumitrescu, D., Chira, C., Lung, R. (eds.) Nature Inspired Cooperative Strategies for Optimization (NICSO 2011), Studies in Computational Intelligence, vol. 387, pp. 21–40. Springer, Berlin Heidelberg (2012). doi:10.1007/978-3-642-24094-2_2
6. al-Rifaie, M.M., Bishop, M., Caines, S.: Creativity and autonomy in swarm intelligence systems. J. Cognit. Comput. **4**(3), 320–331 (2012). doi:10.1007/s12559-012-9130-y
7. al-Rifaie, M.M., Bishop, M.J., Blackwell, T.: An investigation into the merger of stochastic diffusion search and particle swarm optimisation. In: Proceedings of the 13th Annual Conference on Genetic and Evolutionary Computation, GECCO'11, pp. 37–44. ACM, New York (2011). doi:10.1145/2001576.2001583
8. Aupetit, S., Bordeau, V., Monmarche, N., Slimane, M., Venturini, G.: Interactive evolution of ant paintings. In: The 2003 Congress on Evolutionary Computation, 2003. CEC'03, vol. 2, pp. 1376–1383 (2004)
9. Boden, M.: Creativity and Art: Three Roads to Surprise. Oxford University Press, Oxford (2010)
10. Borgia, G.: Complex male display and female choice in the spotted bowerbird: specialized functions for different bower decorations. Anim. Behav. **49**, 1291–1301 (1995)
11. Bown, O.: Generative and adaptive creativity. In: McCormack, J., d'Inverno, M. (eds.) Computers and Creativity. Springer, Berlin (2011)
12. Callan, R.: Artificial Intelligence. Palgrave Macmillan (2003)
13. Clark, A.: Natural-Born Cyborgs: Minds, Technologies, and the Future of Human Intelligence. Oxford University Press, Oxford (2003)
14. Collomosse, J., Hall, P.: Genetic paint: a search for salient paintings. Appl. Evol. Comput. 437–447 (2005)
15. Curtis, C.J.: Loose and sketchy animation. In: ACM SIGGRAPH 98 Electronic Art and Animation Catalog, p. 145 (1998)
16. Dorin, A., Korb, K.: Creativity refined. in computers and creativity. In: McCormack, J., d'Inverno, M. (eds.) Computers and Creativity. Springer, Berlin (2011)
17. Eberhart, R., Kennedy, J.: A new optimizer using particle swarm theory. In: Proceedings of the Sixth International Symposium on Micro Machine and Human Science, vol. 43. IEEE, New York (1995)
18. Etzioni, A., Ben-Barak, A., Peron, S., Durandy, A.: Ataxia-telangiectasia in twins presenting as autosomal recessive hyper-immunoglobulin m syndrome. IMAJ **9**(5), 406 (2007)
19. Galanter, P.: Computational aesthetic evaluation: past and future. In: McCormack, J., d'Inverno, M. (eds.) Computers and Creativity. Springer, Berlin (2011)
20. Gradinaru, V., Thompson, K.R., Zhang, F., Mogri, M., Kay, K., Schneider, M.B., Deisseroth, K.: Targeting and readout strategies for fast optical neural control in vitro and in vivo. J. Neurosci. **26:27**(52), 14,231–14,238 (2007)
21. Greenfield, G.: Evolutionary methods for ant colony paintings. APPLICATIONS OF EVOLUTIONARY COMPUTING, PROCEEDINGS **3449**, 478–487 (2005)

22. Hertzmann, A.: Paint by relaxation. In: Computer Graphics International 2001. Proceedings, pp. 47–54. IEEE (2001)
23. Johnson-Laird, P.N.: Freedom and constraint in creativity. In: Sternberg, R.J. (ed.) The Nature of Creativity: Contemporary Psychological Perspectives, pp. 202–219. Cambridge University Press, Cambridge (1988)
24. Kennedy, J., Eberhart, R.C.: Particle swarm optimization. In: Proceedings of the IEEE International Conference on Neural Networks, vol. IV, pp. 1942–1948. IEEE Service Center, Piscataway, NJ (1995)
25. Monmarche, N., Aupetit, S., Bordeau, V., Slimane, M., Venturini, G.: Interactive evolution of ant paintings. In: B. McKay et al. (eds.) 2003 Congress on Evolutionary Computation, vol. 2, pp. 1376–1383. IEEE Press (2003)
26. Moura, L., Ramos, V.: Swarm paintings-nonhuman art. ARCHITOPIA Book, Art, Architecture and Science, pp. 5–24 (2007)
27. Nagel, T.: What is it like to be a bat? Philos. Rev. **83**(4), 435–450 (1974)
28. Nasuto, S., Bishop, J., Roesch, E., Spencer, M.: Zombie mouse in a chinese room. Philos. Technol. 1–15 (2014). doi:10.1007/s13347-014-0150-2
29. Preston, J., Bishop, M.: Views into the Chinese Room: New Essays on Searle and Artificial Intelligence. Oxford University Press, Oxford (2002)
30. Restany, P.: Hundertwasser: The Painter-King with the Five Skins: The Power of Art. Taschen America Llc (2001)
31. Rothenberg, A., Hausman, C.: The Creativity Question. Duke University Press, Durham (1976)
32. Schlechtweg, S., Germer, T., Strothotte, T.: Renderbots-multi-agent systems for direct image generation. Comput. Graph. Forum **24**, 137–148 (2005)
33. Searle, J.: Minds, brains, and programs. Behav. Brain Sci. **3**(3), 417–457 (1980)
34. Semet, Y., O'Reilly, U.M., Durand, F.: An interactive artificial ant approach to non-photorealistic rendering. In: Genetic and Evolutionary Computation-GECCO 2004, pp. 188–200 (2004)
35. Sternberg, R.: The Nature of Creativity: Contemporary Psychological Perspectives. Cambridge University Press, Cambridge (1988)
36. Turk, G., Banks, D.: Image-guided streamline placement. In: Proceedings of the 23rd Annual Conference on Computer Graphics and Interactive Techniques, pp. 453–460. ACM (1996)
37. Urbano, P.: Playing in the pheromone playground: experiences in swarm painting. Appl. Evol. Comput. 527–532 (2005)
38. Urbano, P.: Consensual paintings. Appl. Evol. Comput. 622–632 (2006)
39. Watanabe, S.: Pigeons can discriminate "good" and "bad" paintings by children. Anim. Cognit. **13**(1) (2009)
40. Weesatchanam, A.M.: Are Paintings by Elephants Really Art? The Elephant Art Gallery (2006)
41. Zhao, M., Zhu, S.: Customizing painterly rendering styles using stroke processes. In: Proceedings of the ACM SIGGRAPH/Eurographics Symposium on Non-Photorealistic Animation and Rendering, pp. 137–146. ACM (2011).

Chapter 3
Theorem: General Intelligence Entails Creativity, Assuming . . .

Selmer Bringsjord

3.1 Apparent Evidence Against the Claim

There appears to be considerable evidence against the claim that general intelligence entails creativity. For example, with this claim unpacked as the proposition that any general-intelligent agent must be creative, the field of AI declares the claim to be false. To see this, we need but note that the dominant and encyclopedic AI textbook [16] defines an intelligent agent as one that computes a mapping from its percepts to its actions. The range of potential mappings explored in this volume is non-trivial, ranging from simple arithmetic functions to functions that leverage declarative knowledge, and beyond. But never is creativity discussed in connection with any of these functions; indeed creativity is nowhere discussed in the book, period; nor for that matter is any synonym (e.g., 'innovative') discussed. In short, as far as this highly influential and comprehensive volume is concerned, general-intelligent agents needn't be creative.

Of course, *AIMA*, as it's known, is a textbook, at the end of the day; a masterful one, yes, but certainly a textbook. It's on the bookshelf of nearly every single AI researcher and engineer on our planet, but the tome doesn't purport to provide a novel account of general machine intelligence. Yet it seems to me that we observe the same lack-of-entailment result if we examine "research-grade" proposals for what abstract machine intelligence is. One example is Hutter's [12] theory of "universal artificial intelligence." Whatever virtues this theory may have (and I do think it has some significant ones), an explanation of creativity isn't one of them. Hutter's formal foundations are avowedly and indeed proudly in sequential decision theory and algorithmic information theory; but such things, if the scientific literature on creativity is any guide, would be top candidates for being *in tension* with creativity. Part of the reason for this is presumably that if we know anything about creativity in the human case, and from that know something about the abstract concept of creativity that can cover information-processing machines and extraterrestrial lifeforms, we know that creativity leverages declarative knowledge to produce new concepts, from

S. Bringsjord (✉)
Rensselaer Polytechnic Institute, Troy, NY, USA
e-mail: selmer@rpi.edu

T.R. Besold et al. (eds.), *Computational Creativity Research: Towards Creative Machines*,
Atlantis Thinking Machines 7, DOI 10.2991/978-94-6239-085-0_3

which new declarative knowledge is generated. The paradigm of this "creativity engine" at work is the evolution of mathematics and mathematical knowledge. If Leibniz hadn't used what he did know to create the concept of an infinitesimal, what we know in knowing analysis via knowing the theorems that constitute it, might never have arrived. So in Hutter's work we have a proposal for what the nature of intelligence is, in the abstract—but nothing in that proposal yields that general intelligence entails creativity.

The somewhat odd thing, though, is that Hutter [12] does *mention* creativity, and indeed does so in a context that seems quite relevant to the present essay. For we read:

> The science of [AI] might be defined as the construction of intelligent systems and their analysis. A natural definition of a *system* is anything that has an input and an output stream. Intelligence is more complicated. It can have many faces like **creativity**, solving problems, pattern recognition, classification, learning, induction, deduction, building analogies, optimization, surviving in an environment, language processing, knowledge, and many more. A formal definition incorporating every aspect of intelligence, however, seems difficult. Further, intelligence is graded ... So, the best we can expect to find is a partial or total order relation on the set of systems, which orders them w.r.t. their degree of intelligence (like intelligence tests do for human systems, but for a limited class of problems). Having this order we are, of course, interested in large elements, i.e., highly intelligent systems ([12], 2–3; bold text from me).

The contrast between this and AIMA is quite interesting. AIMA defines an agent, as we've noted, as an input-output device, with inputs as percepts and outputs as actions. So what Hutter says about systems fits the AIMA framework well. But then the list of "faces" that he gives, and casts aside as infeasible targets for targeted formalization, include many things that AIMA in fact provides computational definitions of (save, as we've noted, for creativity!). I see no reason to despair of formalizing all of these parts of human cognition, and for the life of me don't understand why Hutter rules such a project out as too difficult. The problem that *I* see, from the standpoint of truly general intelligence, abstracted away from us and our machines to cognizers in general, is that many of these parts of human cognition aren't necessarily part of highly intelligent cognizers in the abstract case. I take up this problem below (Sect. 3.2.1), and suggest a solution.

It's also interesting to note that Hutter is to this point roughly in line with what I shall propose, which is a hierarchy of intelligence (and one inspired by my psychometric tendencies, which renders Hutter's comment about human intelligence tests welcome)—but for reasons that remain utterly mysterious to me, he takes maximization of some utility function to be the essence of intelligence, to which all the "faces" he lists are supposed to be reducible. He writes: "Most, if not all, known facets of intelligence can be formulated as goal driven or, more precisely, as maximizing some utility function." [12, p.3]. But no proof or argument is offered in support of this credo. Any concern regarding the absence of such a systematic case is perhaps magnified by the fact that on some accounts of creativity, for instance on some interpretations of what Boden [2] calls *P-creativity*, to be creative is to somehow produce something which cannot be understood from, let alone derived from, antecedents e.g., see [5].[1] In the case of Leibniz, Zeno's paradoxes of motion stood iron-strong for century upon century, and then suddenly new infinitary concepts arrive on the scene, and soon thereafter ordinary physical motion makes perfect mathematical sense. At any rate, whether or not Hutter is right, the fact remains that while his *Universal Artificial Intelligence* is a certified research-grade proposal for what general intelligence, in man or machine, is, the book's Index contains no entry for creativity; and therefore at the very least we have no reason to think, on the basis of Hutter's book, that what I'm declaring to be a theorem is one.

Let's try a third tack. In keeping with so-called *Psychometric AI* (PAI, rhymes with 'π'; [3, 6, 7]), according to which AI consists in the engineering of artificial agents capable of high performance on well-defined tests of various vaunted mental powers in the human sphere, we can quickly see that, once again, general intelligence doesn't seem to entail creativity: Let a be an agent able to perfectly answer every question on every established, psychometrically validated test of general (human) intelligence.[2] And now pull off the shelf every single established test of creativity used by psychometricians and psychologists.[3] Next, does our assumption of a's prowess enable us to deduce that a will score at a high level on the selected test of creativity? No. Indeed, the negative here is so obvious, and so firm, that I will not trouble the reader with any details, and will instead sum up the psychometric chasm between tests of general human intelligence and tests of human creativity by giving this telling, representative fact: It doesn't follow from the proposition that some agent is able to achieve perfection at digit recall[4] that that agent can quickly invent new things to do with tin cans.[5]

3.2 The Setup

Despite the foregoing, in point of fact it *is* possible to show that high general intelligence, whether of the human, alien, or machine variety, *does* entail creativity—as long as certain assumptions are made. I find these assumptions to be eminently reasonable, but lay no claim that they in fact are: Readers are invited to judge for themselves. As this "theorem" is a conditional, a natural way to construct the proof of it is to assume, with respect to arbitrary instances of all the categories over which quantifiers

---

[1] This point should not be interpreted so as to deny the brute fact that contemporary AI, viewed pluralistically, is agent-based, where the agents often have goals/aims, and attempt to reach them by straightforward processing of antecedents. Confirmation is provided by the widely-used [16].

[2] Two ideal picks would be Raven's Progressive Matrices [15], and the WAIS discussed in connection with AI in [7].

[3] One good choice would be the Torrance Tests of Creative Thinking (TTCT; [19]), targeted in the AI work reported in [4].

[4] Digit recall is a sub-test on the WAIS (see note 2). On this sub-test, the test-taker attempts to repeat back a string of digits given to him by the tester.

[5] A typical question on TTCT. See note 3.

in the theorem range, the relevant antecedent, and then derive the consequent. This is the route I take. But please note that as the present essay is intended to be digestible by a general-educated audience, I provide herein only an informal proof-*sketch*, not a full-blown proof. And I seek to make the presentation here largely self-contained: I assume only that my readers have had but a bit of elementary mathematical logic and recursion theory. If you had this, but have forgotten it, your memory will soon be refreshed.

We turn now to setting out our preliminaries. This will be a review for many readers; for others it will serve to secure the self-contained nature of the essay.

3.2.1 A Focus on Arithmetic

I begin by setting some context: Since we are operating under the framework of PAI, the objective is a proof that if some agent a is general-intelligent under an "alien-fair" test $\mathscr{T}_{gi}$ of general intelligence, then a, under some "alien-fair" test $\mathscr{T}_c$ of creativity, is creative.

What is $\mathscr{T}_{gi}$, the alien-fair test of general intelligence (= gi)? It should be obvious that if our test of general intelligence is to be not only culture-fair relative to the cultures in place on Earth, but also a fair test of gi for any spot in the universe, and indeed for any place in any universe, we can't base $\mathscr{T}_{gi}$ on anything that is clearly just a part of our local environment as human beings. The solution is to restrict the test of gi to something that every single class or race of general-intelligent agents must to an appreciable degree master: arithmetic.[6]

Notice that I don't say 'mathematics.' Rather, I specifically refer to arithmetic. This is because clearly parts of the vast edifice of human-discovered mathematics might not be tackled by general-intelligent aliens. For example, extraterrestrials on Alpha Centauri (assuming they are there, for the sake of exposition), however brilliant they may be, might never take up geometry. But there would seem to be absolutely no way these aliens can avoid seeking and securing arithmetical competence. A genuinely general-intelligent alien agent, as well as an information-processing agent that we or such an alien brings into existence, couldn't dodge arithmetic, and the search for substantial knowledge of it.[7]

---

[6] This may be the spot to politely register my astonishment that cognitive modeling, the subfield devoted to erecting computational frameworks that captures the fundamental invariants and capacities of human intelligence, seems to have steered clear of the formal nature of arithmetic, and the nature of what it takes to understand that formal nature, to varying degrees e.g., see [1]. As far as I can tell, most cognitive architectures are provably unable to represent the axioms, theorems, and proofs that correspond to arithmetic, and facility with it. In short, cognitive architectures should probably be presented in such a way that it's clear what circle in Fig. 3.1, if any, they can handle.

[7] I think it's reasonable to assume that bright, accomplished aliens would need *both* natural-number and real-number arithmetic. Under this assumption, that I leave aside consideration of axiomatic treatments of the latter kind of arithmetic isn't a defect. But what about the possibility of an alien race whose intellectual and technological prowess is based *exclusively* on real-number arithmetic?

But we must be a bit more systematic about what arithmetic is, and how much of it must be mastered if an agent is to earn the right to be classified as gi. The next step is to quickly review some standard formal machinery from mathematical logic and computability.[8]

3.2.2 Basic Machinery

Let **A** be some axiomatic theory of arithmetic based on some corresponding formal language $\mathscr{L}_{\mathbf{A}}$. Let $\mathscr{I}_{\mathscr{A}}$ be the natural and received model-theoretic interpretation of ordinary arithmetic with which you and I are intimately familiar. (Warning: I use **A** as a variable for a given axiom system, instances to be visited below; but I use $\mathscr{A}$ to refer to arithmetic, period.) **A** is simply a set of formulae from $\mathscr{L}_{\mathbf{A}}$: viz., a certain set of **axioms** of arithmetic. Now let $\alpha_{\mathbf{A}} \in \mathscr{L}_{\mathbf{A}}$ be some arbitrary formula about **A** arithmetic. (When the context is clear, we shall sometimes drop the subscript $_{\mathbf{A}}$ and refer to a given arithmetical formula as simply α.) To say that α is true on some interpretation $\mathscr{I}$, we write the customary:

$$\mathscr{I} \models \alpha$$

where $\phi_{\mathbf{A}}$ (here too we sometimes omit the subscript $_{\mathbf{A}}$) is a set of formulae based on $\mathscr{L}_{\mathbf{A}}$, α is a **consequence** of ϕ iff

$$\text{For every } \mathscr{I}\text{, if all of } \phi \text{ are true on } \mathscr{I}\text{, then } \mathscr{I} \models \alpha.$$

$\mathbf{A}^{\models}$ denotes the set of all formulae that are consequences of **A**. We assume a standard finitary proof-theory τ based in first-order logic (e.g., resolution-, natural deduction-, or equational-based), and write the usual

$$\phi \vdash_{\tau} \alpha$$

(Footnote 7 continued)
I find this scenario unlikely, but that will hardly impress skeptics; and the fact of the matter is that such a scenario is mathematically and, it would seem, even technologically, possible. However, there is clearly an analogue to my theorem-and-proof based in axioms, theories, and theorems cast in systems that yield *both* natural-number and real-number arithmetics: viz., axiomatic set theory. I recommend that any readers interested in pursuing this route, or the narrow route of a real-number version of what the case I give herein, consult [17], which works backwards from all of mathematics, to axioms systems.

[8] My notation and focus is devised for the purposes at hand, but in general nice coverage is provided in the venerable [8], which I have long used in classroom teaching of intermediate mathematical logic. But there is an especially good background provided in [18], which has the added benefit of a learned discussion of potential ways of distinguishing between an understanding of basic arithmetic, versus understanding more. Unlike [18], I have high-ish standards: I interpret basic arithmetic to include truths of arithmetic beyond ordinary, mechanical proof in first-order logic.

to indicate that α is provable from ϕ in this theory. $\mathbf{A}^{\vdash_\tau}$ denotes the set of all theorems that can be proved from $\mathbf{A}$ in the proof. Finally, we set

$$\text{TRUE}_{A/X} = \{\alpha \in \mathscr{L}_{\mathbf{A}} : \mathscr{I}_{\mathscr{A}} \models \alpha\}$$

where X is a placeholder index enabling reference to modifications in the underlying language $\mathscr{L}_{\mathbf{A}}$ and axiom system $\mathbf{A}$. For instance, writing $\text{TRUE}_{A/I}$ will denote the set of all first-order arithmetical formulae true on the standard first-order interpretation $\mathscr{I}_{\mathscr{A}}$ of arithmetic.

3.2.3 Context for the Theorem

Our context is set out diagrammatically in Fig. 3.1. The reader will be able to make sense of much of this figure given his/her assimilation of the review provided in Sect. 3.2.2. Here's the remaining explanation that is needed:

First, a circle C inside another circle C' indicates a proper subset relation; that is, $C \subset C'$. The contents of each circle is just a set of formulae.

Four axiom systems of arithmetic appear in the diagram. At the inner core, in the smallest circle, the system **EA**, "elementary arithmetic," appears. More precisely, every theorem (consequence) of the axioms **EA** is what composes the innermost circle. Smith [18] refers to this axiom system as "Baby Arithmetic," and I follow suit, and so deploy '**BA**,' as can be seen the diagram. No one could take **BA** seriously as an axiom system that captures what even moderately intelligent pre-teen humans know about arithmetic. For example, while any true instance of an equation of the form $n + m = k$ is deducible from **BA**, and likewise any true instance of an equation of the form $n \times m = k$ as well, **BA** is severely limited, since for instance it doesn't even allow formulae and deduction with the quantifiers $\exists$ and $\forall$. There is thus little point here in saying anything further about **BA**, since even moderately intelligent schoolchildren know truths of arithmetic that make use of variables and (implicit) quantifiers (e.g., $\forall x(x \times 1 = x)$, which such children would recognize as $x \times 1 = x$). In other words, an alien-fair $\mathscr{T}_{gi}$ would have to include questions like the following, which aren't in the innermost circle.

Q $\forall x(x \times 1 = x)$?

In Fig. 3.1, I indicate that an agent in the innermost circle understands everything within this circle. I do this by depicting the "face" of the agent inside this circle. But notice that there is an arrow flowing from this picture of the agent that leaves the innermost circle and travels to the immediate superset, that is, to the next circle. And notice that this arrow has a check on it. What this says is that any agent of moderate intelligence who has reached pre-college development will not only understand **BA**, but also, if they have an understanding of basic arithmetic, **Q**, or—as it is sometimes known—**Robinson Arithmetic**.

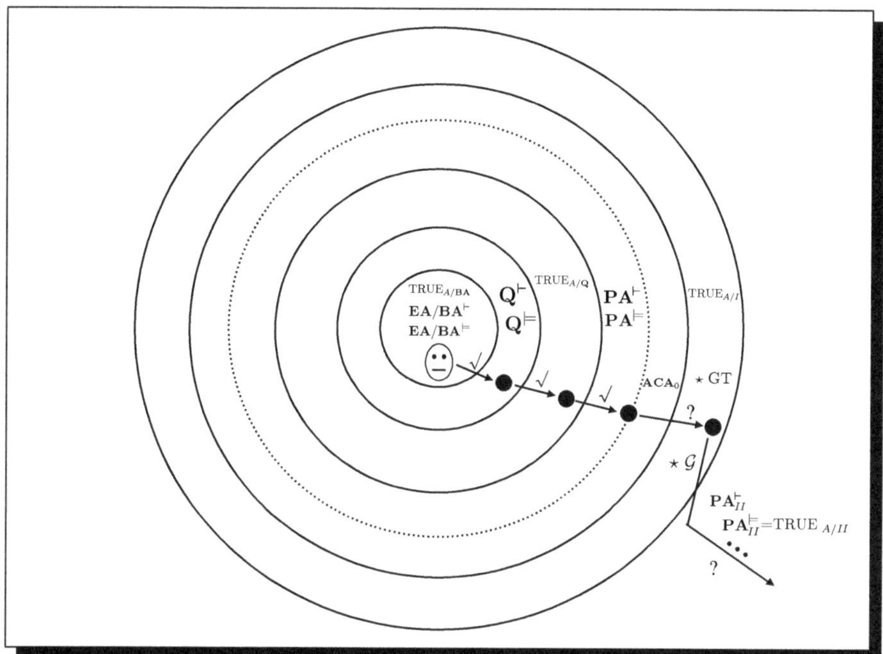

Fig. 3.1 The context for the theorem

The axioms of **Q** do allow quantification, and can be associated with the standard proof theory τ that I invoked above.[9] While the above question Q is settled by **Q**, this axiom system isn't powerful enough to support the test $\mathscr{T}_{gi}$. This is clear from the fact that for example the following question would be easy for a young student, but features a theorem that isn't in $\mathbf{Q}^{\vdash_\tau}$.

Q' $\forall x(0 + x = x)$?

Obviously, then, we can't identify alien-fair gi with the second circle in Fig. 3.1. In light of this, note that I give the agent in the diagram a free pass to the next circle, which is labeled with 'TRUE$_{A/\mathbf{Q}}$, the set of all formulae that are true on the standard interpretation of arithmetic, relativized to **Q**. Obviously $\forall x(0 + x = x)$ is in TRUE$_{A/\mathbf{Q}}$.

[9] **Q** is composed of seven axioms (where s is the successor or "increment-by-one" function):

Axiom 1 $\forall x(0 \neq s(x))$
Axiom 2 $\forall x \forall y(s(x) = s(y) \rightarrow x = y)$
Axiom 3 $\forall x(x \neq 0 \rightarrow \exists y(x = s(y)))$
Axiom 4 $\forall x(+(x, 0) = x)$
Axiom 5 $\forall x \forall y(+(x, s(y)) = s(+(x, y)))$
Axiom 6 $\forall x(\times(x, 0) = 0)$
Axiom 7 $\forall x \forall y(\times(x, s(y)) = +(\times(x, y), x))$

We come now to the third axiom system: $\mathbf{PA}_I$. This is standard Peano Arithmetic, which most readers will at least have heard something about. It includes the first six of the seven axioms composing $\mathbf{Q}$ (see note 9), plus one additional axiom schema:

Induction Schema Every sentence that is the universal closure of an instance of this schema:

$$[\phi(0) \wedge \forall x(\phi(x) \rightarrow \phi(s(x))] \rightarrow \forall x \phi(x)$$

$\mathbf{PA}_I$ is what I shall take as the springboard from which to launch to a completion of the definition of alien-fair gi.[10] We can jump from the circle containing all the theorems provable from $\mathbf{PA}_I$ (which is the same set as all the consequences of $\mathbf{PA}_I$) to a larger class: namely, all the truths of standard first-order arithmetic. (Circles beyond this one involve second-order logic (as indicated by the subscript $_{II}$), and are left for future expansions of my case for the gi-implies-creativity theorem.) But notice that here the arrow designed to reflect the "travel" of our agent is not labeled with a check, but rather with a question-mark. This is so because making this jump requires some impressive intelligence. What do you reach if you make this jump, specifically? I give two examples in the diagram, each one marked with a $\star$. And what are the examples? The first is $\mathscr{G}$, and is a label for the formula that Gödel pointed to via his first incompleteness theorem. Each of these formulae is of course such that neither it nor its negation can be proved from $\mathbf{PA}_I$—but each such formula is true on $\mathscr{I}_{\mathscr{A}}$. The second $\star$ proposition I call out here is a particular number-theoretic fact: Goodstein's Theorem [11], and is indicated by 'GT.' While GT and instances of $\mathscr{G}$ are all true on the standard interpretation of arithmetic, they are beyond the theorems of $\mathbf{PA}_I$, a nice result first proved by Kirby and Paris [13].[11]

3.2.4 Key Definitions

I define a truly gi agent, whether human, alien, or machine, to be one that understands[12] not only basic arithmetic (i.e., that understands $\mathbf{PA}_I^{\vdash_\tau}$ and below, to include $\therefore$ $\mathbf{Q}^{\vdash_\tau}$ and $\mathbf{EA}^{\vdash}/\mathbf{BA}^{\vdash_\tau}$), but also at least one $\star$ truth.

---

[10] Alert readers will note that in jumping from this board I pass straight through $\mathbf{ACA}_0$ without comment, and they will have already have noticed the dotted circle I drew for this axiom system. What gives? Ultimately, $\mathbf{ACA}_0$ supports a class of first-order theorems that doesn't exceed those provable from $\mathbf{PA}_I$; hence the dotted circle rather than a solid one. As to what $\mathbf{ACA}_0$ is, we shall have to rest content with the highly informal piece of information that it's a restricted form of second-order arithmetic. A philosophically rich presentation of $\mathbf{ACA}_0$ is provided in [18].

[11] GT is simply the fact that a particular sequence of natural numbers, the **Goodstein sequence**, starting with any natural number n, eventually terminates at zero. But many folks who first understand the sequence are utterly convinced that it's both astonishingly fast-growing and never terminates, and simply returns larger and larger numbers as the sequence progresses, forever. See [14] for a nice version of the proof, which makes use of infinitary concepts and techniques, and turns these intuitions upside down to yield a result that a truly general-intelligent agent can appreciate.

[12] As alert readers will have noted, multiple times above I've made use of the concept of *understanding*, and here I sustain the practice. This concept does carry a lot of baggage, yes; but on

The test $\mathscr{T}_{gi}$ is composed of truths of arithmetic; that is, of members of the set $\text{TRUE}_{A/I}$, with an accompanying request for a supporting proof. In the case of $\mathbf{PA}_I$, there can be a supporting formal proof in a standard, finitary, mechanizable proof theory (our τ). This will not be possible for members of the set

$$\text{TRUE}_{A/I} - \mathbf{PA}_I^{\vdash_\tau}$$

I further define creativity as passing beyond such proofs in basic arithmetic in order to reach at least one $\star$ truth. Again, Goodstein's Theorem is currently an ideal example.

3.3 The Proof-Sketch Itself

Theorem: Suppose that an agent a is alien-fair gi. Then a is creative.

Proof-Sketch: Trivial, given our setup. Assume the hypothesis of the theorem. By definition, a, since its gi includes command of all of basic arithmetic, knows at least one $\star \in \text{TRUE}_{A/I}$ on the strength of a proof π discovered and confirmed by a. But since π as a matter of mathematical fact exceeds the mechanical type of proof that characterizes our τ, a has left behind mere mechanical, first-order techniques, and is by definition creative. **QED**

3.4 Objections

Some objections can be anticipated; I discuss two.

3.4.1 Begging the Question?

Objection: "As you yourself note, given your setup, the theorem is easily established. So you have simply begged the question. Why would anyone accept your setup in the first place?"

My reply: Well, every theorem presupposes background machinery, and some of it will be objectionable to some; the present situation is no exception. I cheerfully admit that anyone unwilling to accept that alien-fair gi must include significant command

(Footnote 12 continued)
the other hand it's undeniable that it stands ineliminably at the very heart of mathematics, and the teaching thereof, for us. Specifically, we find it impossible to forego language in which we speak of whether or not a young student *understands* some aspect of arithmetic, or whether or not a college student *understands* (differential and integral) calculus. Not only is this linguistic practice standard and unavoidable, but so is (at least in pedagogy) the need to try to confirm that understanding is in place via giving tests. In the present paper, my use of the concept of understanding, and recourse to tests designed to confirm its presence, simply follows a suit played in the real world.

over basic arithmetic at the level of the truths of arithmetic will not be persuaded. But I maintain that at the very least it's undeniable that it's not unreasonable to construe alien-fair gi in such a manner that understanding of one or more $\star$ is included. After all, Earth-bound empirical evidence is on my side, given the remarkable creativity it has taken to reach some $\star$ truths. I also maintain that it's not unreasonable to identify creativity with a process of coming to know some $\star$, since invariably this hard-won knowledge comes via reasoning that is beyond the rigid, mechanistic construction of standard formal proofs in first-order logic.

3.4.2 A Non-creative Route to a $\star$?

Objection: "But here's a *non*-creative way to reach the performance you say is creative, which serves as a counter-example to your so-called theorem: We know that the set of all formulae reachable from the grammar and alphabet of $\mathscr{L}_{\mathbf{A}}$ by standard recursive rules for well-formedness is countably infinite. Even the extensions of the language to make room for the moves to second-order logic of course stay within the bound of countably infinite. Hence there is a machine $\mathscr{M}$ which prints out (in accordance with some lexicographic ordering) the first such formula, then the second, and so on. In addition, $\mathscr{M}$ is assumed to be equipped with a random "formula picker" P such that, given a formula in the relevant class, it returns either TRUE or FALSE randomly. Clearly, if $\mathscr{M}$-plus-P is lucky, it will declare all the $\star$'s you're talking about to be true—and yet clearly this 'agent' is operating in purely mechanistic, naïve fashion, indeed more so than the searching for proofs in the proof theory τ."

My reply: Multiple problems are fatal to this objection; I mention two here. First, if this objection worked, then basic incompleteness results like $\mathscr{G}$ would in some sense be surmounted as well, by stunningly "dumb" means. But no one thinks there's a shortcut here to establishing formulae that are beyond $\mathbf{PA}^{\vdash_\tau}$. Second, my tests of both gi and creativity (in the realm of arithmetic) are such that to pass requires *understanding*, and the behavioral correlate to understanding, taken to confirm its presence, is *justification*.[13] In other words, and this repeats what has been said above, to pass $\mathscr{T}_{gi}$, an agent must prove that their answers to basic arithmetic are correct; and to pass $\mathscr{T}_c$ must prove at least one $\star$ truth about arithmetic. The dim contraption $\mathscr{M}$-plus-P does nothing of the sort.

3.4.3 Mere Mechanical via Ordinal-Logics?

Objection: "You are ignoring perhaps the very earliest systematic attempt to 'surmount' Gödelian incompleteness, one seminally inaugurated by none other than Turing, in his doctoral dissertation, and soon thereafter published in [20]: viz., *ordinal*

[13] See footnote 12.

logics. Apparently Church, Turing's advisor, who already had on hand his system $\mathcal{O}$, a set of constructive notations a for ordinals α, encouraged Turing to pursue the ordinal-logic direction.[14] Turing acceded, specifically by exploring sequences Λ composed of sequences of logics L_a where a is in $\mathcal{O}$. Mechanisms for generating larger and larger elements in such sequences are known as *reflection principles* [9]. Today we know that such principles, which are pretty much just mechanical, do overcome the key boundary in your case (and specifically in your Fig. 3.1), and hence we have a counter-example to your so-called theorem."

Reply: This objection is impaled by either of two unavoidable horns in a dilemma.

Horn 1: It should be pretty obvious that while all of the creativity in this approach is at the meta-theoretical, not the object-theoretical, level, there *is* certainly a lot of creative, far-from-obviously-mechanical thinking and reasoning going on. This holds true starting at the genesis of the approach, when Church, armed with $\mathcal{O}$ (and other machinery, e.g., his λ-calculus), encouraged to Turing to "take the plunge," and then as Turing ran into negative, not positive, results. While Turing showed that Λ_p is Π^0_1-complete, Π^0_2 formulae were out of reach. But notice that this result is completely independent of any object-level results: that is, of any proof or disproof of a particular Π^0_1 or Π^0_2 theorem of natural and specific interest to mathematicians. The situation is no different with respect to more recent work on ordinal logics: elegance, insight, and creativity abounds at the meta-theoretical level; but results at the object level isn't part of the game. Therefore, my case is firmly intact.

Horn 2: The skeptic could offer a rebuttal here in which he concedes my observation, but points out that he is in fact talking exclusively about the *object*-level, and therefore the merely mechanical is in operation, only. But now the second horn rears up. For the fact of the matter is that [11]'s proof, like, as a matter of brute empirical fact, all proofs of particular theorems at this point or above, is stunningly *non*-mechanical. In fact, the reasoning is patently *infinitary* in nature, and hence the idea, which is needed here by my critic, that the reasoning is mechanical in the sense of being in line with a Turing-equivalent process, is no more than an article of faith.[15]

3.4.4 Additional Objections

Life, including specifically intellectual life, being what it is, I'm quite sure that additional objections will occur to those wishing to resist the entailment I've defended herein. For example, some may urge me to consider alien beings who directly apprehend via intuition, without proof, all arithmetic truth. But the point of my essay is to defend my "theorem" in a context short of the divine! I'm concerned with gi,

---

[14] A marvelous non-technical and historical overview is provided in [10].

[15] I'm not making here any such claim as that the reasoning in Goodstein's case is irreducibly infinitary (though I do happen to believe that). On the other hand, I do remind readers that a way to surmount Gödelian incompleteness at the object level is via the ω-rule. And for all we know from the alien-fair perspective, there may be entirely new proof theories (including perhaps diagrammatic ones) that allow aliens or future humans to move beyond incompleteness.

which means a form of intelligence *short* of God, who is traditionally taken to be omniscient. The alien beings here imagined by my critic are just gods, and would likely be called such for instance in a science-fiction story. Inevitably, any number of additional gymnastic moves will be made, and pressed against me. But I claim to have articulated a serious case at this point, one suitable as a springboard for future dialectic. And that, I think, for now, is enough.

Acknowledgments I'm deeply grateful for the opportunity to speak in Montpelier at C3GI 2012, a lecture that marked my pursuit, seemingly a pipe dream then, for a rigorous rationale in favor of what was then confessedly only an inchoate suspicion: that general intelligence (when both high and "alien-fair") entails creativity. I have benefitted from conversations with Naveen Sundar G., Matthias Scheutz, and Simon Ellis, and am grateful to them as well. Thanks are due to anonymous reviewers of the paper, for insightful analysis; and also to Tarek Besold for the same, and first-rate editorial guidance as well. Finally, the support of AFOSR has been crucial. Needless to say, any errors appearing herein are mine alone, unrelated to any of the kind minds who have interacted with me on matters treated herein.

References

1. Anderson, J., Lebiere, C.: The Newell test for a theory of cognition. Behav. Brain Sci. **26**, 587–640 (2003)
2. Boden, M.: The Creative Mind: Myths and Mechanisms. Basic Books, New York (1991)
3. Bringsjord, S.: Psychometric artificial intelligence. J. Exp. Theor. Artif. Intell. **23**(3), 271–277 (2011)
4. Bringsjord, S., Ferrucci, D.: Artificial Intelligence and Literary Creativity: Inside the Mind of Brutus, A Storytelling Machine. Lawrence Erlbaum, Mahwah (2000)
5. Bringsjord, S., Ferrucci, D., Bello, P.: Creativity, the turing test, and the (better) lovelace test. Mind. Mach. **11**, 3–27 (2001)
6. Bringsjord, S., Licato, J.: Psychometric artificial general intelligence: the Piaget-MacGuyver room. In: Wang, P., Goertzel, B. (eds.) Foundations of Artificial General Intelligence, pp. 25–47. Atlantis Press, Amsterdam, The Netherlands (2012). http://kryten.mm.rpi.edu/Bringsjord_Licato_PAGI_071512.pdf. This url is to a preprint only
7. Bringsjord, S., Schimanski, B.: What is artificial intelligence? Psychometric AI as an answer. In: Proceedings of the 18th International Joint Conference on Artificial Intelligence (IJCAI-03), pp. 887–893. Morgan Kaufmann, San Francisco (2003). http://kryten.mm.rpi.edu/scb.bs.pai.ijcai03.pdf
8. Ebbinghaus, H.D., Flum, J., Thomas, W.: Mathematical Logic, 2nd edn. Springer, New York (1994)
9. Feferman, S.: Reflecting on incompleteness. J. Symb. Logic **56**(1), 1–49 (1991)
10. Feferman, S.: Turing in the land of O(Z). In: Herken, R. (ed.) The Universal Turing Machine, 2nd edn. Springer, Secaucus, pp 103–134 (1995)
11. Goodstein, R.: On the restricted ordinal theorem. J. Symb. Logic **9**(31), 33–41 (1944)
12. Hutter, M.: Universal Artificial Intelligence: Sequential Decisions Based on Algorithmic Probability. Springer, New York (2005)
13. Kirby, L., Paris, J.: Accessible independence results for peano arithmetic. Bull. Lond. Math. Soc. **14**, 285–293 (1982)
14. Potter, M.: Set Theory and its Philosophy: A Critical Introduction. Oxford University Press, Oxford (2004)
15. Raven, J.C.: Advanced Progressive Matrices Set II. H. K. Lewis, London (1962). Distributed in the United States by The Psychological Corporation Inc., San Antonio, Texas

16. Russell, S., Norvig, P.: Artificial Intelligence: A Modern Approach, 3rd edn. Prentice Hall, Upper Saddle River (2009)
17. Simpson, S.: Subsystems of Second Order Arithmetic, 2nd edn. Cambridge University Press, Cambridge (2010)
18. Smith, P.: An Introduction to Gödel's Theorems, 2nd edn. Cambridge University Press, Cambridge (2013)
19. Torrance, E.P.: The Torrance Tests of Creative Thinking. Scholastic Testing Service, Bensenville (1990)
20. Turing, A.: Systems of logic based on ordinals. Proc. Lond. Math. Soc. (Series 2), **45**, 161–228 (1939)

Chapter 4
The Computational Creativity Complex

Dan Ventura

Abstract We briefly examine the subject of computational creativity through the lenses of three different systems for producing creative artifacts in three different domains: music, visual art and cookery. From these we attempt to abstract something of a "general-purpose" computationally creative agent and examine how this agent might behave in the context of an archetypical "algorithm" for creativity. Finally, we consider this agent's behavior from the point of view of the theory of (Turing) computability and suggest that computational creativity as a field provides an interesting opportunity for looking at computational complexity in new ways.

4.1 Inception

The question of computational creativity can be approached from several different angles. Here we assume that creativity is possible in computational systems and examine the idea of a general, abstract mechanism for computational creativity that exists somehow independently from any particular domain. This is a difficult question, and it is clear that much of creativity is not a domain independent construct. However, we posit that there does exist some core abstraction or creativity "algorithm" that can be applied to any domain (with suitable domain-specific augmentation, of course). Given this hypothesis, we attempt an inductive approach to the problem by first examining three computationally creative systems in three different domains and then attempt to generalize from these specific examples an abstract model of a creative system.

As a complementary view of the problem, we also discuss an abstract "algorithm" for creativity and relate this "algorithm" to the abstract system, asking whether this "algorithm" could, in fact, become an *algorithm* in the formal sense, and attempt to reason about the answer to that question. In other words, we try to ascertain whether computational creativity is actually computable.

D. Ventura (✉)
Computer Science Department, Brigham Young University, Provo, UT 84602, USA
e-mail: ventura@cs.byu.edu

T.R. Besold et al. (eds.), *Computational Creativity Research: Towards Creative Machines*,
Atlantis Thinking Machines 7, DOI 10.2991/978-94-6239-085-0_4

4.2 Three Blind Mice

Our three specific examples, from which we will try to generalize an abstract system are DARCI, a computational artist that creates visualizations for communicating concepts [1–5]; CARL, a computational composer that discovers musical motifs in non-musical sources and composes music around them [6, 7]; and PIERRE a computational chef that creates original slow cooker recipes [8]. Each of these systems has been presented in more detail elsewhere, and we give only enough detail here to support the generalization that is our goal.

4.2.1 DARCI

DARCI is a system for generating original images that convey intention and is inspired by other artistic image generating systems such as AARON [9] and The Painting Fool [10]. Central to the design philosophy of DARCI is the notion that the communication of meaning in art is a necessary part of eliciting an aesthetic experience in the viewer [11], and it is unique in that it creates images that explicitly express a given concept using *visual metaphor*. This is currently done at two levels: using iconic nouns as surrogates for the target concept and using image filters to convey associated adjectival content.

DARCI is composed of two major subsystems, an *image analysis* component, and an *image generation* component. The image analysis component learns how to associate images with concepts in the forms of nouns and adjectives. The image generation component composes an original source image as a collage of iconic noun concepts and then uses a genetic algorithm, governed by the analysis component, to render this source image to visually convey an adjective. Figure 4.1 outlines this process of creating artifacts.

4.2.1.1 Image Analysis

DARCI's understanding of images is derived from two sources: a mapping from low-level image features to descriptive adjectives and semantic associations between linguistic concepts.

Visuo-Linguistic Association In order for DARCI to make associations between images and associated adjectives, the system learns a mapping from low-level computer vision features [12–17] to words using images that are hand-labeled with adjective tags. The use of WordNet's [18] database of adjective synsets allows images to be described by their affect, most of their aesthetic qualities, many of their possible associations, and even, to some extent, by their subject.

To collect training data we have created a public website for training DARCI (http://darci.cs.byu.edu), where users are presented with a random image and asked

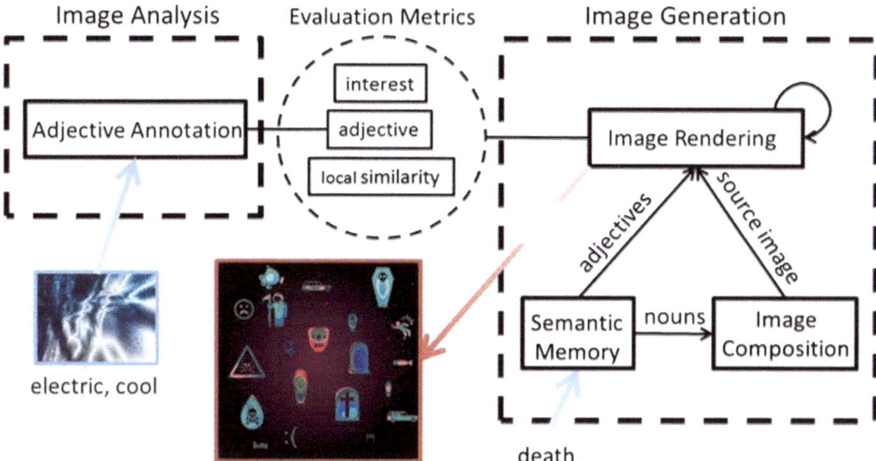

Fig. 4.1 A diagram outlining the two major components of DARCI. *Image analysis* learns how to annotate new images with adjectives using a series of *appreciation networks* trained with labeled images (outlined in *blue*). *Image generation* uses a *semantic memory* model to identify nouns and adjectives associated with a given concept. The nouns are composed into a source image that is rendered to reflect the adjectives, using a genetic algorithm that is governed by a set of evaluation metrics. The final product (outlined in *red*) is an image that communicates the given concept

to provide adjectives that describe the image. When users input a word with multiple senses, they are presented with a list of the available senses, along with the WordNet gloss, and asked to select the most appropriate one. Additionally, for each image presented to the user, DARCI lists seven adjectives that it associates with the image. The user is then allowed to flag those labels that are not accurate.

Learning image to synset associations is a multi-label classification problem [19], meaning each image can be associated with more than one synset. To handle this, we use a collection of artificial neural networks (ANNs) that we call *appreciation networks*, each of which outputs a single real value, between 0 and 1, indicating the degree to which a given image can be described by the network's corresponding synset (adjective). An appreciation network is created for each synset that has a sufficient number of training data, and as data are incrementally accumulated, new neural networks are dynamically added to the collection to accommodate any new synsets. There are currently close to 300 appreciation networks in the system.

Semantic Memory Model The system also contains a simple cognitive model, built as a semantic network forming a graph of associations between words [20, 21]. These word associations are acquired in one of two ways: from people and by automatic inference from a corpus, with the idea being to use the human word associations to capture general knowledge and then to fill in the gaps using the corpus associations.

For the human word associations, we use two pre-existing databases of *free association norms* (FANs): the Edinburgh Associative Thesaurus [22] and the University of Florida's Word Association Norms [23]. These word associations were acquired by asking hundreds of human volunteers to provide the first word that comes to mind when given a cue word. This technique is able to capture many different types of word associations including word co-ordination (pepper, salt), collocation (trash, can), super-ordination (insect, butterfly), synonymy (starving, hungry), and antonymy (good, bad). The association strength between two words is simply a count of the number of volunteers that said the second word given the first word. FANs are considered to be one of the best methods for understanding how people, in general, associate words in their own minds [23].

For the corpus-based associations, we build a (term $\times$ term) co-occurrence matrix from a large corpus, in a manner similar to that employed in the *Hyperspace Analog to Language* (HAL) model [24]. For our corpus, we use the entire (English) text of Wikipedia, as it is large, easily accessible, and covers a wide range of human knowledge [25]. Once the co-occurrence matrix is built, we use the co-occurrence values themselves as association strengths between words. This approach works, since we only care about the strongest associations between words, and it allows us to reduce the number of irrelevant associations by ignoring any word pairs with a co-occurrence count less than some threshold.

Our final semantic network is a composition of the human- and corpus-based associations, which essentially merges the two separate graphs into a single network before querying it for associations. This method assumes that the human data contains more valuable word associations than the corpus data because such human data is typically used as the gold standard in the literature. However, the corpus data does contain some valuable associations not present in the human data. To combine the graphs, we add the top n associations for each word from the corpus data to the human data but weight the corpus-based association strengths lower than the human-based associations. This is beneficial for two reasons. First, if there are any associations that overlap, adding them again will strengthen the association in the combined network. Second, corpus-based associations not present in the human data will be added to the combined network and provide a greater variety of word associations. We keep the association strength low because we want the corpus data to reinforce, but not dominate, the human data.

4.2.1.2 Image Generation

DARCI generates images in two stages: the creation of a source image composed of a collage of concept icons and the rendering of this source image using various parameterized image filters. The collage generation is driven by the semantic network, while the filtered rendering is achieved using an evolutionary mechanism whose fitness function is defined in terms of the outputs of the visuo-linguistic association networks.

Image Composition. The semantic memory model can be considered to represent the meaning of a word as a (weighted) collection of other words. DARCI effectively makes use of this collection as a decomposition of a (high-level) concept into simpler concepts that together represent the whole, the idea being that in many cases, if a (sub)concept is simple enough, it can be represented visually with a single icon. To represent these "simple enough" concepts, DARCI makes use of a collection of icons provided by *The Noun Project* [26]. Given such a collection of iconic concepts, DARCI composes their visual representations (icons) into a single image.

When given a concept, DARCI first uses the semantic memory model to retrieve all words associated with the given concept, including itself. These word associations are filtered by returning only nouns for which DARCI has icons and adjectives for which DARCI has appreciation networks. The nouns are sorted by association strength and the top few are chosen as a collective iconic representation of the concept. These icons are scaled to between 25 and 100 % of their original size according to their association strength rank.

An initial blank white image is created, and the set of scaled icons are drawn onto the blank image at random locations, the only constraints being that no icons are allowed to overlap and no icons are allowed to extend beyond the border of the image. The result is a collage of icons that collectively represent the original concept.

DARCI then probabilistically (weighted by each adjective's association strength) selects an adjective from the set returned by the semantic memory model and then uses its image rendering component to render the collage image according to the selected adjective. The final image will both be artistic and in some way communicate the concept to the viewer.

Image Rendering. DARCI uses an evolutionary mechanism to render images so that they visually express the meaning of given synsets. The genotypes that comprise each gene pool are lists of filters (and their accompanying parameters) for processing a source image, similar to those found in Adobe Photoshop and other image editing software. The processed image is the phenotype.

Every generation of the evolution, each phenotype is created from the same source image. The function used to evaluate the fitness of each phenotype created during the evolutionary process can be expressed by the following equation:

$$\text{Fitness}(f^P) = \lambda_A A(f^P) + \lambda_S S(P) \tag{4.1}$$

where P is the phenotype image and f^P is the vector of image features for a given phenotype, and $A : F^P \rightarrow [0, 1]$ and $S : I \rightarrow [0, 1]$ are two functions for modeling appreciation and similarity, respectively. These functions compute a real-valued score for a given phenotype (here, F^P represents the set of all phenotype feature vectors, I represents the set of all images and $\lambda_A + \lambda_S = 1$).

The appreciation function A is computed as the weighted sum of the output(s) of the appropriate appreciation network(s). The similarity function S borrows from research on *bag-of-visual-word* models [27, 28] to analyze local image features (in contrast to the global features represented by f^P). A bag-of-visual-words is created for the source image and for the phenotype image, and the Euclidean distance between their vector representations is calculated. This effectively measures the number of interest points shared by the two images and can be translated into a measure of image similarity.

Fitness-based tournament selection determines those genotypes that propagate to the next generation and those genotypes that participate in crossover. One-point crossover is used to allow for variable length offspring. Crossover is accomplished in two stages: the first occurs at the filter level, so that the two genomes swap an integer number of filters; the second occurs at the parameter level, so that filters on either side of the cut point swap an integer number of parameters. By necessity, parameter list length is preserved for each filter.

Mutation occurs at both the filter and parameter levels. Filter mutation is a wholesale change of a single filter (discrete values), while parameter mutation is a change in one or more parameter values for a filter (continuous values).

Because there are potentially many ideal filter configurations for modeling any given synset, sub-populations are used within each gene pool, allowing the evolutionary mechanism to converge to multiple solutions. Limited migration between sub-populations is allowed, with the exception that the most fit genotype per sub-population is not allowed to migrate, and sub-population size balancing is enforced.

Figure 4.2 shows an example image produced by DARCI when given the concept *war* as the stimulus.

Fig. 4.2 Image produced by DARCI to express the concept of *war*

4.2.2 CARL

CARL is a computational composition system designed to incorporate two competing compositional ideas: (1) to produce "acceptable" music and (2) to produce innovative music. Many compositional systems accomplish one or the other of these goals very well. In the case of the former goal, approaches tend to learn a model from a set of training examples and probabilistically generate new music based on the learned model. These approaches effectively produce artifacts that mimic classical music literature, but little thought is directed toward expansion and transformation of the music domain. For example, David Cope [41] and Dubnov et al. [29] seek to mimic the style of other composers in their systems. In the case of the latter goal, approaches utilize devices such as genetic algorithms [30, 31] and swarms [32]. While these approaches can theoretically expand the music domain, they often have little grounding in accepted musicality, and their output often receives little acclaim from either music scholars or average listeners.

In order to serve both goals, CARL couples machine learning (ML) techniques with an inspirational component. The ML component maintains grounding in music literature and effects innovation by employing the strengths of generative models. It embraces the compositional approach found in the period of common practice and the early 20th century. The inspirational component introduces non-musical ideas and enables innovation beyond the musical training data. The system focuses on the composition of motifs, an atomic level of musical structure, defined as "the smallest structural unit possessing thematic identity" [33].

First, an ML model is trained on a set of monophonic MIDI themes retrieved from The Electronic Dictionary of Musical Themes.[1] Then, a set of candidate motifs is extracted from an inspirational media file (e.g., pitch detection is performed on an audio file or edge detection is performed on an image file). Members of the set of candidate motifs that are most probable according to the ML model are selected as the building blocks for a composition. A high-level system pipeline is shown graphically in Fig. 4.3.

4.2.2.1 Machine Learning Models

Two different ML models are trained, one over a set of 128 possible pitches and one over 32 possible rhythmic durations (32nd note multiples up to a whole note). A variety of ML approaches, including HMMs, variable order Markov models and recurrent neural networks have been used as generative (and discriminitive) models for music composition. In this context, the model is used discriminatively—to identify motifs that conform (to some degree) to known musical conventions (or, more accurately, to such conventions as might be extracted from the training data), and two classes of model, Prediction by Partial Match [34] and Context Tree Weighting [35], have proven particularly effective at this task.

[1] http://www.multimedialibrary.com/barlow/all_barlow.asp.

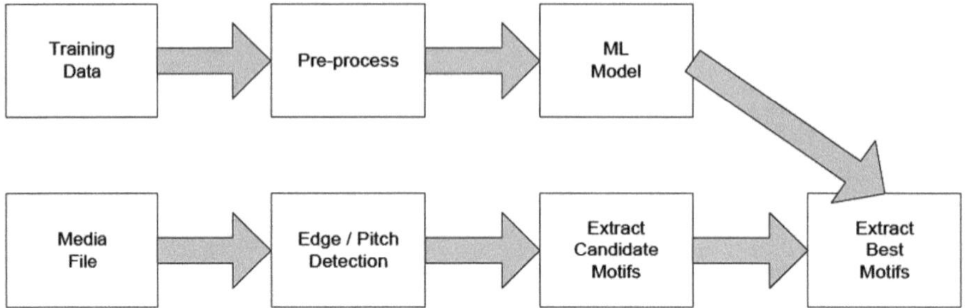

Fig. 4.3 A high-level system pipeline for motif discovery. An ML model is trained on pre-processed music themes. Pitch detection is performed on an audio file or edge detection is performed on an image file in order to extract a sequence of notes. The sequence of notes is segmented into a set of candidate motifs, and only the most probable motifs according to the ML model are selected

4.2.2.2 Audio Pitch Detection

Pitch detection is performed on the audio file using an open source command line utility called Aubio,[2] which combines note onset detection and pitch detection to output a string of notes (each comprised of a pitch and duration). The string of detected notes is post-processed to make the sequence more manageable: each duration is quantized to the nearest 32nd note value, and pitch intervals that are larger than an octave are modified to the equivalent interval that is less than an octave.

4.2.2.3 Image Edge Detection

Edge detection is performed on an image using a Canny edge detector,[3] which returns a new image comprised of black and white pixels. The original image is also converted to a greyscale image. To extract strings of notes analogous to those extracted from audio, both images are iterated over one pixel at a time using a spiral pattern starting from the outside and working inward. For each sequence of b contiguous black pixels (delimited by white pixels) in the edge-detected image, a single note is created. The pitch of the note is the average intensity of the corresponding b pixels in the greyscale image, and the duration of the note is proportional to b.

4.2.2.4 Motif Discovery

After the string of notes is detected and processed, candidate motifs are extracted (see Algorithm 1). All contiguous motifs of length greater than or equal to l_min

[2] http://www.aubio.org.

[3] http://www.tomgibara.com/computer-vision/canny-edge-detector.

Algorithm 1 *extract_candidate_motifs*

```
Input: notes, l_min, l_max
candidate_motifs ← {}
for l_min ≤ l ≤ l_max do
    for 0 ≤ i ≤ |notes| − l do
        motif ← (notes_i, notes_{i+1}, ..., notes_{i+l−1})
        candidate_motifs ← candidate_motifs ∪ motif
    end for
end for
return candidate_motifs
```

and less than or equal to l_max are stored. After the candidate motifs are gathered, the motifs with the highest probability according to the model of the training data are selected (see Algorithm 2). Pitches and rhythms are learned separately, weighted, and combined to form a single "probability" defined as

$$P_{p+r}(m) = Pr(m_p){N_p}^{|m|} + Pr(m_r){N_r}^{|m|} \tag{4.2}$$

where $Pr()$ is computed with the appropriate model (pitch or rhythm), m is a motif, m_p is the motif pitch sequence, m_r is the motif rhythm sequence, N_p and N_r are constants with $N_p > N_r$. The resulting value is not a true probability because it can be greater than 1.0, but this is not significant because we are only interested in the relative probability of motifs.

Since shorter motifs are naturally more probable than longer motifs, an additional normalization step is taken in Algorithm 2 to ensure that each motif length has equal probability:

$$P_{equal} = \frac{1}{(l\_max - l\_min + 1)} \tag{4.3}$$

Since the probability of a generative model emitting a motif of length l is

$$P(l) = \sum_{m \in C, |m|=l} Pr(m|model) \tag{4.4}$$

a length-dependent normalization term is introduced to equalize the probability of selecting motifs of various lengths (step 5 of Algorithm 2):

$$norm(l) = \frac{P_{equal}}{P(l)} \tag{4.5}$$

Algorithm 2 *discover_best_motifs*

```
Input: notes, model, num_motifs, l_min, l_max
C ← extract_candidate_motifs(notes, l_min, l_max)
best_motifs ← {}
while |best_motifs| < num_motifs do
    m* ← argmax_{m∈C} [norm(|m|) Pr(m|model)]
    best_motifs ← best_motifs ∪ m*
end while
return best_motifs
```

4.2.2.5 Lead Sheet Composition

Using the motif discovery system as an entry point, a simple, hybrid composition system for creating lead sheets can be built. This system composes from the bottom up because it uses motifs as a starting point. It also utilises a top down approach because it imposes formal global constraints on the final piece of music.

After motifs are collected from an inspirational file, they are randomly concatenated together to form an initial population for a genetic algorithm. The genetic algorithm outputs a section of monophonic music which is harmonized using a hidden Markov model trained on a lead sheet database. The monophonic melody and the chords from the hidden Markov model comprise a lead sheet.

From a set of motifs, we create an initial population of 100 themes by repeatedly choosing four random motifs and concatenating them. We run a genetic algorithm with the initial population for 100 iterations without a stopping criterion, assuming a 4/4 meter. The fitness function, crossover, and mutation operations are taken from the NormalDistributionFE, OnePointCrossover, and ComplexMutater classes in jMusic,[4] respectively.

The genetic algorithm above is executed twice in order to create an A theme and a B theme. The A and B themes are then concatenated to create a melody line in ABAB form, a typical form in pop music and jazz.

After the melody line for the entire piece of music is created, chords are added to the melody by following the process described in [36]. Chord transition probabilities are accumulated in a square matrix by counting chord transitions in a chords-only lead sheet database called *The Imaginary Book*, which can be downloaded at the Impro-Visor Yahoo Group.[5] Melodies are generated for *The Imaginary Book* with Impro-Visor's improvisation tool using the Bill Evans grammar, and melody emission probabilities are accumulated in a matrix by counting the melody note durations for each accompaniment chord.

Next, the composed theme (A or B) is segmented into measures, where each measure becomes an observation in a hidden Markov model, and the chord transition matrix represents the state transitions. The measure observations are calculated by

---

[4] http://explodingart.com/jmusic.

[5] https://groups.yahoo.com/neo/groups/impro-visor/info.

Fig. 4.4 A portion of a lead sheet created by CARL. The full melodic line was created using a genetic algorithm on an initial population of concatenated motifs discovered from the input file *Bioplazm2.jpg*. Harmonization was accomplished using the Viterbi algorithm on an HMM which was trained on a lead sheet database

taking the dot product of the appropriate row in the melody emission matrix with the pitch vector that represents the measure, and this constitutes the emission matrix of a hidden Markov model. The Viterbi algorithm is used to calculate the most likely chord sequence that accompanies the given melody.

Figure 4.4 shows part of a lead sheet created by CARL when the inspirational source was an abstract image file called *Bioplazm2.jpg*.

4.2.3 PIERRE

PIERRE is a system for culinary creation, which, given access to existing recipes, learns to produce new crockpot recipes. PIERRE is composed primarily of two modules, for handling evaluation and generation, respectively. Each of these components takes input from an inspiring set and each is involved in producing recipes to send to the presentation module, as shown in Fig. 4.5. In addition, the system interacts with the web, both acquiring knowledge from online databases and (potentially) publishing created recipes.

Recipe generation is a complicated task that requires not only precise amounts of ingredients but also explicit directions for preparing, combining, and cooking the ingredients. To focus on the foundational task of the type and amount of ingredients,

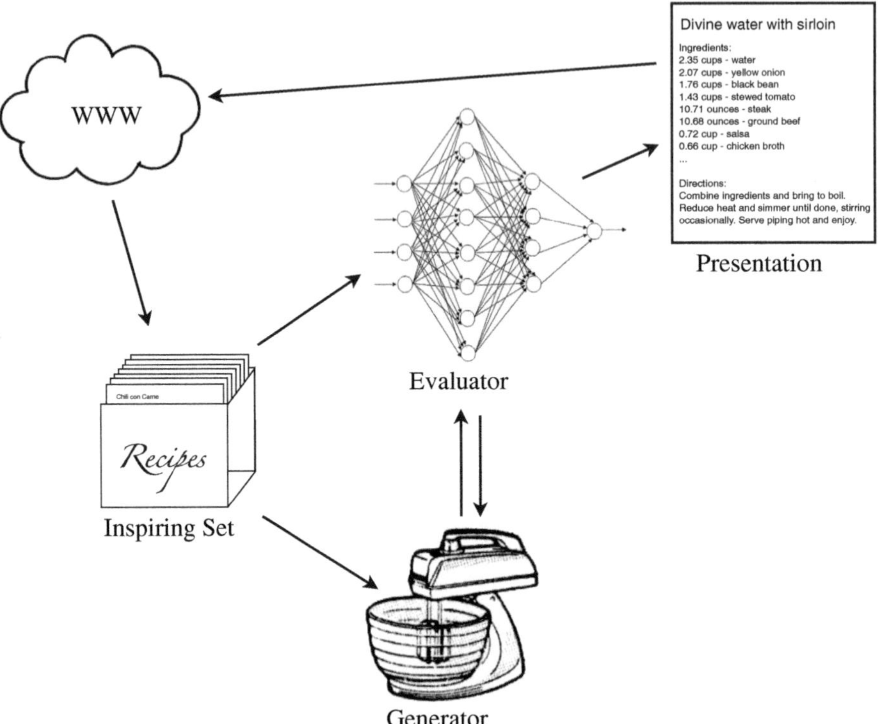

Fig. 4.5 High-level view of the system architecture. Inspiring set recipes are taken from online sources and inform the evaluator and generator. Recipes are created through an iterative process involving both generation and evaluation. Eventually, generated recipes with the highest evaluation are fed to the presentation module for rendering and may be published online

we restrict our focus to recipes (specifically soups, stews, and chilis) that can be cooked in a crockpot and thus require no special preparation instructions—crockpot recipes simplify the cooking process to essentially determining a set of ingredients (and their amounts) to be cooked together. Varshney et al. have recently presented a system that uses a cognitive-based approach to produce additional types of recipes that do require a nontrivial preparation component, but they have also avoided this issue, by taking a similar simplifying approach—producing, instead of the complete recipe, an interesting set of ingredients and leaving the preparation instructions to be invented by a human collaborator [37].

4.2.3.1 Inspiring Set

The inspiring set contains soup, stew, and chili recipes gathered from popular online recipe websites.[6] From these recipes we manually create both a list of measurements and ingredients in order to parse recipes into a consistent format. This parsing enables (1) grouping identical ingredients under a common name, (2) grouping similar ingredients at several levels, and (3) gathering statistics about ingredients and ingredient groups across the inspiring set. Recipes in the inspiring set are normalized to 100 ounces.

The database of ingredients is explicitly partitioned into a hierarchy in which similar ingredients are grouped at a *sub*-level and these ingredient groups are further grouped at a *super*-level. For example, as shown in Fig. 4.6, the super-group *Fruits and Vegetables* is composed of the sub-groups *Beans*, *Fruits*, *Leafy Vegetables*, and others. The sub-group of *Beans* includes many different types of beans including *Butter Beans*, *Red Kidney Beans*, *Garbanzo Beans*, and others.

Statistics (minimum, maximum, mean, standard deviation, and frequency) are kept for each ingredient. These statistics are also aggregated at the sub- and super-group levels, enabling comparison and evaluation of recipes at different levels of abstraction. In addition, gathering statistics at the group level provides a mechanism for smoothing amounts for rare ingredients. Each statistic ω for such ingredients is linearly interpolated with the corresponding statistic of the sub-group, according to the following:

$$\omega = \begin{cases} \left(\frac{\alpha}{\alpha+\beta}\right) x + \left(\frac{\beta}{\alpha+\beta}\right) \xi & \text{if } \alpha < \theta \\ x & \text{if } \alpha \geq \theta \end{cases} \tag{4.6}$$

where x is the raw statistic of the ingredient, ξ is the statistic of the sub-group, α is the number of times the ingredient occurs in the inspiring set, β is the number of times any of the sub-group ingredients occur in the inspiring set, and the rarity threshold θ is set to 100.

The inspiring set is used differently for generation than it is for evaluation. During artifact generation (Sect. 4.2.3.2) the inspiring set determines the initial population used for the genetic algorithm. During artifact evaluation (Sect. 4.2.3.3) the inspiring set determines which recipes and ratings are used as training examples. Since the inspiring set is used in multiple ways, employing a different inspiring set for generating artifacts than the one used to evaluate artifacts can have useful effects.

4.2.3.2 Generation

PIERRE generates new recipes using a genetic algorithm acting on a population of recipes, each composed of a list of ingredients. The population is initialized by choosing recipes uniformly at random from the inspiring set, and the fitness of each recipe

[6] http://www.foodnetwork.com and http://www.allrecipes.com.

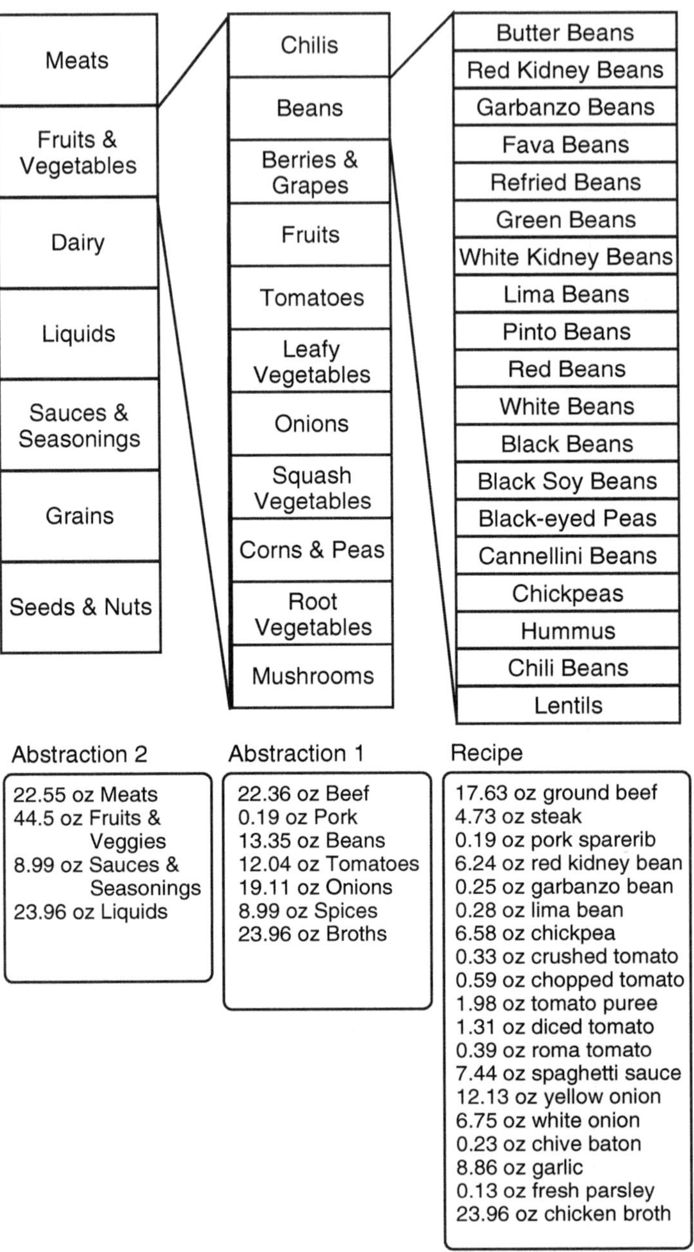

Fig. 4.6 *Above*, a view of the ingredient hierarchy, showing the super-group (*left*), sub-group (*middle*), and ingredient (*right*) levels of abstraction. The *Fruits & Vegetables* super-group is expanded to show its sub-groups, including *Beans*, which is expanded to show its ingredients. *Below*, an example recipe is shown as it would appear at each level of abstraction

is evaluated using the MLP evaluator described in Sect. 4.2.3.3. To produce each generation, a number of new recipes are generated equal to the number of recipes in the population. For each new recipe, two recipes are selected, with probability proportional to their fitness, for genetic crossover. The crossover is performed by randomly selecting a pivot index in the ingredient list of each recipe, thus dividing each recipe into two sub-lists of ingredients. A new recipe is then created by combining the first sub-list of the first recipe with the second sub-list of the second recipe.

After crossover, each recipe is subject to some probability of mutation. If a mutation occurs, the type of mutation is selected uniformly from the following choices:

- *Change of ingredient amount.* An ingredient is selected uniformly at random from the recipe and its quantity is set to a new value drawn from a normal distribution that is parameterized by the mean and standard deviation of that ingredient's amount as determined from the inspiring set.
- *Change of one ingredient to another.* An ingredient is selected uniformly at random from the recipe and is changed to another ingredient from the same super-group, chosen uniformly at random. The amount of the ingredient does not change.
- *Addition of ingredient.* An ingredient is selected uniformly at random from the database and inserted into a random location (chosen uniformly) in the recipe's ingredient list. The amount of the new ingredient is determined by a draw from a normal distribution parameterized by the mean and standard deviation of the ingredient amount as determined from the inspiring set.
- *Deletion of ingredient.* An ingredient is selected uniformly at random and removed from the recipe.

At the completion of each iteration, evolved recipes are re-normalized to 100 ounces for equal comparison to other recipes. The next generation is then selected by taking the top 50 % (highest fitness) of the previous generation and the top 50 % of the newly generated recipes. The rest of the recipes are discarded, keeping the population size constant.

4.2.3.3 Evaluation

To assess the quality of recipes, PIERRE uses an interpolation of two multilayer perceptrons (MLPs). Taking advantage of the (online) public user ratings of the recipes in the inspiring set, these MLPs perform a regression of the user rating using ingredient amounts as inputs. The two MLPs are trained at different levels of abstraction within our ingredient hierarchy, with one operating at the super-group level and the other at the sub-group level. Thus, the model at the higher level of abstraction attempts to learn the proper relationship of major groups (meats, liquid, spices, etc.), and the other model works to model the correct amounts of divisions within those groups.

Because we assume any recipe from the online websites is of relatively good quality, regardless of its user rating, we supplemented the training set with randomly constructed recipes given a rating of 0. These negative examples enabled the learner to discriminate between invalid random recipes and the valid ones, created by actual people.

Each MLP has an input layer consisting of real-valued nodes that encode the amount (in ounces) of each super-group (sub-group), a hidden layer consisting of 16 hidden nodes and a single real-valued output node that encodes the rating (between 0 and 1). The MLP weights are trained until there is no measurable improvement in accuracy on a held out validation data set. The set of weights used for evaluating generated recipes are those that performed the best on the validation data set.

4.2.3.4 Presentation

Colton has suggested that *perception* plays a critical role in the attribution of creativity (or, more precisely, in the attribution of *un*creativity) [38]. In other words, a computationally creative system could (and possibly must) take some responsibility to frame its work to engender a perception of creativity (or at least to avoid being summarily labeled uncreative).

In an attempt to help facilitate such a perception of its artifacts, PIERRE contains a module for recipe presentation. First, the module formats the recipe for human readability. Ingredient quantities are stored internally in ounces, but when recipes are rendered for presentation, the ingredients are sorted by amount and then formatted using more traditional measurements, such as cups, teaspoons, dashes, and drops. Recipes are presented in a familiar way, just as they might appear in a common cookbook.

Second, the presentation module generates a recipe name. Standard recipes always have a name of some sort. While this task could be a complete work by itself, PIERRE employs a simple name generation routine that produces names in the following format: *[prefix] [ingredients] [suffix]*. This simple generation scheme produces names such as "Homestyle broccoli over beef blend" or "Spicy chicken with carrots surprise." The components of the name are based on prominent recipe ingredients and the presence of spicy or sweet ingredients. This simple approach creates names that range from reasonable to humorous.

Recipe 1 is an example of one of PIERRE's creations and was among those served during a computational creativity festival titled *You Can't Know My Mind* [39]. To produce this recipe, a population size of 150 recipes was allowed to evolve for 50 generations with a mutation rate of 40 %.

4.3 The Blind Leading the Blind

We now make two nontrivial conceptual leaps, positing both an abstract model of a domain-independent creative agent, and an abstract creativity "algorithm" for that agent to run. Here we will simply present both in enough detail to facilitate the discussion to follow, in which we consider how such an agent might or might not

Recipe 1 Scrumptious Broth with Bean

Ingredients:

2 3/8 cups - vegetable broth
2 1/3 cups - chicken broth
1 1/3 cups - coconut milk
1 cup - halloumi
1 cup - leek
2 1/2 sticks - cinnamon
7/8 cup - yellow onion
1 1/2 green chile
3 5/8 tablespoons - lemon juice
1/4 cup - white onion
1/4 cup - crushed tomato
1/2 red bell pepper
1/8 cup - green onion
1/8 cup - vegetable juice
1 5/8 tablespoons - green tea
1 1/2 tablespoons - asiago
1 3/8 tablespoons - sherry
1/8 cup - black beans, precooked
1 tablespoon - evaporated milk
1 3/8 teaspoons - baking cocoa
1 1/3 teaspoons - pepper

Directions: Combine ingredients and bring to boil. Reduce heat and simmer until done, stirring occasionally. Serve piping hot and enjoy.

be able to implement different steps of the "algorithm", grounding the discussion to some extent using the specific cases of DARCI, CARL and PIERRE.

4.3.1 An Abstract Creative Agent

A creative agent is composed of many internal mechanisms/processes, some which include *background knowledge*, an ability to *learn*, *intentionality*, an ability to *conceptualize*, a sense of *aesthetic* (that is, an evaluation mechanism) and some method of *generating* artifacts.

In DARCI's case, background knowledge is encoded in the form of visuo-linquistic associations, semantic networks and iconic noun representations; learning happens by building neural and semantic networks; intentionality is effected by the goal of communicating a concept through visual art; an aesthetic is encoded in the fitness function, is tied to what is learned and can change over time as more information is acquired; visual artifacts are generated with an evolutionary mechanism.

In CARL's case, background knowledge is encoded in the model of relevant musical themes; learning happens by building a variable order Markov model; intentionality is effected by the dual goals of innovation and musicality; an aesthetic is encoded to some degree in the model's probability distribution over motifs and weakly in the evolutionary mechanism for generating a lead sheet but is in general less developed than that of DARCI; musical artifacts are generated with a simple generate-and-test mechanism at the motif level and an evolutionary mechanism at the lead sheet level.

In PIERRE's case, background knowledge is encoded in the form of a database of ingredients and their statistical properties and neural network models of recipe quality; learning happens by adding to the ingredient database and building the neural networks; intentionality is effected, to some extent, by the goal of creating edible recipes but is less developed than in CARL and especially in DARCI; an aesthetic is encoded in the neural network models; culinary artifacts are generated with an evolutionary mechanism.

These internal mechanisms interact with each other in multiple ways, and both the mechanisms and their interactions are the subject of ongoing research, with both human and computational subjects. However, we will use them here only as instruments for grounding and guiding the discussion.

In addition to these internal mechanisms, because the agent exists in an environment, it interacts with the environment in multiple ways, including being *taught*, the *presentation* of artifacts, being *inspired*, receiving *feedback* and other *influences*.

DARCI is taught by human-labeled images, human-generated linguistic responses and both structured and unstructured data resources on the web; presents its artifacts as visual images, sometimes with simple titles; is weakly inspired by target concepts and can autonomously discover potentially interesting concepts online to some extent; receives some feedback in the form of negative reinforcement for poor visuo-linguistic associations but none yet directly for its creations, though some of those creations have been indirectly critiqued when appearing in surveys and art exhibitions.

CARL is taught by structured data resources on the web; presents its artifacts as musical motifs or lead sheets; is inspired by various types of non-musical input, including audio, images and even sleep data recordings; doesn't really receive any feedback at this stage.

PIERRE is taught by structured data resources on the web; presents its artifacts as recipes, including appropriate and sometimes amusing titles; is sometimes weakly inspired, when the set of recipes used to drive generation and evaluation differ significantly; receives no direct feedback at this stage, though some of its creations have been indirectly critiqued by users of online recipe repositories and by people actually cooking the recipes and eating the result.

Figure 4.7 offers a gross visualization of a our archetypal agent. Both on the human and computational fronts, there have been significant advances in understanding many of the individual mechanisms shown in the figure. What is still not understood is how these mechanisms, both internal and external, interact to realize the creativity "algorithm", and it is this question that we will try to say something about here.

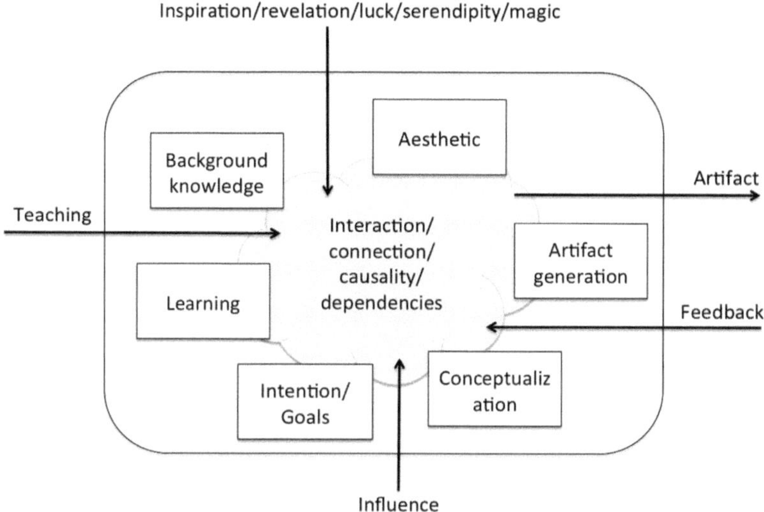

Fig. 4.7 *Possible logical overview of a creative agent.* The component internal mechanisms are meant to be representative rather than exhaustive. In the same spirit, no attempt here is made to accurately visualize the dependencies and communication between these mechanisms. The agent (potentially) communicates with the environment in several ways, represented by labeled arrows entering or leaving the agent

4.3.2 The Creativity "Algorithm"

Because there is not even an agreed upon definition for creativity (and very likely never will be), there does not exist a generally accepted canonical "algorithm" for creativity. As a surrogate for such an "algorithm", we will employ five general steps distilled by Csikszentmihalyi [40]:

1. preparation
2. incubation
3. insight
4. evaluation
5. elaboration

In what follows we will discuss each of these steps in the context of computability. Note that, as Csikszentmihalyi observes, these steps should not be taken as a single-iteration process but rather as parts of a multiple-iteration, possibly recursive process in which the steps may be revisited multiple times, in varying order as necessary. Here, we will ignore this obviously important issue of flow control and focus only on the five steps. In what follows we will consider the archetype agent of Fig. 4.7 whose ambition is creativity, and we will consider how that agent might follow Csikszentmihalyi's archetypal "algorithm" for being so.

4.4 Out of the Frying Pan

We will treat each of the steps of the "algorithm" in turn, positing something about the salient agent mechanisms and their interactions and what the prospects are for its implementation in a computational setting.

4.4.1 Preparation

Preparation is the initial process of learning about the domain in which an agent will attempt creativity. It entails significant interaction with the environment for the acquisition of background knowledge and understanding accepted practices and approaches as well as open problems. In addition, an agent must acquire or develop some aesthetic sense of the domain, where we use aesthetic in the sense of some abstract notion of quality. Initially this sense could be taught to the agent by the environment in just the same way that the background knowledge is. Of course, agents that develop new aesthetic sensibilities (a meta-level creative act?) are likely to be considered more creative in their output. Eventually, an agent may use its acquired background information to learn/develop such novel aesthetics. It is sometimes argued that too much preparation can result in the repression of creative possibility as old, set ideas are assimilated too thoroughly. However, it is certainly the case that a good deal of preparation is necessary to facilitate downstream processes, particularly those of evaluation and elaboration.

Computational challenges inherent in this step include the acquiring, encoding, and understanding of knowledge, ontologies, formalization, etc. as well as methods for learning/developing evaluation strategies. These are nontrivial tasks, to be sure, but many proof-of-concept structured, semi-structured and unstructured projects put the knowledge acquisition aspects squarely in the category of difficult-but-manageable engineering tasks (cf., Wikipedia,[7] WordNet [18], ConceptNet [42], the semantic web[8] and even the World-Wide-Web itself). As for learning/developing an aesthetic, general purpose machine learning techniques exist for inferring structural relations from data. In many respects, this preparation step is not unlike developing pedagogy for human students, and many AI approaches to the problem, from ontologies to machine learning would be recognized to some extent by educational practitioners.

---

[7] http://www.wikipedia.org.

[8] http://www.w3.org/2013/data/.

4.4.2 Incubation

Incubation is the process of "putting it on the back burner"—allowing ideas to simmer in a possibly unconscious way, the development of goals, the exploration of unusual connections, brainstorming, etc. This is often described as an open-ended process without a clear time line or quantifiable goals, other than "finding something interesting". The agent conceptualizes and generates ideas using its knowledge base and additional outside environmental influences. These concepts and ideas are judged against the agent's aesthetic sense and very often discarded immediately. While this step can be performed consciously and intentionally, as in the aforementioned brainstorming session, it is often described as best happening when the conscious mind is otherwise engaged (with another task, while exercising, while in the shower or even while sleeping). It is unclear whether this unconscious aspect is necessary or simply catalyzing and whether intentionality may be uncoupled from consciousness.

Given an effective organization and acquisition of knowledge, it is not difficult to argue that computational systems will actually (eventually) enjoy a significant advantage over human intelligence in this step—speed, lack of bias, nonsusceptibility to fatigue, distraction, boredom, etc. all favor computational approaches to the exploration of potentially interesting connections and the generation of ideas and conceptualizations at scale. Of course, any "intelligent" biases should be formalized and leveraged by computational systems for obvious reasons; however, determining whether a bias is useful or potentially detrimental is likely classifiable as a creative task itself (another meta-level concern?).

4.4.3 Insight

Insight is the step that may be tempting to describe as having nothing explicitly to do with any action or intention of the agent, as originating from outside. Depending on a person's bent, this might be called inspiration or revelation or luck or serendipity or magic or something else. It is often associated with an "Aha!" moment, when things just fall into place, the answer suddenly becomes clear, etc. This presents us, apparently, with something of a Gödelian quandary, which may (or may not be) resolvable in one of several ways.

One possibility is that insight is an agent fabrication that is not really necessary for creativity; a second possibility is that insight, though a necessary part of the "algorithm", does not, in fact, originate outside the agent at all[9]; a third possibility is that insight is somehow necessary for human creativity but may not be for a computational variant,[10] and it is therefore unimportant for the current discussion; a fourth possibility is that, in fact, insight is necessary for creativity in any medium

[9] The agent's belief that it does may be explainable by appeal to the unconscious, insufficient understanding of neural psychological and cognitive processes, etc.

[10] An adaptation of the common airplane versus bird analogy.

and does also, in fact, represent a Gödelian process over which the agent may not have any control.

The computational challenge faced in realizing this step of the "algorithm" depends upon which, if any, of the possibilities above best explains insight. In the first three cases, the simplest solution must involve some variation on a brute force search (in what space? Is identification/construction of the search space another meta-level problem?) Such an approach will (eventually) produce artifacts that satisfy the agent's aesthetic and are potentially considered creative. Of course in any interesting domain, the search space is very likely to be infinite and so the first real computability concern raises its head. Such a search will not be computable in the strong sense of decidability (see more on this in Sect. 4.4.4); however, it will be in the weaker sense of recognizability, and this could be argued to be no more egregious than is the case for human creativity—we can't define or guarantee it, but we know it when we see it. The next obvious solution is to introduce search heuristics to circumvent the complexity/computability issues associated with the brute force approach. These may be learned from the environment[11] or invented by the agent (meta-level process, again) and there will be a tradeoff between computational guarantees and possibility of success.

In the fourth case, we have the possibility that creativity has an analog to Gödel's incompleteness theorem in that something from outside the agent is necessary. This would, of course, preclude any general (closed) computational creative system and will perhaps seem appealing to some who may see creativity as a last bastion of humanity or as something *ex vi termini* impossible computationally. And yet, if the premise were indeed true, the same would have to be said about the human variety as well. Even if this is the case, we still see creative acts, both personal and historical, occurring with regularity, and we might yet simulate this productivity computationally by acting ourselves as the requisite extra-agent component of insight. That is, computational creativity would be effective at least some of the time only with the aid of external intervention, suggesting something of co-creativity, and if that external intervention comes from a person, the possibility yet persists of maintaining a small toe-hold on the precipice of human superiority.

As a last comment, we note that in at least one theory, insight has been equated with re-representation [43]. That is, creativity is difficult (or impossible) when the agent's representation of the problem is not amenable to it—the agent can (figuratively) wander around forever and not discover anything useful until—Aha!—it "lucks into" the right representation (this appears like yet another potential meta-level issue, with at least the outside possibility that there may be no access to the meta-level by the agent).

[11] And may, in fact, simulate some unconscious cognitive or sub-cognitive process.

4.4.4 Evaluation

Evaluation is the application of the aesthetic measurement process to the product of the generation process. Both of these processes may be learned during preparation or they may be the product of a (meta)creative process themselves. This is an internal evaluation, not to be confused with the external appraisal and feedback from the environment to which all potentially creative acts must be subject.[12] A result that passes the aesthetic test will be elaborated and eventually presented to the environment for that external assessment.

Though the high-level process description is deceptively simple, the computational challenges posed at this step are non-trivial. Assume that evaluation is computable in principle, so we have an algorithm E that computes it.

What we want is another algorithm F that can tell us whether an artifact a is accepted by E; that is, we are interested in the language $L(F) = \{a|E \text{ accepts } a\}$. Initially, let's optimistically assume that E is even computable in the strong Turing sense, that is, it is *decidable*. Then, we have an easy algorithm for F (run E on input a), and, thus the rudimentary makings of an algorithm C for using F to solve whatever the problem is (that is, to be creative):

```
C()
  do
    choose a
  until a in L(F)
```

Of course, in any interesting case, the space to explore is infinite and E may be very selective, so this algorithm may be worthless, but at least it is computable—it has been reduced to a "simple" question of complexity. To make it useful, we need a good exploration strategy. It is possible that this might be learned from the environment during the preparation step, but if such a search strategy is already known, then the problem to which the search strategy is to be applied is likely already (nearly) solved. So, for non-trivial problems, it is likely that the agent must discover a search strategy. This is yet again a meta-level problem, and one we'll examine in a bit more detail.

We now have a new (meta)space to explore (this one containing exploration strategies for the original space containing the artifacts a) in which we are looking for a new (meta)artifact (the exploration strategy), so we have to reconsider the five steps in that context. Of course this meta-problem immediately suggests yet another meta-meta-problem—how do we search the space of search strategies?[13]

---

[12] In a very real sense, creativity is a social phenomenon. It is impossible to attribute creativity in a vacuum—both a creator and one or more receivers are necessary for creative attribution. The creator produces an artifact, or engages in a process, and the receiver(s) experience the result and attribute creativity based upon their perception of the artifact's, the process' and/or the creator's characteristics.

[13] It is possible that this third-level question is still related to the base domain in a non-trivial way, so that perhaps we don't have a really complete abstraction.

4.4.4.1 (meta)Preparation

Is this a new domain with its own background knowledge, etc? How should knowledge be represented at this abstract level? Does (already) knowing the base domain suggest the strategy? Is there some abstract exploration strategy domain that an agent must master before it can reasonably expect any success at this level? Or, perhaps there is often not much to consider here, and one just hops between meta- and base-level steps 2–4…

4.4.4.2 (meta)Incubation

How does the agent make connections at this level? How are abstract connections related to base-level connections? Another way to think about this is that the agent is looking for ways to structure the base space so that it is easy to explore. So, the dual problem is one of representation rather than one of exploration strategy—if the agent can re-represent the base domain so that, for example, it is (approximately) concave, the exploration problem becomes trivial.

4.4.4.3 (meta)Insight

This is still an “Aha!” moment. Or not. The same arguments apply as were given for the base level.

4.4.4.4 (meta)Evaluation

The agent now has some (meta)aesthetic for recognizing a good search strategy/representation, which suggests the following interesting philosophical question: Can an agent recognize this (meta)aesthetic without recognizing where it points in the base search (and thus already solving the base problem)? A more concrete version of this question is the following: is it possible to recognize a good fitness function without knowing what objects score well under that function?

4.4.4.5 (meta)Elaboration

In many cases, this likely degenerates to simply applying the exploration strategy (or the re-representation) back in the base domain. There may be situations in which the search strategy or re-representation itself is of general interest and perhaps even supersedes anything discovered in the base domain to which it applies. In such cases, consideration must be given to communicating the (meta)discovery and its import.

Returning to our base-level discussion, we first note the potential difficulty this apparent recursion introduces—it is not clear that there is a base case for terminating the recursion. Perhaps there exists a level of abstraction sufficient so that no further meta-level issues can arise. Or perhaps there will always be a point at which an "Aha" moment must be provided (by a human?) that will serve the purpose of tipping the process out of the recursion.

Finally, we will mention that it is very probably unrealistic to suppose that the evaluation function E is decidable rather, it is likely more realistic to suggest that E is at best *semi-decidable*—a quality artifact can be recognized, but it is not possible to recognize an artifact that does *not* measure up to the aesthetic.[14]

Now, the algorithm for F cannot simply consist of running E on a[15] because E may not halt. In this case, we need $F(E, a)$ to be decidable in some other way. Unfortunately, the obvious trivial reduction from the classical Halting Problem[16] means that this is not possible. So, in the absence of a decidable aesthetic, the problem of computational creativity is not computable in the strong sense, independent of whether the insight problem is real and independent of any difficulties (or lack thereof) due to meta-level recursion issues.

4.4.5 Elaboration

The elaboration step is often described as the "99 % perspiration" that complements the "1 % inspiration" of insight. The process is deliberate and intentional—the artifact is situated relative to the background knowledge, additional variations and details are generated and evaluated against the aesthetic, feedback from the environment may drive additional iterations and local refinement (or, even potentially major revisions). Herein lies all the hard work of development and polishing ideas, framing results and selling the finished product, and these processes may themselves require additional creativity, both large and small—iterating or recursing on some or all of the five "algorithmic" steps. Edison and his lightbulb are the perfect example here, not only for the oft-cited methodical search for the right filament material but also for the equally important development of the necessary infrastructure, marketing and educational programs that facilitated his version of the technology being adopted over all previous and contemporary competitors.

The computational challenges here are in many ways similar to those at the preparation stage, only in the reverse. Now, the system, rather than needing to acquire knowledge must dispense it, communicating both results and their import. The hard

---

[14] Perhaps the environment itself accepts those artifacts that everyone appreciates and rejects those that no one appreciates but isn't sure about those with mixed reception. Any aesthetic that accurately models such a scenario will not be decidable given the existence of all three types of artifact.

[15] Unless it is acceptable to have programs that may not terminate. If the insight issue is the sticky fourth case, this will be unavoidable, in which case F may remain a simple simulation of E without incurring any additional computational penalty for the overall "algorithm".

[16] Actually, the most obvious reduction is from the related Acceptance Problem.

work of filling in details, exploring possible processes, etc. may again be argued to be advantage computer for the same reasons cited above. The difficulty of framing or marketing the result is a more complex consideration, and may be regarded as a creative act itself: what story to tell, how to write the research paper, how to market a product, how to explain a piece of art.

4.5 Coda

It is unlikely that the creativity "algorithm" is computable in the strong Turing sense of decidability. It is somewhat more likely that creativity is weakly Turing computable in the sense of semi-decidability (recognizability), though there are some non-trivial hurdles to overcome before this might be demonstrated.

Still, Turing computability is a very strong result, and it is not surprising that a creativity "algorithm" might resist this level of constraint; indeed, most of human intelligence, if held to the strict standards of the current theory of computability, is a failure. That is not to say that efforts at computationally simulating it are failures but that humans themselves fail to "compute" by such strict standards. Also, it is certainly true that other uncomputable problems of interest, in many instances, do yield themselves to computational attacks of varying efficacy, so it is not unreasonable to expect that similar inroads might be forged into computational creativity.

Of course, there is also the (remote) possibility that in fact all the assumptions necessary to render creativity strongly computable will prove true, and we will discover that we can, simply, brute force an "Aha!" moment, no luck required. However, it is much more likely that this will not prove true. Some have begun to suggest that our traditional approach to computational complexity is not sufficient for the current state of the field of computer science [44], and it seems quite possible that computational creativity is making that same argument for our idea of what it means to compute.

References

1. Heath, D., Norton, D., Ventura, D.: Autonomously communicating conceptual knowledge through visual art. In: Proceedings of the 4th International Conference on Computational Creativity, pp. 97–104 (2013)
2. Heath, D., Norton, D., Ventura, D.: Conveying semantics through visual metaphor. ACM Trans. Intell. Syst. Technol. **5**(2), 31:1–31:17 (2014)
3. Norton, D., Heath, D., Ventura, D.: Establishing appreciation in a creative system. In: Proceedings of the 1st International Conference on Computational Creativity, pp. 26–35 (2010)
4. Norton, D., Heath, D., Ventura, D.: Autonomously creating quality images. In: Proceedings of the 2nd International Conference on Computational Creativity, pp. 10–15 (2011)
5. Norton, D., Heath, D., Ventura, D.: Finding creativity in an artificial artist. J. Creat. Behav. **47**(2), 106–124 (2013)

6. Johnson, D., Ventura, D.: Musical motif discovery from non-musical inspiration sources. J. New Music Res. in submission (2014)
7. Johnson, D., Ventura, D.: Musical motif discovery in non-musical media. In: Proceedings of the 5th International Conference on Computational Creativity, pp. 91–99 (2014)
8. Morris, R., Burton, S., Bodily, P., Ventura, D.: Soup over bean of pure joy: culinary ruminations of an artificial chef. In: Proceedings of the 3rd International Conference on Computational Creativity, pp. 119–125 (2012)
9. McCorduck, P.: AARON's Code: Meta-Art, Artificial Intelligence, and the Work of Harold Cohen. W. H. Freeman & Co. (1991)
10. Colton, S.: The painting fool: stories from building an automated painter. In: McCormack, J., d'Inverno, M. (eds.) Computers and Creativity. Springer, Berlin (2011)
11. Csíkzentmihályi, M., Robinson, R.E.: The Art of Seeing. The J Paul Getty Trust Office of Publications (1990)
12. Datta, R., Joshi, D., Li, J., Wang, J.Z.: Studying aesthetics in photographic images using a computational approach. Lect. Notes Comput. Sci. **3953**, 288–301 (2006)
13. Gevers, T., Smeulders, A.: Combining color and shape invariant features for image retrieval. IEEE Trans. Image Process. **9**, 102–119 (2000)
14. King, I.: Distributed content-based visual information retrieval system on peer-to-peer(p2p) network. http://appsrv.cse.cuhk.edu.hk/~miplab/discovir/ (2002)
15. Li, C., Chen, T.: Aesthetic visual quality assessment of paintings. IEEE J. Sel. Top. Signal Process. **3**, 236–252 (2009)
16. Wang, W.N., He, Q.: A survey on emotional semantic image retrieval. In: Proceedings of the International Conference on Image Processing pp. 117–120 (2008)
17. Wang, W.N., Yu, Y.L., Jiang, S.M.: Image retrieval by emotional semantics: a study of emotional space and feature extraction. IEEE Int. Conf. Syst. Man Cybern. **4**, 3534–3539 (2006)
18. Fellbaum, C. (ed.): WordNet: An Electronic Lexical Database. The MIT Press, Cambridge (1998)
19. Tsoumakas, G., Katakis, I.: Multi-label classification: an overview. Int. J. Data Warehous. Min. **3**(3), 1–13 (2007)
20. De Deyne, S., Storms, G.: Word associations: norms for 1,424 Dutch words in a continuous task. Behav. Res. Methods **40**(1), 198–205 (2008)
21. Sun, R.: The Cambridge Handbook of Computational Psychology, 1st edn. Cambridge University Press, New York (2008)
22. Kiss, G.R., Armstrong, C., Milroy, R., Piper, J.: An associative thesaurus of English and its computer analysis. In: Aitkin, A.J., Bailey, R.W., Hamilton-Smith, N. (eds.) The Computer and Literary Studies. University Press, Edinburgh (1973)
23. Nelson, D.L., McEvoy, C.L., Schreiber, T.A.: The University of South Florida word association, rhyme, and word fragment norms. http://www.usf.edu/FreeAssociation/ (1998)
24. Lund, K., Burgess, C.: Producing high-dimensional semantic spaces from lexical co-occurrence. Behav. Res. Methods Instrum. Comput. **28**, 203–208 (1996)
25. Denoyer, L., Gallinari, P.: The Wikipedia XML corpus. In: INEX Workshop Pre-Proceedings, pp. 367–372 (2006)
26. Thomas, S., Boatman, E., Polyakov, S., Mumenthaler, J., Wolff, C.: The noun project. http://thenounproject.com (2013)
27. Csurka, G., Dance, C.R., Fan, L., Willamowski, J., Bray, C.: Visual categorization with bags of keypoints. In: Proceedings of the Workshop on Statistical Learning in Computer Vision, pp. 1–22 (2004)
28. Sivic, J., Russell, B.C., Efros, A.A., Zisserman, A., Freeman, W.T.: Discovering objects and their location in images. Int. J. Comput. Vis. **1**, 370–377 (2005)
29. Cope, D.: Experiments in Musical Intelligence, AR Editions Madison, WI (1996)
30. Dubnov, S., Assayag, G., Lartillot, O., Bejerano, G.: Using machine-learning methods for musical style modeling. Computer **36**(10), 73–80 (2003)
31. Biles, J.: GenJam: a genetic algorithm for generating jazz solos. In: Proceedings of the International Computer Music Conference, pp. 131–137 (1994)

32. Burton, A.R., Vladimirova, T.: Generation of musical sequences with genetic techniques. Comput. Music J. **23**(4), 59–73 (1999)
33. Blackwell, T.: Swarm music: improvised music with multi-swarms. In: Proceedings of the AISB Symposium on Artificial Intelligence and Creativity in Arts and Science, pp. 41–49 (2003)
34. White, J.D.: The Analysis of Music. Prentice-Hall, New Jersey (1976)
35. Cleary, J., Witten, I.: Data compression using adaptive coding and partial string matching. IEEE Trans. Commun. **32**(4), 396–402 (1984)
36. Willems, F., Shtarkov, Y., Tjalkens, T.: The context-tree weighting method: basic properties. IEEE Trans. Inf. Theory **41**(3), 653–664 (1995)
37. Simon, I., Morris, D., Basu, S.: Mysong: automatic accompaniment generation for vocal melodies. In: Proceedings of the SIGCHI Conference on Human Factors in Computing Systems, pp. 725–734 (2008)
38. Varshney, L., Pinel, F., Varshney, K., Schorgendorfer, A., Chee, Y.M.: Cognition as a part of computational creativity. In: Proceedings of the 12th IEEE International Conference on Cognitive Informatics and Cognitive Computing, pp. 36–43 (2013)
39. Colton, S.: Creativity versus the perception of creativity in computational systems. In: Proceedings of the AAAI Spring Symposium on Creative Systems (2008)
40. Colton, S., Ventura, D.: You can't know my mind: a festival of computational creativity. In: Proceedings of the 5th International Conference on Computational Creativity, pp. 351–354 (2014)
41. Csikszentmihalyi, M.: Creativity: Flow and the Psychology of Discovery and Invention. Harper Perennial, New York (1996)
42. Liu, H., Singh, P.: ConceptNet—a practical commonsense reasoning tool-kit. BT Technol. J. **22**, 211–226 (2004)
43. Ohlsson, S.: Information-processing explanations of insight and related phenomenon. Adv. Psychol. Think. **1**, 1–44 (1992)
44. Vardi, M.Y.: On P, NP, and computational complexity. Commun. ACM **53**(11), 5 (2010)

Chapter 5
How Models of Creativity and Analogy Need to Answer the Tailorability Concern

John Licato, Selmer Bringsjord and Naveen Sundar Govindarajulu

Abstract Analogy is a major component of human creativity. Tasks from the ability to generate new stories to the ability to create new and insightful mathematical theorems can be shown to at least partially be explainable in terms of analogical processes. Artificial creativity and AGI systems, then, require powerful analogical subsystems—or so we will soon briefly argue. It quickly becomes obvious that a roadblock to such a use for analogical systems is a common critique that currently applies to every one in existence: the so-called "Tailorability Concern" (TC). Unfortunately, TC currently lacks a canonical formalization, and as a result the precise conditions that must be satisfied by an analogical system intended to answer TC are unclear. We remedy this problem by developing a still-informal but clear formulation of what it means to successfully answer TC, and offer guidelines for analogical systems that hope to progress further toward AGI.

5.1 Introduction

Creativity and the ability to reason analogically have a strong relationship [18, 19, 23, 28]. For example, many of the greatest lines in literature make use of familiar analogical processes: "A rose by any other name would smell as sweet" [41] makes sense if that rose and its name are understood to correspond to Romeo and his given name (analogical mapping). The subsequent analogical inference is that just as the nature of the rose is independent of its name, so is Romeo's nature independent of his.

Even if one believes that overuse of analogy can be harmful to creative thought [39], many researchers argue that the ability to determine analogical similarity is

J. Licato (✉) · S. Bringsjord · N.S. Govindarajulu
Rensselaer Polytechnic Institute, Troy, NY, USA
e-mail: licatj@rpi.edu

S. Bringsjord
e-mail: selmer@rpi.edu

N.S. Govindarajulu
e-mail: govinn@rpi.edu

T.R. Besold et al. (eds.), *Computational Creativity Research: Towards Creative Machines*,
Atlantis Thinking Machines 7, DOI 10.2991/978-94-6239-085-0_5

important at least for combinatorial creativity, which seems to be very easy for humans but very difficult for AI systems [5, 6]. Furthermore, the importance of analogy is of course not limited to creativity, as analogical ability has been proposed as an indispensable component of artificial general intelligence as well [11, 26].

With the above in mind, it makes sense to develop models of analogy, both computational and theoretical. Much work has been done in this direction; a few implementations include SME [17], LISA [29], HDTP [40], and recently our own META-R [32]. On the surface, it seems that the current generation of analogical systems sufficiently capture and explain all of the phenomena commonly associated with analogical reasoning, and that they will eventually reach levels characteristic of human cognition. It may well be the case that the most important principles underlying the nature of analogy have been expressed. But a serious objection has been raised recently which, as will be argued, should be the primary focus of analogical researchers over the next few years—at least if any significant further progress is to be made in the direction of creativity and AGI.

The objection is raised by Gentner and Forbus [24]. They call it the 'tailorability concern' (TC), and the objection echoes a common criticism of cognitive systems in general: that they operate on toy examples manually constructed in such a way as to guarantee the desired solution. However, though this concern may have been stated in many forms throughout the years [36], it lacks, to our knowledge, a formulation clear enough to anchor productive scientific discussion. And this ambiguity in turn negatively impacts not only the relevant science, but AI engineering as well: absent a definition of TC, it is difficult to understand precisely what an analogical system must do in order to successfully answer TC. In the remainder, we take steps toward addressing this problem as it applies to analogical systems.

5.2 The Tailorability Concern

A frequently appearing criticism of cognitive systems in general is that they are only applied to manually constructed 'toy examples', a problem many researchers in the field themselves acknowledge. Gentner and Forbus [24] refer to the problem as the *tailorability concern* (TC): "that is, that (whether knowingly or not) the researchers have encoded the items in such a way as to give them the desired results" [24].

Of course, nothing is wrong with toy examples per se. They can be extremely useful in demonstrating key concepts, helping to illustrate particular qualitative strengths or weaknesses of computational models, or helping to get a new model off the ground. Indeed, the present authors plead guilty to using toy examples in these ways; but properly done, carefully chosen microcosmic cases can be very useful as demonstrations-of-concept. But we should be careful not to treat such examples as the final proof of a system's worth, since in most of these examples it is not clear that the principles used to solve them generalize to other problems, nor is it clear that such principles can be used to mechanically find useful solutions just as effectively in the absence of human assistance.

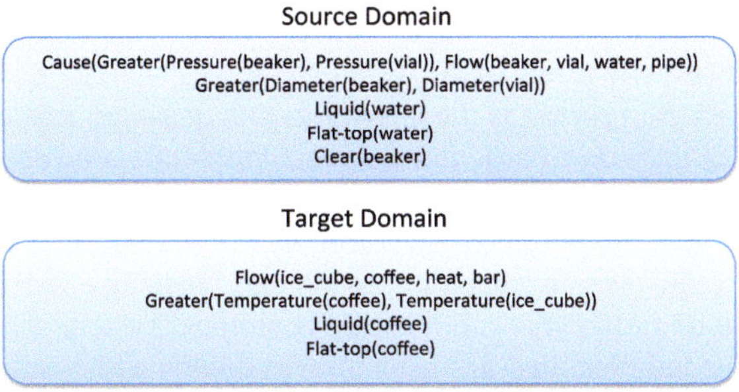

Fig. 5.1 One version of the popular "heat flow" example from Falkenhainer et al. [17] used to test many analogical systems

For example, the well-known water-flow/heat-flow analogy (Fig. 5.1) has been used as a demonstration of many models of analogy [17, 27, 31, 43]. But little to nothing is written about how the structural representations used in examples such as these are acquired in the first place. One approach is to model the acquisition of structured representations through sensory data (e.g., see [16]), and another is to presume the existence of a large database of already-structured data (such as that to which a neurobiologically normal adult might be expected to have access), and some sort of filtering process that ensures that from this database, proper representations are selected and any unnecessary data that would produce incorrect matching results are excluded.[1] Yet even when such filtering processes are proposed, they are not put to the test and proven to perform well with a large database containing enough knowledge to match that of a child's, much less an adult's. The TC rightfully attempts to refocus efforts on these filtering processes, by requiring that they demonstrate the ability to produce clean source and target analogs as required by the analogical mappers.

The field of case-based reasoning (CBR), which overlaps quite heavily with that of analogical reasoning (AR), also deals with some of the issues raised by the TC. There are differing opinions on what features distinguish the CBR and AR approaches (see [1, 14, 34]), but two common themes are that CBR tends to deal with source and target cases that come from the same domain, and cases are selected and adapted with some clear pragmatic goal in mind. AR approaches, on the other hand, try to be more generally applicable across different domains, and tend to focus more on the mapping process that actually determines analogical similarity. CBR approaches, then, deal with the TC by trading generality for effectiveness, so that a program designed to work well in one domain (medical diagnosis, for example, is a popular field) may not work so well in another without a significant amount of human assistance.

Unfortunately, the CBR approach of restricting generality does not sufficiently answer the TC. Analogy research can be seen as centering around fundamental

[1] It would be less appropriate to urge a careful treatment of the TC and tie it so closely to large semantic databases if they weren't available. But over the past few years, natural-language processing and semantic-web technologies have been progressing to the point where we now have access to large collections of semantic databases containing wide-ranging general knowledge. These include Cyc [35], Freebase [7], and DBPedia [3]. Many of these have easy-to-use interfaces.

questions, one of them being: How can we find good analogies? The TC is especially problematic because it forces analogy researchers to prove that their theoretical process is the answer to this question, and although it can be promising to see that some particular approach produces good analogies in some limited domain, no approach can constitute a completely satisfying answer to this question unless it is versatile enough to perform well in *many* domains.

Any system that can answer the challenge of the TC will instantly distinguish itself from every other extant analogical system, since (at least, to our knowledge) the only one that has been able to do this with some degree of success is the SME-based family of systems [20, 24, 33]. Later in this chapter we will briefly discuss SME (particularly, the CaseMapper software) and point out several of its features which we believe constitute partial answers to the TC. But first, it is important to clarify what it means to answer this challenge and discuss why it is such a non-trivial feat.

5.2.1 Answering the TC

Gentner and Forbus [24] suggest that there are two possible ways to answer the TC. One applies to visual domains, and involves using automatic encodings of visual representations. The other more generally applicable direction involves two key features: first, the use of pre-existing databases; second, an automated or semi-automated parsing process that goes from input text to a sufficiently rich semantic representation. A first attempt at a precise statement of what it means to answer the TC is as follows:

> **$\mathbf{TCA_1}$** A computational system of analogy answers the TC if, given no more than a pre-existing database and an unparsed input text, it is able to consistently produce *good* analogies across many domains.

At least one general intent behind **$\mathbf{TCA_1}$** is clear: it attempts to place emphasis on the filtering process (whose job is, as we said, to either select some subset of available source analogs from a large database and recommend only some of them for the more computationally expensive step of analogical matching, or to automatically construct structured representations from sensory data). By removing the reliance on human intervention, TC ensures that the filtering is not done manually in such a way that guarantees desired results. However, in order to truly answer the TC in a satisfactory way, we must be precise about its purpose and motivations: What concerns are behind the TC in the first place? Furthermore, **$\mathbf{TCA_1}$** is hopelessly vague: the words ‘unparsed’ and ‘good’, if left open to interpretation, make it too easy for anyone to prematurely claim victory over TC. Also: Why is it important that the database be pre-existing? What degree of separation must there be between the creators of the database and the designers of the analogical system? For example, does the database’s underlying knowledge-representation philosophy need to overlap with that of the analogical system?

5.2.1.1 What Is a 'Good' Analogy?

Though the widely influential Structure-Mapping Theory [22] offers the systematicity principle and the one-to-one constraint as indicative features of a good analogy, it does not provide a clear quantitative measure of analogy quality. SME evaluates match scores by combining the scores of the evidence provided by its match rules; this allows for a comparison between different matches of the same problem. But the resulting match score is not normalized, and as a result match quality between different problems cannot be compared [21]. Other analogical models do not help much in this regard. Holyoak and Thagard's [27] Multiconstraint Theory, for example, introduces additional criteria to evaluate what makes an analogical match a good one, making cross-domain analogical match quality more difficult to assess.

This is especially problematic when considering TC. Given a set of predefined problems and desired answers for all of them, the good analogy is simply the one that performs as intended. But when applied to problem sets where the best analogy, or even the existence of a good analogy, is not clear even to the persons running the experiment, do the guidelines currently available still apply?

Paul Bartha [4] offers a comprehensive theory of analogy that can be useful here. His goal is to produce a *normative* account of analogy, and in the process he sheds light on the proper role of analogy within a larger context of reasoning. His Articulation Model is based on the idea that there are two features common to all good analogical arguments: a prior association (a relationship in the source analog that is transferred to the target), and a potential for generalization. But perhaps most relevant to our present discussion is his claim that by virtue of the analogical argument, the resulting hypothesis inferred through an analogical argument contains no more than *prima facie* plausibility, which can be understood as something like a suggestion that the inferred hypothesis is worth exploring further. If there is an independent reason to reject the hypothesis, such as a deductive argument showing it leads to a contradiction, or contrary empirical evidence, then it can be abandoned.

The idea that the proper role of an analogical argument is to do no more than provide *prima facie* plausible hypotheses [37, 38] suggests that the relationship between an analogy's match quality and its tendency to produce hypotheses which can be independently verified may not be as simple as it might seem. In the end, a *model* of analogy is a good one only if it produces good analogies, and an analogy is a good one only if it produces plausible hypotheses.[2]

This complicates things further. In toy examples, the information available to the analogical matcher is very limited; a hypothesis in these cases is plausible if the information available to the analogical matcher does not provide a reason to reject that hypothesis. But if we are suggesting that the information available to the analogical system (or more specifically, the filtering process) is actually a large

[2] We might leave room here to exclude models of analogy that have psychological or neurological plausibility as their primary end goals. In these cases, it might be the goal of the model to replicate poor analogical reasoning as well, if it matches human performance. But it is our assumption (at least in the present inquiry) that the ultimate goal of AGI research is not to model poor human reasoning.

database conceivably encompassing, say, all of the knowledge on the World Wide Web, then how are we to realistically evaluate the system's performance?

In light of these complications, we propose the following alterations of **TCA$_1$**. If the word 'good' is replaced with 'useful,' then we connote an evaluation method that is not based in the particular metrics as defined by the analogical theory itself (which can be controversial), but rather based in an intuitive notion that can and should be evaluated independently of the model. In other words, researchers might disagree on how to measure an analogical match, but whether the resulting analogically inferred hypothesis is useful can be evaluated without any knowledge of the analogical matcher used. Of course, since what is 'useful' can be very domain-dependent, we do not claim that this word is completely unambiguous. Later in this chapter, a better replacement for the word 'useful' will be suggested. For now, the aim behind this move is to divorce the metric which determines the quality of an analogical match's results (which may be very domain-dependent) from the theory-specific metric that the matcher is specifically designed to optimize. That (deceptively) small change gives us **TCA$_2$**:

> **TCA$_2$** A computational system of analogy answers the TC if, given no more than a pre-existing database and an unparsed input text, it is able to consistently produce *useful* analogies across many domains.

5.2.1.2 What Are Acceptable Databases?

In **TCA$_2$**, it is clear that the knowledge available in the database used is a limiting factor in how informative the inferences produced by the analogical system can be. The suggestion phrased by Gentner and Forbus [24] as "pre-existing databases" requires more clarification. The implication (at least as we interpret it) is that the dataset and the structures within cannot have been constructed for the purpose of solving the particular toy examples that are of interest. Otherwise this introduces bias and tailorability concerns, in spite of the best intentions of the designers. Two issues immediately come to mind. First, what is the proper level of separation between the database and the analogical system? Secondly, how large does the database have to be?

The first question is at least partially answered by considering the level of *representational agreement* between the database and the analogical system. For example, if the database is a purely distributed one with no localist concepts whatsoever (which is, we acknowledge, an unlikely possibility), and the analogical system is one that uses only localist, explicitly structured data, then a significant level of work will be needed to first extract the information from the database and put it into the form that the analogical reasoner requires (this can be considered a *re-representation* step [15, 30, 44]). The choice of database becomes important for this reason, and if no database exists that does not require a high level of re-representation, then it suggests a problem: Consider that although proponents of localist, distributed, and hybrid representation styles make claims all the time about the scalability of their assumptions of knowledge representation, the researchers who have to design and work with large semantic databases actually have to put these ideas to the test. If the state-of-the-art research in the database field has significant difficulty with some style of representation as required by an analogical matcher, then perhaps that matcher needs to carefully consider its assumptions about the nature of knowledge representation, or else be able to easily extract the necessary data.[3] The choice of database is therefore a sensitive one.

It may help to go back to the criticisms which may have motivated this requirement in the first place. Hofstadter's group criticizes SME by saying that "the analogy is already effectively given in the representations" [36]. The information provided to the matcher is selected in such a way that it does not include a significant amount of extra data that would lead to false analogies, and the choice of predicates and objects is done in such a way that presupposes the desired matching. This is a criticism composed of two points: the amount of information provided as input (which is not too little, or too much, but *just right*), and the nature of the information (the corresponding predicates on both sides just happen to have the right amount of arguments in the necessary order). The system is not robust enough to produce good analogies when given the same input data in a variety of formats.

Two possible changes can begin to answer these critiques. One is to require that the databases are large enough, and the input minimal enough, to introduce a significant amount of false matches that would confuse less robust analogical matchers. Critics claiming that the input is too carefully chosen could be pointed to the fact that the analogical matcher must plumb through an answer space that is large (at least relative to the input problem). The larger the search space, the more impressive the ability of the analogical matcher to select only a few potentially relevant source analogs. Furthermore, an inability to demonstrate scalability to large datasets weakens any architecture's claim to psychological plausibility: if the architecture can't handle a dataset large enough to produce non-trivial answers, how can it be an accurate model of a process used by human-level reasoners?[4]

Secondly, we could require robust and consistent performance on a variety of input forms. For example, in the heat-flow problem (Fig. 5.1) Mitchell and Hofstadter [36] note that there are many possible ways to structure the input: heat could be described as an object, or as an attribute of coffee, or *heat flow* could be a relation with three rather than four arguments [36]. Consistent performance across various input forms puts more pressure on the analogical matcher's re-representation algorithm(s), rather than relying on a separate NLP module. This also allows for a leveling of the playing field across different systems: In order to show that a given example adheres to this requirement, a localist, structured analogical system would have to demonstrate two things with regard to that particular example:

---

[3] We do not mean here to say that what works best for large artificial databases is the same as what is employed by the human brain. But if a researcher declares that the structure of human knowledge has certain properties, and large datasets cannot be created that do not have those properties for reasons of scalability, then it should be at least a weak hint that perhaps the assumption of those properties is not practicable.

[4] This is a common criticism of Hummel and Holyoak's LISA system; see [24].

- Across multiple, more-or-less equivalent structural representations of the same input data, and a wide variety of domains, the matcher still produces the desired results.
- The desired results are still produced when the input is minimal; meaning any redundant information or structural constructs which might be identified by critics as being used only to aid the matcher can be removed.

Of course, if a system did happen to require no more than unstructured, natural-language descriptions as input, or direct sensory data from visual representations, it would satisfy both of these conditions. This allows our criteria to encompass the alternate route to answering the TC mentioned by Gentner and Forbus [24]—a route which seeks to answer TC by not necessarily having a large database, but by having one that at least attempts to directly construct structured representations from low-level sensory or natural-language data. The generality of these conditions allows us to claim the converse of **TCA$_2$**, leading us to our next iteration:

TCA$_3$ A computational system of analogy answers the TC if and only if given no more than either

- unstructured textural and/or visual data, or
- a large, pre-existing database,

and minimal input, it is able to consistently produce *useful* analogies and demonstrate stability through a variety of input forms and domains.

One might be satisfied with this set of criteria, which draws its strength from its lack of commitment to any particular theory of analogy, and its emphasis on large sets of non-tailored input. But **TCA$_3$** is undeniably ambiguous, and may not be focused enough to guide any research program. We encourage the reader to take **TCA$_3$** and develop it further, but first, to close out this paper we will take some steps of our own to reduce some of its weaknesses.

5.2.1.3 Strengthening the TC with Psychometric AI

We will make two important moves to sharpen **TCA$_3$**. One, we turn to *Psychometric AI* (PAI) [9, 13], according to which, in a nutshell, commendable AI systems are those that demonstrate prowess on tests of various mental abilities from psychometrics. Our second move is to embed analogy-generation systems within broader AI problem-solving systems that make use of additional forms of fundamental reasoning in human-level intelligence; e.g., deduction. In particular, we place TC within the context of the integration of analogical reasoning with deduction, which we dub *analogico-deductive reasoning* (ADR), and which we have explained and demonstrated elsewhere [11]. An ADR system does use analogy generation, but analogies are used to guide solutions that can be rigorously verified by proof. The architecture-sketch of an ADR system that accords with our pair of moves is shown in Fig. 5.2. While we don't have the space to provide details here, this system receives puzzles that are part linguistic and part visual in nature (e.g., so-called *seating puzzles*), and

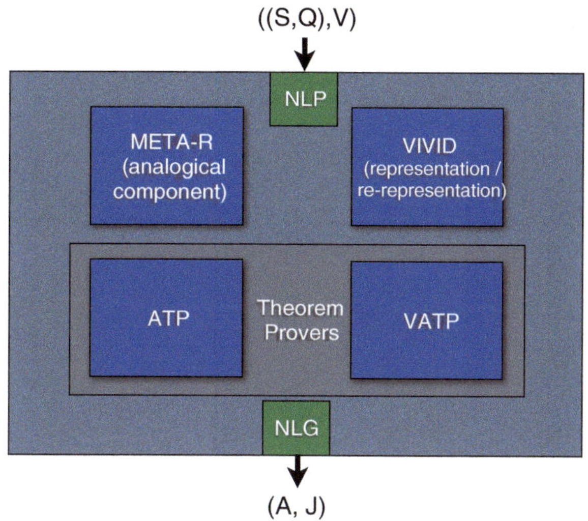

Fig. 5.2 Architecture-Sketch of an ADR System that Answers TC. Input includes a **S**tory and a **Q**uestion, along with **V**isual content. Output includes an **A**nswer and—in the form of a proof or at least a rigorous argument—a **J**ustification

harnesses not only our own analogical system (META-R), and not only ATP technology, but *visual* theorem-proving; for formal details see [2]. This architecture-sketch is inspired by, but abstracts and extends beyond, AI systems able to solve analogy-relevant problems. One such inspiring system is by Lovett et al. [33]; it can solve items on the Raven's Progressive Matrices. But it cannot for example prove that its answers are correct, which is part of what the architecture-sketch in Fig. 5.2 demands.

Given the context we have now set, we can articulate a new biconditional:

> **TCA$_4$** A computational system $\mathscr{A}$ for analogy generation answers the TC if and only if, given as input no more than either
>
> - unstructured textual and/or visual data, or
> - a vast, pre-existing database not significantly pre-engineered ahead of time by humans for any particular tests of $\mathscr{A}$,
>
> is—in keeping with aforementioned *Psychometric AI*—able to consistently generate analogies that enable $\mathscr{A}$ to perform *provably well* on precisely defined tests of cognitive ability and skill.

To comment briefly on **TCA$_4$**, first note that we remove the explicit requirement that the ability to find useful analogies be stable across a variety of input forms and domains. This is subsumed by the requirement of good performance on precisely defined tests; it is assumed that a sufficiently difficult psychometric test would provide questions that are both varied in their form (e.g., word problems, puzzle solving, story comprehension) and in their domains. The implicit requirement of domain variety rules out the possibility of an artificial reasoning agent that can only process, for example, certain types of math problems, as an acceptable answer to the TC.

Some might also see **TCA$_4$** and the use of PAI as too restrictive in that it relies too heavily on problem-solving and not enough on either creative thinking or the kind of everyday analogical thinking that may not be goal-oriented in nature. PAI, however, provides a tool for measuring those abilities which, at least at the surface, don't rely on directed problem-solving, such as reading comprehension. Additionally, it is difficult to imagine that any research program in AGI would be able to demonstrate clear progress without showing increased performance in an ability that can be measured according to some psychometric test. Non-goal-oriented analogical reasoning is a good example of this principle: If the cognitive processes underlying normal analogical reasoning when it is non-goal-oriented (as in everyday reasoning) and when it is goal-oriented (as during psychometric testing) are largely the same, then an artificial system capable of performing the latter has a strong claim to performing the former. A system that only has sporadic performance on psychometrically-measurable tasks is difficult to defend as generally intelligent.

One might ask: Can a system do well on psychometric tests and still be subject to claims of tailorability? The answer, if the requirements in **TCA$_4$** are not met, is certainly *Yes*. PAI is not meant to be a replacement for the input format and large database requirements we have been developing in this paper; rather, it is only one possible sharpening of the ambiguous concepts in **TCA$_3$**. Other possibilities may exist, but we do not at present know of any satisfying alternatives.

5.3 Case Mapper: A Step in the Right Direction?

TCA$_3$ and **TCA$_4$** reflect a growing realization among the AI community that if some cognitive system is to ever move closer to AGI, it needs to be able to demonstrate its performance with a large knowledge base. *Case Mapper*, from Northwestern University's Qualitative Reasoning Group, is a system currently in development that allows users to test SME [17], MAC/FAC, and other analogical reasoning tools, along with large databases that are included. Among these databases is a version of OpenCyc [35], which comes with millions of facts connecting tens of thousands of concepts and relations. Although in our opinion the database is not yet large enough to match the performance of an 8-year-old child on analogy problems that require recollection of arbitrary source cases, the way it gives the user access to such a collection of powerful tools suggests it can become very useful to anyone interested in the field.

Case Mapper works with *cases*, which are collections of facts that can be constructed automatically by the program given some concept in the database. After mapping two cases, a set of candidate inferences can be drawn from the source to the target case. We can explore these candidate inferences manually in order to have some idea of the system's current ability. For example, having Case Mapper construct cases for *Dog* and *Cat* and setting them as source and target cases respectively, among the candidate inferences produced are:

```
(Collection
  (SubcollectionOfWithRelationToTypeFn
    (:skolem Shedding)
    (:skolem bodilyDoer)
    Cat
  )
)
```

This essentially means, "Cats shed", which is an hypothesis that could easily be confirmed by observation, allowing an ADR-based system to add to its own knowledge base autonomously. Turning now back to **TCA$_4$**, we note that a similar approach can be used to solve Psychometric-styled analogy questions,[5] such as the following:

Baseball – Ball:BaseballGame::TennisBall: ?

In order to solve a problem like this, a system might find a mapping between a source case constructed around the concepts *Baseball-Ball* and *BaseballGame*, and a target case from *TennisBall*, where the mapping is instructed beforehand to map the two ball concepts together. In the resulting mapping, whichever element in the target case corresponds to *BaseballGame* should be the correct answer. All of these settings are available in the current version of Case Mapper, but it does not output the correct answer because there is not enough information in the cases to draw a mapping with confidence. The relevant facts in the provided OpenCyc database are as follows:

```
(isa
  Baseball-Ball
  (EquipmentTypeForEventTypeFn
    BaseballGame
  )
)

(isa
  TennisBall
  (EquipmentTypeForStructuredActivityFn
    Tennis-TheGame
  )
)
```

Note that although a human would be expected to realize that in these contexts the symbols *EquipmentTypeForEventTypeFn* and *EquipmentTypeForStructuredActivityFn* are close together enough semantically to map for the purposes of this problem,

[5] We must mention here that standard proportion-type analogy questions seem to be falling out of favor recently; they have for example been removed from the SAT, and it is unclear at this time whether they will be brought back anytime soon.

they do not register as a match here, and as a result the concepts *BaseballGame* and *Tennis-TheGame* are not linked. Nevertheless, were the database more complete (as one would hope it will be eventually), such a psychometric problem would likely have been trivial.[6]

5.4 Conclusion and Future Work

The concerns reflected by TC have been more or less echoed in many criticisms of analogical systems over the past 30 years or so, but little to no discussion has attempted to precisely state the conditions under which an analogical system would no longer be subject to this common complaint. We humbly acknowledge the limitations of our formulation here, and remind the reader that even a system with a strong claim to meeting the conditions of $\mathbf{TCA_4}$ (e.g., an implementation of Fig. 5.2) may not be completely immune to TC-based criticisms. But it seems that such criticisms would be somewhat difficult to justify. We hope that this paper inspires further discussion, and invite researchers to pick up from and sharpen $\mathbf{TCA_4}$, or to steer away from Psychometric AI and therefore develop $\mathbf{TCA_3}$ in a different direction.

Those who strive for something deeper than a test-based anchor for a response to TC will presumably be more sanguine than Bringsjord about the ultimate reach of computing machines. Bringsjord holds that it is impossible for any mere computing machine to *originate* anything [10], or even to genuinely understand anything [12]. But he holds on the other hand that for any determinate behavioral test, a computing machine can be programmed to pass it—even if the test is one which, if passed, suggests to most human observers that the passing machine is not simply the product of human intelligence, harnessed [8]. In short, the turn to $\mathbf{TCA_4}$ as a way of fleshing out $\mathbf{TCA_3}$ reflects deep pessimism about computing machines, and specifically a deep affirmation of the penetrating and profound nature of TC. Licato, in contrast, is more optimistic about the potential of analogical-hybrid systems, but is not prepared to take any philosophical positions on whether a system that passes TC demonstrates true understanding or creativity.

What about AI work devoted to building creative systems on the strength of analogical reasoning? Our work continues along this path. We are currently investigating the potential of using our analogical matching system META-R, which is designed to allow for a flexible application of heuristics across a wide variety of domains, with the community-built dataset Freebase [7]. We are also investigating the use of META-R in modeling the automatic generation of logical theorems and high-level proofs from semi-formal and formal domains [25, 32]. Future work in this direction includes the investigation of Analogico-Deductive Moral Reasoning (ADMR) using a hybrid of

[6] Of course, this discussion does not even mention the difficulty artificial systems have in finding an overlap between symbols on the basis of the semantics of the referents of those symbols. Such a problem is somewhat out of this chapter's scope, but is certainly something that researchers interested in transcending the TC should eventually tackle.

CLARION [42] and LISA [29]. And finally, we are seeking to leverage analogical reasoning in order to engineer systems capable of automatic programming.[7]

With profound gratitude, the support of both the John Templeton Foundation and AFOSR is acknowledged.

References

1. Aamodt, A., Plaza, E.: Case-based reasoning: foundational issues, methodological variations, and system approaches. AI Commun. **7**(1), 39–59 (1994)
2. Arkoudas, K., Bringsjord, S.: Vivid: an ai framework for heterogeneous problem solving. Artificial Intelligence 173(15), 1367–1405 (2009). http://kryten.mm.rpi.edu/vivid_030205.pdf. The http://kryten.mm.rpi.edu/vivid/vivid.pdf provides a preprint of the penultimate draft only. If for some reason it is not working, please contact either author directly by email
3. Auer, S., Bizer, C., Kobilarov, G., Lehmann, J., Cyganiak, R., Ives, Z.: DBpedia: a nucleus for a web of open data. In: Proceedings of the 6th International Semantic Web Conference (2007)
4. Bartha, P.F.: By Parallel Reasoning: The Construction and Evaluation of Analogical Arguments. Oxford University Press, New York (2010)
5. Boden, M.A.: Creativity and unpredictability. Stanford Humanit. Rev. **4**(2), 123–139 (1995)
6. Boden, M.A.: Computer models of creativity. AI Mag. **30**(3), 23–34 (2009)
7. Bollacker, K., Evans, C., Paritosh, P., Sturge, T., Taylor, J.: Freebase: a collaboratively created graph database for structuring human knowledge. In: Proceedings of the 2008 ACM SIGMOD International Conference on Management of Data (SIGMOD'08), ACM (2008)
8. Bringsjord, S.: Could, how could we tell if, and why should-androids have inner lives? In: Ford, K., Glymour, C., Hayes, P. (eds.) android epistemology, pp. 93–122. MIT Press, Cambridge (1995)
9. Bringsjord, S.: Psychometric artificial intelligence. J. Exp. Theor. Artif. Intell. **23**(3), 271–277 (2011)
10. Bringsjord, S., Ferrucci, D., Bello, P.: Creativity, the turing test, and the (better) lovelace test. Mind. Mach. **11**, 3–27 (2001)
11. Bringsjord, S., Licato, J.: Psychometric Artificial General Intelligence: The Piaget-Macgyver Room. In: Wang, P., Goertzel, B. (eds.) Theoretical Foundations of Artificial General Intelligence. Atlantis Press (2012). http://kryten.mm.rpi.edu/Bringsjord_Licato_PAGI_071512.pdf
12. Bringsjord, S., Noel, R.: Real robots and the missing thought experiment in the chinese room dialectic. In: Preston, J., Bishop, M. (eds.) Views Into The Chinese Room: New Essays on Searle and Artificial Intelligence, pp. 144–166. Oxford University Press, Oxford (2002)
13. Bringsjord, S., Schimanski, B.: What is artificial intelligence? psychometric ai as an answer. In: Proceedings of the 18th International Joint Conference on Artificial Intelligence (IJCAI-03), pp. 887–893. Morgan Kaufmann, San Francisco, CA (2003). http://kryten.mm.rpi.edu/scb.bs.pai.ijcai03.pdf
14. Burstein, M.H.: Analogy vs CBR: the purpose of mapping. In: Proceedings from the Case-based Reasoning Workshop, pp. 133–136. Pensacola Beach, Florida (1989)

---

[7] In the case of automatic programming, the input shown in Fig. 5.2 would be instantiated as the informal definition of a number-theoretic function (where that definition can be partly linguistic and partly visual), and the answer is code in some conventional programming language, accompanied by a proof of the correctness of this code relative to the input. Automatic-programming systems seemingly require the ability to judge two programs analogous. More precisely, such systems seemingly would need to be able to answer this question: Given a set of programs $\mathscr{P}$ in some programming language, can the system produce a similarity metric $\rho\colon \mathscr{P} \times \mathscr{P} \rightarrow \mathbb{R}$ capturing which pairs of programs are semantically analogous?

15. Chi, M.T., Feltovich, P.J., Glaser, R.: Categorization and representation of physics problems by experts and novices. Cogn. Sci. **5**(2), 121–152 (1981)
16. Doumas, L.A., Hummel, J.E., Sandhofer, C.: A theory of the discovery and predication of relational concepts. Psychol. Rev. **115**, 1–43 (2008)
17. Falkenhainer, B., Forbus, K.D., Gentner, D.: The structure-mapping engine: algorithm and examples. Artif. Intell. **41**, 1–63 (1989)
18. Finke, R.: Creative Imagery: Discoveries and Inventions in Visualization, pp. 239–260. Lawrence Erlbaum Associates, Mahwah (1990)
19. Finke, R., Ward, T., Smith, S.: Creative Cognition: Theory, Research, and Applications. MIT Press, Cambridge (1992)
20. Forbus, K., Klenk, M., Hinrichs, T.: Compaion cognitive systems: design goals and lessons learned so far. IEEE Intell. Syst. **24**(4), 36–46 (2009)
21. Forbus, K.D., Gentner, D.: Structural evaluation of analogies: what counts. In: Proceedings of the Eleventh Annual Conference of the Cognitive Science Society, vol. 34 (1989)
22. Gentner, D.: Structure-mapping: a theoretical framework for analogy. Cogn. Sci. **7**, 155–170 (1983)
23. Gentner, D., Brem, S., Ferguson, R., Wolff, P., Markman, A., Forbus, K.: Analogy and creativity in the works of Johannes Kepler. In: Ward, T., Smith, S., Vaid, J. (eds.) Creative Thought: An Investigation of Conceptual Structures and Processes, pp. 403–459. American Psychological Association (1997)
24. Gentner, D., Forbus, K.: Computational models of analogy. Wiley Interdisc. Rev.: Cogn. Sci. **2**(3), 266–276 (2011)
25. Govindarajulu, N.S., Licato, J., Bringsjord, S.: Small steps toward hypercomputation via infinitary machine proof verification and proof generation. In: Unconventional Computation and Natural Computation—12th International Conference (UCNC 2013) (2013)
26. Hofstadter, D.R.: The analogical mind: perspectives from cognitive science. In: Gentner, D., Holyoak, K.J., Kokinov, B.N. (eds.) Epilogue: Analogy as the Core of Cognition, vol. 15, pp. 331–336. The MIT Press, Massachusetts (2001)
27. Holyoak, K.J., Thagard, P.: Analogical mapping by constraint satisfaction. Cogn. Sci. **13**(3), 295–355 (1989)
28. Holyoak, K.J., Thagard, P.: Mental Leaps: Analogy in Creative Thought. MIT Press, Cambridge (1995)
29. Hummel, J.E., Holyoak, K.J.: A symbolic-connectionist theory of relational inference and generalization. Psychol. Rev. **110**, 220–264 (2003)
30. Krumnack, U., Gust, H., Kühnberger, K.U., Schwering, A.: Re-representation in a logic-based model for analogy-making. In: Proceedings of the 21st Australasian Joint Conference on Artificial Intelligence. Aukland, New Zealand (2008)
31. Krumnack, U., Schwering, A., Gust, H., Kühnberger, K.U.: Restricted higher-order anti-unification for analogy making. In: Proceedings of Twenties Australian Joint Conference on Artificial Intelligence, LNAI 4830, pp. 273–282. Springer (2007)
32. Licato, J., Govindarajulu, N.S., Bringsjord, S., Pomeranz, M., Gittelson, L.: Analogico-deductive generation of gödel's first incompleteness theorem from the liar paradox. In: Proceedings of the 23rd Annual International Joint Conference on Artificial Intelligence (IJCAI-13) (2013)
33. Lovett, A., Forbus, K., Usher, J.: A structure mapping model of raven's progressive matrices. In: Proceedings of CogSci, vol. 10 (2010)
34. Mantaras, RLd, Mcsherry, D., Bridge, D., Leake, D., Smyth, B., Craw, S., Faltings, B., Maher, M.L., Cox, M.T., Forbus, K., Keane, M., Aamodt, A., Watson, I.: Retrieval, reuse, revision and retention in case-based reasoning. Knowl. Eng. Rev. **20**(3), 215–240 (2005)
35. Matuszek, C., Cabral, J., Witbrock, M., DeOliveira, J.: An introduction to the syntax and content of cyc. In: Proceedings of the 2006 AAAI Spring Symposium on Formalizing and Compiling Background Knowledge and its Applications to Knowledge Representation and Question Answering, vol. 3864 (2006)

36. Mitchell, M., Hofstadter, D.R.: Perspectives on copycat: comparisons with recent work. In: Hofstadter, D.R. (eds.) The Fluid Analogies Research Group. Fluid Concepts and Creative Analogies: Computer Models of the Fundamental Mechanisms of Thought, Chap. 6. Basic Books, Inc (1995)
37. Norton, J.D.: Formal and Material Approaches to Analogical Inference. Forthcoming (Forthcoming)
38. Owen, S.: Analogy for Automated Reasoning. Academic Press (1990)
39. Riesbeck, C.K.: Inside Case-based Reasoning. Lawrence Erlbaum Associates, Hillsdale (1989)
40. Schmidt, M., Krumnack, U., Gust, H., Kühnberger, K.U.: Heuristic-driven theory projection: An overview. Computational Approaches to Analogical Reasoning: Current Trends. Studies in Computational Intelligence 548 (2014)
41. Shakespeare, W.: Romeo and juliet. In: The Complete Works of William Shakespeare, p. 254. Wordsworth Editions Limited (2007)
42. Sun, R.: Duality of the Mind: A Bottom Up Approach Toward Cognition. Lawrence Erlbaum Associates, Hillsdale (2002)
43. Wilson, W.H., Halford, G.S., Gray, B., Phillips, S.: The STAR-2 model for mapping hierarchically structured analogs. In: Gentner, D., Holyoak, K.J., Boicho, K.K. (eds.) The Analogical Mind: Perspectives from Cognitive Science, chap. 4. The MIT Press, Cambridge, MA (2001)
44. Yan, J., Forbus, K., Gentner, D.: A theory of rerepresentation in analogical matching. In: Proceedings of the 25th Annual Conference of the Cognitive Science Society (2003)

Chapter 6
On the Role of Computers in Creativity-Support Systems

Bipin Indurkhya

Abstract We report here on our experiences with designing computer-based creativity-support systems over several years. In particular, we present the design of three different systems incorporating different mechanisms of creativity. One of them uses an idea proposed by Rodari to stimulate imagination of the children in writing a picture-based story. The second one is aimed to model creativity in legal reasoning, and the third one uses low-level perceptual similarities to stimulate creation of novel conceptual associations in unrelated pictures. We discuss lessons learnt from these approaches, and address their implications for the question of how far creativity can be tamed by algorithmic approaches.

6.1 Introduction

Even though the last few decades have seen a steady progress in the development of computer systems that produce artifacts in the domain of visual art [8, 43], music [7, 40, 44], literature [39, 51]; and so on, generally they have received a negative press as regard to their creativity: computers cannot have emotions, programs do not have intents, creativity cannot be algorithmic, and so on [4, 57]. Even designers of computational creativity systems seem to take an apologetic tone when it comes to ascribing creativity to their systems. For example, Colton [9] argues that it is not enough to generate an interesting or creative artifact, but one must also take into account the process by which the artifact was generated. Krzeczkowska et al. [38] took pains to project some notion of purpose in their painting tool so that it might be perceived as creative. Such views blatantly expose the implicit assumptions underlying creativity: namely that it crucially needs a creator with emotions, intentions, and such. A consequence of this view is that creativity is considered an essentially human trait, and cannot be ascribed to computer programs or AI systems (or to animals like elephants and gorillas).

B. Indurkhya (✉)
Department of Computer Science, AGH University of Science and Technology, Cracow, Poland
e-mail: bipin@agh.edu.pl

T.R. Besold et al. (eds.), *Computational Creativity Research: Towards Creative Machines*,
Atlantis Thinking Machines 7, DOI 10.2991/978-94-6239-085-0_6

We critically examine this traditional view in the light of our previous experiences in designing creativity-support systems and modeling creativity. We present three very different case studies here, each of which incorporates a different mechanism of creativity. These systems are based on our previous research, and we will mention here only the main ideas behind each of the systems and the results. After a brief discussion of these systems, we will present our views on the role of computers in supporting and modeling creativity.

6.2 Some Case Studies of Computer-Based Creativity-Support Systems

We have been studying and modeling different aspects of creativity for over twenty years [23–34]. During this time, we have also explored various computational approaches to creativity, and have developed some computational systems that stimulate imagination and emergence of novel ideas and associations in the users, or model such processes. We present here three such systems in order to provide some concrete examples of how computers can play a crucial role in supporting creativity.

6.2.1 Stimulating Creativity in Generating Stories

We implemented a system *Story Telling from Pictures* [34] inspired by an idea *Little Red Riding Hood in a helicopter* from Rodari's fascinating book *The Grammar of Fantasy* [54]. In this technique, children are given a list of five or six words, and are asked to make a story that involves all of them. Rodari's idea was that if all but one of the words in the list are chosen so that they remind the children of some familiar story, and one unrelated but familiar word is thrown in with them, children's imagination is stimulated in incorporating the unrelated concept in the familiar story. Children enjoy this activity, and produce a great many imaginative variations of the original story.

For example, suppose children are given the words 'grandmother', 'wolf', 'forest', 'cape', and 'helicopter'. The first four words remind the children of the story *Little Red Riding Hood.* However, the last word is completely unrelated. We must emphasize here two necessary conditions for this technique to work. The first is that the children must be familiar with the story *Little Red Riding Hood.* If they do not know the story, or if the words used in the list do not remind them of the story, for there are several versions of the story, then this technique is not so effective in terms of stimulating imagination. The second condition is that the children must be familiar with the unrelated word as well. If they have no idea what a helicopter is, then the technique does not work either.

In this technique, the children find it interesting and challenging to make a story that incorporates the strange but familiar element (the helicopter) in the familiar story. Each child tackles this task in her or his own way. Their imagination is stimulated, and they enjoy the activity. They listen to each other's stories, and react to them enthusiastically.

Our creativity-support system based on this technique was implemented in three stages. In the first stage, we showed a number of pictures to the children and asked them to describe each picture. Our aim here was to find out the concepts with which our target user group (the children) was familiar. In the second stage, we created a library of *picture elements*, where each picture element depicted an object familiar to the children. Picture elements were also organized in a semantic association network to reflect which ones of them are related and which ones are not. In the third stage, we composed a picture by combining picture elements such that all the picture elements were associated except one that was semantically distant or unrelated. So, for example, the system might generate a picture of a cow in a classroom by combining some classroom-related picture elements like desks, blackboard, notebooks, children and teacher, and add the unrelated picture element cow. On evaluating the system, we found that the children found writing about such pictures more interesting, and they wrote longer stories.

This system was a straightforward implementation of Rodari's technique. The point we would like to emphasize is that it is not so difficult for a computer program to add an unrelated object in a scene. However, this task is harder for people, for as soon as a concept or word is given, all the associated concepts and their corresponding words get automatically activated—it is difficult to suppress these activations and to look beyond them to find an unrelated word or concept.

The T-puzzle (Fig. 6.1) provides an excellent example of how our past experiences, and perceptual and conceptual associations constrain us. The puzzle has four simple wooden pieces, and the objective is to arrange them in the form of the letter T. However, people have a very hard time solving this puzzle because their prior perceptual experiences keep them trying the same combinations over and over again [62].

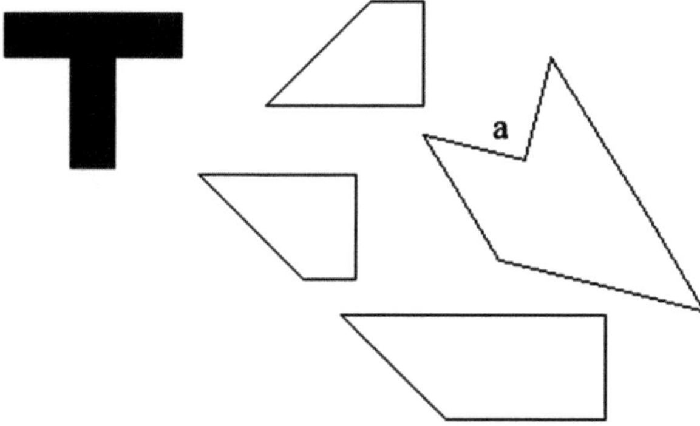

Fig. 6.1 The T-puzzle

On analyzing many of Rodari's techniques to stimulate creativity and imagination, we find that they are essentially elaborate ways to get some unrelated combination of words or concepts. When people try to connect these unrelated concepts in their imagination, their creativity is stimulated. For example, in one method, two children are asked to go to different rooms, close their eyes, open a dictionary at a random page, and put their finger at some point on the open page. This produces two random words, and then the children make a story connecting these two words. In another activity, each child is asked to bring a picture from some newspaper or magazine. The children sit in a circle, and all the pictures are put face down in the middle. Children take turns at random (by drawing lots), and the child whose turn it is turns over one of the pictures, and starts a story based on that picture. Each child (in random order) repeats this process, except that all the subsequent children have to continue the story generated so far by incorporating the picture they just turned over. This process is akin to generating novel and creative metaphors by combining unrelated words or concepts together [29].

In terms of computational modeling, at least this aspect of creativity is easily modeled algorithmically [31]. This is because it is very difficult to model commonsense conceptual associations on the computer, and it has been a challenging research areas since the advent of Artificial Intelligence. This implies that it not difficult for a computer program to break or ignore these conceptual associations and generate a combination of two unrelated words or concepts. For us humans, on the contrary, conceptual associations are an unalienable part of us, and so elaborate methods have to be devised to look past those existing associations.

6.2.2 Modeling Creativity in Legal Reasoning Computationally

We now move to a completely different topic and present another piece of previous research where we studied creativity in legal reasoning, and sought to model it computationally [23, 25]. The main idea behind this approach was that creative insights often come from applying the high-level structure (or gestalt) of one situation to the low-level details of another situation. The distinction between the levels (high and low levels) is important, for if both the situations are considered from a high level, then only traditional analogy can result, which, as far as creativity is concerned, is counterproductive (see [24, 29] for detailed arguments with examples). We explain this approach with an example below.

We focused on situations where new categories are brought in to analogize or distinguish between prior cases and a new case in order to argue for a particular resolution of the new case. Our domain was a particular tax law in the US, which allowed taxpayers to deduct their home-office expenses from their taxable income under certain conditions. These exception conditions were based on factors such as whether the home office was the *principal place of business* (*PPB*) (e.g. when a doctor saw patients regularly at a home clinic), whether the employer provided office space for the employee, and so on.

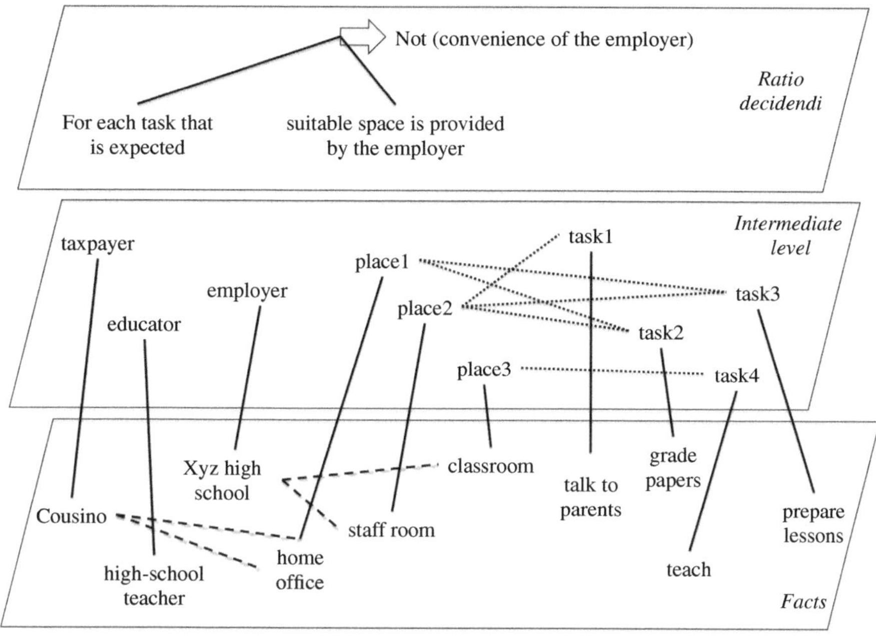

Fig. 6.2 A partial representation of the *Cousino case* (*Dashed lines* means *provided by* and *dotted lines* mean *designated space for*.)

In particular, among the precedents, there was a case of a high-school teacher Cousino, who claimed home-office deduction, but the courts denied him because the school provided him a *suitable space:* a classroom where he could teach, and an office equipped with a phone and office supplies. Let us refer to this as the *Cousino case*. A partial representation of the *Cousino case* is shown in Fig. 6.2.

In this figure, three levels of representations are shown. The *Facts* level corresponds to the low-level or perceptual features and the *Ratio decidendi* level corresponds to the high-level conceptual or the gestalt features. At the *Ratio decidendi* level, the justification for the decision is presented in abstract legal terms. An *intermediate* level of representation is also shown, which mediates between and connects the high-level representation to the low-level representation. (See [25] for more details.)

Another case among the precedents concerned a concert violinist Drucker. He claimed tax deduction for a studio he maintained at home where he practiced regularly. The courts allowed him the deduction arguing that for musicians, their principal activity is rehearsal, and the employer did not provide any space where the musician could rehearse. Let us call this the *Drucker case*. A partial representation of the *Drucker case* is shown in Fig. 6.3.

Now consider the case of a college professor Weissman, who claims tax deduction for an office he maintained at home. In the *Weissman case,* the college provided him a shared office. But the taxpayer argued that because many staff members shared the

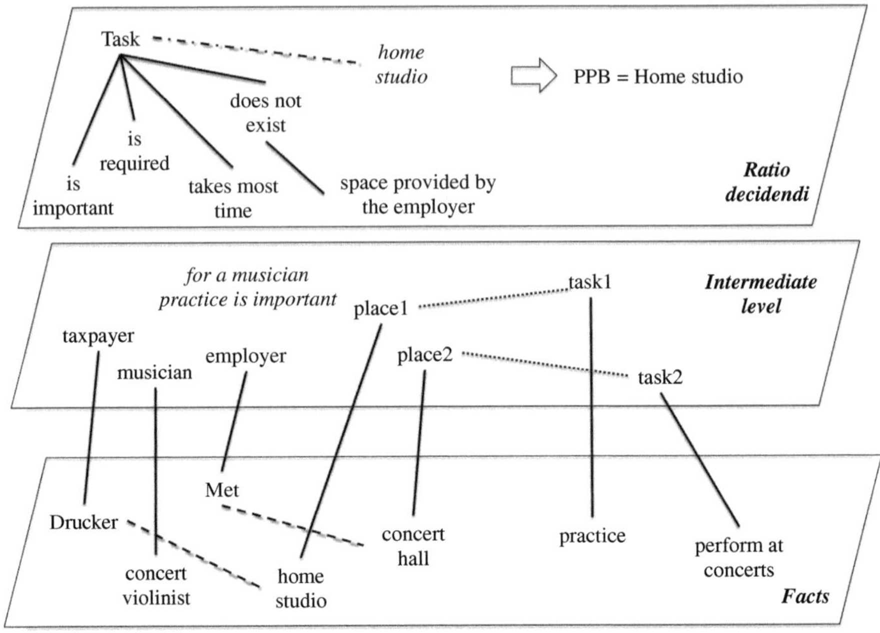

Fig. 6.3 A partial representation of the *Drucker case* (*Dashed lines* means *provided by, dotted lines* mean *designated space for,* and *dot-dash lines* means *performed at.* PPB is *principal place of business.*)

office, he could not leave his books and other material safely there. Let us consider this to be the new case.

With this background, the *Cousino case* is very similar to the *Weissman case* (the new case) and supports a decision against Weismann. When the *Drucker case* is applied to the facts of the *Weissman case,* we also get a decision against Weissman, because for each task that the college professor had to perform as part of his duties, there was some place (the shared office) provided by the employer. This situation is shown in Fig. 6.4.

However, when the *Cousino case* is activated, the category *suitable space* comes into play. Now we can reinterpret the *Drucker* using this category to argue that the decision favored the taxpayer because the employer did not provide any *suitable space* to carry out the activity. This reinterpretation of the *Drucker case* is shown in Fig. 6.5.

With this reinterpretation, the *Drucker case* is rendered similar to the *Weissman case* (the new case), and supports a decision in favor of Weissman as shown in Fig. 6.6. (See [25] for details, and also [23].)

It may seem a small semantic quibble to some readers, but legal arguments often hinge on such quibbles. There is another example discussed in [25] that hinges on introducing the term *substantial.* We should also add that all the analyses and

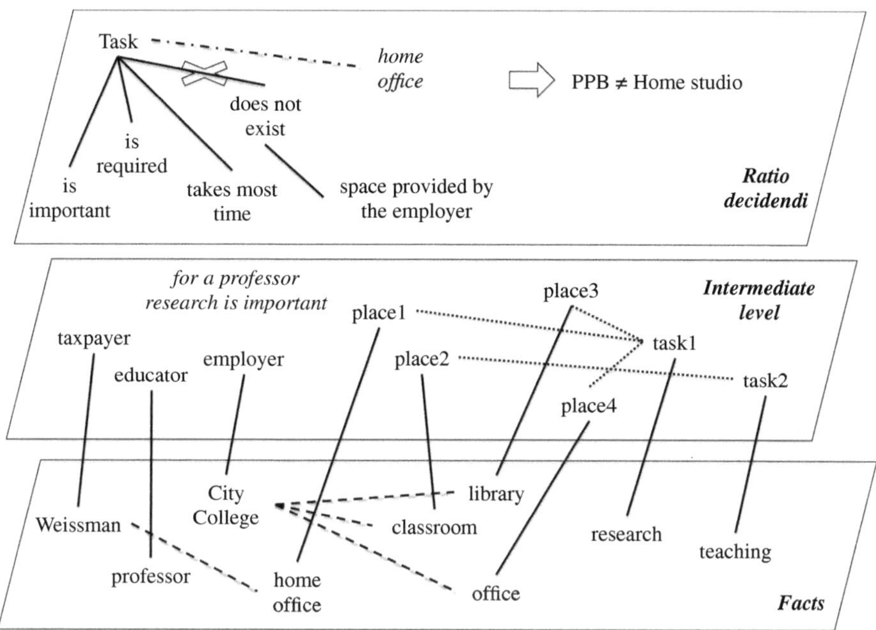

Fig. 6.4 *Drucker case* applied to the *Weissman case* (*Dashed lines* means *provided by*, *dotted lines* mean *designated space for*, and *dot-dash lines* means *performed at*. PPB is *principal place of business*.)

representations used in these examples were derived from actual opinions written by the judges when rendering their decisions.

The implications of this for computational systems is that we need to be able to model the process of reinterpretation, by which concepts and categories are applied to different data (for which they may not have been intended) in novel ways. This is consistent with recent research that suggests *psychological distance* as a mechanism for enhancing creativity [60]. Moreover, it has been demonstrated that psychological distance can be induced by such simple devices as taking another person's perspective or thinking of the problem as if it is unreal [35].

It is interesting to point out that the ability to get a new insight or perspective was one of the advantages claimed for case-based reasoning when it was promoted by Riesbeck and Schank ([53, pp. 9–14]). They compared and contrasted three modes of reasoning: (1) reasoning with ossified cases (rules or abstract principles), (2) reasoning with paradigmatic cases (cases with a given interpretation), and (3) reasoning with stories (cases with many possible interpretations and capable of re-interpretations). They argued that it is the third mode of reasoning that displays the most flexibility and power of having a knowledge base containing cases. However, most approaches to case-based reasoning in the 1990s and early part of the 2000s worked largely with indexed cases, which precludes this reinterpretation step. (See, for instance [6].) But in recent years, advances in data mining and unsupervised learning techniques pro-

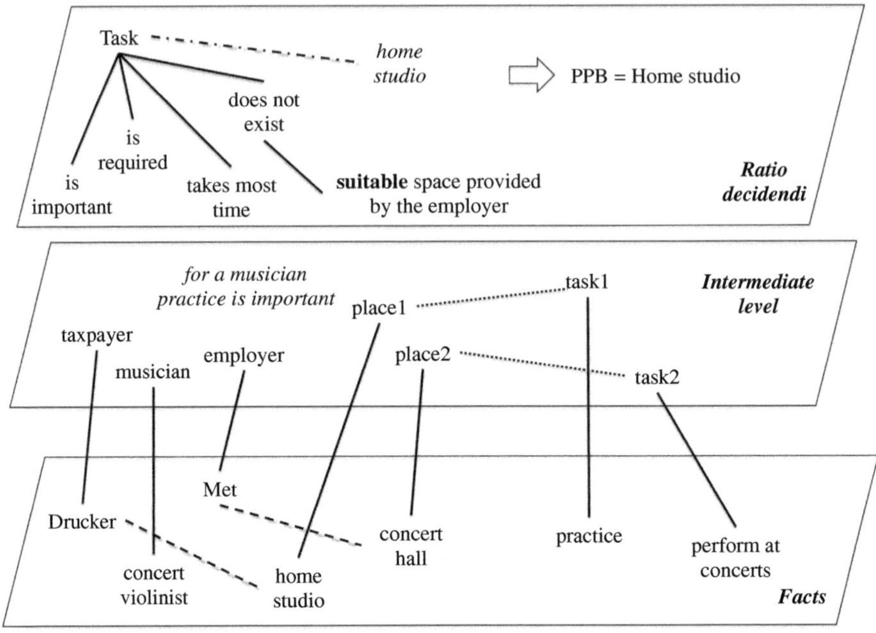

Fig. 6.5 *Cousino case* applied to reinterpret the *Drucker case* by introducing the category *suitable space* (*Dashed lines* means *provided by, dotted lines* mean *designated space for,* and *dot-dash lines* means *performed at* PPB is *principal place of business.*)

vide us with many new approaches to model the reinterpretation process. (See, for example [21, 65, 66].)

6.2.3 Role of Low-Level Perceptual Similarities in Stimulating Novel Conceptual Associations

Finally, we present a third approach to creativity in which we assessed the role of low-level perceptual similarities—namely similarities with respect to shape, color, texture, etc.—on emergent features when two images are juxtaposed. A feature related to a metaphor is considered *emergent* if it is not normally related to either of the two terms of the metaphor alone. For example, in "Her gaze, a flash of diamond", 'seduction' is an emergent feature as it is not normally related to 'gaze' or 'diamond' [16]. A major methodological problem in working with images is in determining the degree of low-level perceptual similarities between two given pictures. One alternative is to ask the participants to rate the degree of perceptual similarities between pairs of pictures, but the drawback is that when we look at a picture, conceptual and perceptual features interact heavily, and it is difficult to be certain that only perceptual

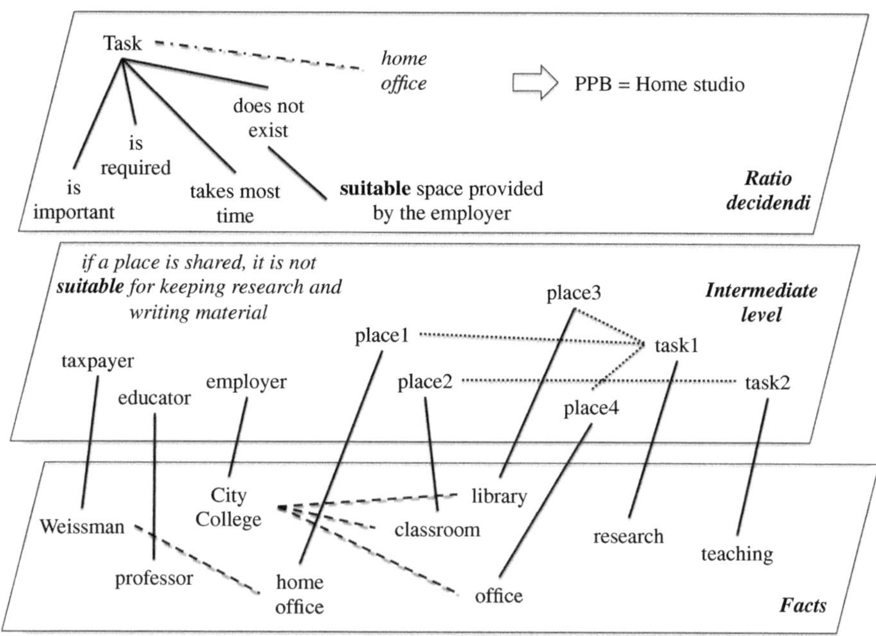

Fig. 6.6 Reinterpreted *Drucker case* applied to the *Weissman case* to support a decision in favor of Weissman (*Dashed lines* means *provided by, dotted lines* mean *designated space for,* and *dot-dash lines* means *performed at* PPB is *principal place of business.*)

features were used in determining the degree of similarity. To address this problem, we turned to image-processing programs.

In the field of machine vision, a number of algorithms have been developed for low-level visual processing. These algorithms extract features (like color, shape, texture, and so on) of images, which are analogous to features found in the early stages of visual processing in humans. So a similarity measure based on these features would reflect perceptual similarity.

We used one such image-based search system called Fast Image Search in Huge Database (FISH), which compares two images based on low-level perceptual features like color, shapes, texture, etc., to get a similarity index for them [64]. We refer to this as *algorithmic perceptual similarity*. For example, consider the pair of images shown in Fig. 6.7. The image on the left is of the world-famous marble mausoleum *Taj Mahal* that was built by the Moghul emperor Shah Jahan in the 17th century. The image on the right is of wine bottles. These two images were given a high perceptual similarity index by the FISH system. In fact, the wine bottles image was retrieved by the system as a *similar* image when queried by the *Taj Mahal* image. If we examine them carefully, we can see the perceptual similarities: the tall slender minarets of the *Taj Mahal* are analogous to the shape of the wine bottles. However, when people

Fig. 6.7 An example of algorithmic perceptual similarity

look at these two images, they tend to focus on conceptual aspects, and fail to find any significant similarities, if they find them similar at all.

Using such stimuli, we experimentally studied how perceptual similarities correlate with people's ability to interpret pairs of images metaphorically, and with emergence of new features that are not a part of either image [46]. Our results show that a pair of perceptually similar images (in terms of color, shape, etc.) is more likely to be given a metaphorical interpretation. Here are some examples of the interpretations given to the pair of images in Fig. 6.7 by the participants: 'Becomes better as it grows old', 'Standing pillars of tradition', 'Beauty in taste', 'Taste of history', 'Taj for eyes, wine for tongue', 'What a waste of time.' We also found that perceptual similarity correlates positively with emergent features.

An implication of these results is that they provide yet another way in which computational approaches can aid creativity. If this hypothesis—namely that low-level perceptual similarities facilitate novel conceptual associations (among people)—is correct then a computational system based on an algorithmic approach to perceptual similarity will be quite effective in stimulating creative imagination in the viewer. Systems based on such approaches can be used for creating persuasive ads, intuitive educational material, aesthetic pleasing art, and so on [32].

6.3 Is Creativity Computational?

We now return back to the question we raised in the introduction: Can creativity be algorithmic? Or is it essentially a human quality? To examine such questions, let us consider two different characterizations of creativity. The first one focuses on the process by which a human being engages in a creative pursuit. If we try to think of creative people, who comes to mind? Perhaps Einstein, Mozart, Michelangelo or Leonardo da Vinci. In the modern times, we might think of Steve Jobs. But what do we mean when we say that they are creative?

Perhaps music came naturally to Mozart. In a letter to his father on Nov. 8, 1777, he wrote: "I cannot write in verse, for I am no poet. I cannot arrange the parts of speech with such art as to produce effects of light and shade, for I am no painter. Even by signs and gestures I cannot express my thoughts and feelings, for I am no dancer. But I can do so by means of sounds, for I am a musician." Perhaps one could say that his brain was structured in a certain way that generated musical patterns naturally. Of course, what makes his work great is because of the way people have responded to his music over more than two centuries. (See also [37, 47].)

Or consider mentally different people, like the case of Stephen Wiltshire, discussed in [55]. Sir Wiltshire has an amazing ability to draw a landscape from memory after seeing it only once. Though he is diagnosed with autism, his work is highly regarded both by critics and general population. He was awarded *Member of the Order of the British Empire* for services to art in 2006. So he is no doubt a very creative person, no matter which criterion one chooses to apply.

But let us think about it a minute. What do we mean by saying that he is creative? His work has a certain style, level of details that most people cannot reach, aesthetic appeal, and all that. As with Mozart, we can go further and say that perhaps this is the way he expresses himself naturally: just like you and I might describe what we did on our last summer vacation, he draws fantastic landscapes.

We can now throw in here examples of people with schizophrenia or brain damage, savants or manic-depressive people, and so on [57]. When these people produce work that is considered creative, often this is their mode of being, and it could not have been otherwise. (See also [1, 17].) Many times the intention is missing as well.

Einstein's brain was preserved after his death so that people can study it to get any clues about the biological basis for creativity. But it is not like he was creative every day of his life. It is the impact of his theory of relativity, and its eventual acceptance by the scientific community that was a key factor in him becoming an icon of scientific creativity of the twentieth century. Moreover, Einstein was also dogmatic at times, perhaps the most famous case being his rejection of Alexander Friedmann's expanding universe hypothesis [61].

If we were to model Einstein's creative process, what would we model? There have been some computational models of scientific discovery, but they almost always greatly simplify the process by putting a number of assumptions in place as to what is significant and what is not. At that point, it is not clear at all if they are modeling the actual mental process of the creative person at the time of the creative act. (See also [5].)

Such examples suggest that the so-called creative humans use a variety of heuristics, some of them consciously and some subconsciously, for creating artifacts or for problem solving. Many of these heuristics can be mechanized, and in principle there seems to be no reason to consider any of them non-algorithmic.

The second characterization of creativity focuses on the nature of creative artifacts. It takes only the audience's perspective, so the creator is not even mentioned. We refer to Barthes' [3] articulation: "We know that to restore to writing its future, we must reverse its myth: the birth of the reader must be ransomed by the death of the Author," though he traced this view to even earlier scholars. Though not everyone subscribes to this extreme position, most accounts of creativity do acknowledge the role of the audience [12, 13, 22, 41].

In the audience-based view of creativity, it is generally accepted that in order for an artifact to be deemed creative, it must be *novel* and *useful*. We have argued above (Sec. 6.2.1) that *novelty* is cognitively difficult for people because we are constrained by our previous conceptual associations. Researchers who study creativity have come up with various techniques to overcome this difficulty. However, computers and AI systems, which do not have any such associations, have a great advantage here, for they can search the unchartered areas of novel concepts and conceptual combination more systematically [19, 31].

However, to automatically assess the *usefulness* of created artifacts is a different cup of tea altogether. As the usefulness is necessarily from a human point of view, the question becomes: Can an algorithm capture usefulness to humans? Here, we can distinguish two different aspects of usefulness. One is aesthetics, which relates to artistic creativity. In this regard, there has been some research to suggest that at least some of our aesthetic values are hardwired in the structure of the brain [52, 68]. Moreover, machine-learning techniques have been applied to *learn* about the cultural preferences of an audience based on the past data. For instance, Ni et al. [45] trained their program with the official UK top-40 singles chart over the past 50 years to learn as to what makes a song popular. A program like this might successfully predict, for instance, the winner of the future Eurovision competitions. However, a limitation of these approaches is that they cannot predict drastic changes in the aesthetic values and tastes: for example atonal music, or abstract art. Moreover, creativity is not the same as popularity. So to be able to predict whether a song, or a book, or a video will become popular [63] is not the same thing as evaluating their creativity.

This problem becomes more severe when we move beyond arts, and consider creativity in problem solving, and in science and technology. Here the usefulness of a novel and creative idea comes down to simply whether it works. This clearly has an objective component, for in a sense it is the reality that determines whether the idea works or not. History of science and technology is full of many interesting and novel ideas that did not work. Prehistory of flight [20] is a rich domain of examples where many novel ideas that were based on numerous observations, experimentations, and in which inventors had complete faith, did not work at all. The following examples provide further support for this argument (see also [28, 48]):

1. Schön [58, 59] described the case of a product-development team in the 1940s, which was working to develop a synthetic-fiber paintbrush that would yield smoothly painted surfaces like the natural-fiber paintbrush did. They came up with an innovative and successful design using the *painting-as-pumping* metaphor. However, during the problem-solving phase, they also considered *painting as masking-a-surface* metaphor, which was quite a novel idea, but it led to no useful insight.
2. Yolanda Baie, a food stand operator and owner, petitioned to have the kitchen of her house, where she prepared food sold at the food stand, qualify as a *home office* for the purpose of tax deduction (Baie vs. C.I.R., 74 T.C. 105), using the argument

that her kitchen was a *manufacturing facility* for her business. The judges, while finding the argument 'ingenious and appealing,' ruled it 'insufficient' nonetheless.
3. John Casti ([5, pp. 7–10]) comments on the fate of Immanuel Velikovsky's theory as outlined in Worlds in Collision, which hypothesized Earth's encounters with a large comet expelled from Jupiter and provided explanations for many biblical events. Velikovsky's theory proposes a novel understanding of the Solar system, but the scientific community has not accepted it.

Considering such examples, we suggest that this *usefulness* aspect of creativity remains essentially non-algorithmic, but not because humans are special, and cognitive processes cannot be computational, but because nature is not bounded by the limits of our cognitive models. To elaborate this further, whether an object flies or not (which is the *usefulness* of the idea) does not depend on how beautiful or elegant the theory is, or how much effort and emotional energy the creator has invested in the object, and so on. In other words, usefulness cannot be addressed from within the cognitive model, but it must be applied and tested in the real world.

Thus, creativity, in our view, represents the open-endedness of our interaction with the environment, and cannot be captured in a cognitive or computational model. Nonetheless, we can have computational models of creativity in limited domains, and computational systems can be designed to stimulate and enhance general creativity in people.

We can contrast our position with some other approaches to model creativity, notably among them being the FACE and IDEA models proposed Colton and his colleagues [10, 11, 49, 50]. The FACE model formalizes novelty by explicitly identifying eight dimensions along which an object can be considered novel. To complement this, the IDEA model formalizes the impact of the artifact (so the *usefulness*) by assuming how it affects an ideal audience. Two dimensions are identified to measure the impact on the audience: one refers to how the well being of the audience has changed in response to the artifact or the work, and the other refers to the cognitive effort required to appreciate the artifact.

Our position is consistent with the FACE model, except that the FACE model is more detailed in its explicit identification of various ways in which novelty can be generated. However, the position we have articulated here with respect to *usefulness* essentially implies that the goal of the IDEA model is not attainable. First of all, it is very difficult to characterize an *ideal audience*. Moreover, as the character of the audience—and here we include both the nature of the individual members of the audience and the membership of the audience—changes as a result of interacting with the artifact, and changes in quite unpredictable ways, it is nearly impossible to measure the two parameters posited in the IDEA model. One only has to consider the history of art genre like atonal music, minimalist music, abstract visual art, conceptual art, and how they gradually became accepted by the audience to appreciate this point. Finally, for scientific creativity and real-world problem solving, the audience is the nature or the real world, which ultimately accepts or rejects the artifact, and this response cannot be modeled, as we have argued above. Nevertheless, in restricted domains, it may be possible to make certain assumptions about the audience, and so IDEA model can be useful in a limited way.

6.4 Conclusions

To summarize the main arguments of this chapter, we would like to rearticulate them in another way. It is generally accepted that the two main characteristics of creativity are originality and intelligibility: the product must be novel or the process must generate a new perspective; and the product or the generated perspective must be intelligible in order to be useful for at least some audience [2, 56, 67]. For novelty, research on real-world creativity shows that it is difficult for people to step out of their conventional and habitual conceptual associations. To overcome this inertia, several methods like making the familiar strange [18], concept displacement [58], bisociation [36], lateral thinking [14], estrangement [54], conceptual blending [15], and so on, have been proposed in the literature. However, computers do not have this inertia, and so they can be very effectively used to generate novel ideas. This argument has been presented in more detail elsewhere [31]. Our experience in developing creativity-assistive systems (reviewed in Sect. 6.2) lends supports to this hypothesis.

However, when it comes to incorporating usefulness of the generated perspective or idea, we have argued that, in general, it is not possible to capture this aspect of creativity algorithmically. The reason is simply that when a new object or style is introduced, people react to it in different ways. Sometimes they adapt to it right away; at other times they do not find it interesting or useful at first, but the same object or style introduced at a later time becomes a big success; and sometimes they do not find it useful at all, in spite of the efforts made by the creators to convince them otherwise.

Nevertheless, one cannot rule out the possibility that in limited domains we might be able to characterize usefulness algorithmically, and to design and implement computer systems that can generate statistically a larger number of useful and interesting artifacts and ideas. So combining this with novelty-generating systems, we can have computer systems that are creative. Systems like Aaron exemplify this approach.

However, even in a limited domain, once usefulness is characterized algorithmically, it loses its novelty, and gradually ceases to be creative. (See, for instance, the model of literary style change proposed by Martindale [42].) So while, we may be able to model some aspect of creativity within a style (with respect to usefulness), it remains doubtful whether creative changes in styles can be modeled successfully in a universal way. Again, to emphasize, novelty can be modeled—it is relatively easy to computationally generate new styles, but the problem is to incorporate which styles will be successful (meaning people will adapt to them and find them useful). Therefore, we claim that this usefulness aspect of creativity will always remain the last frontier for computational modeling techniques.

Acknowledgments I am grateful to Georgi Stojanov and an anonymous reviewer for carefully going through an earlier draft of this chapter and making many helpful comments.

References

1. Abraham, A., Windmann, S., McKenna, P., Güntürkün, O.: Creative thinking in schizophrenia: the role of executive dysfunction and symptom severity. Cognit. Neuropsychiatry. **12**(3), 235–2581 (2007)
2. Amabile, T.M.: Creativity in Context. Westview Press, Boulder (1996)
3. Barthes, R.: The Death of the Author. In: Barthes, R. (ed.) Image Music Text (trans. S. Heath). London: Fontana (1977). Original work published 1967
4. Boden, M.A.: Computer models of creativity. AAAI. AI. Mag. **30**(3), 23–33 (2009)
5. Casti, J.L.: Paradigms Lost: Images of Man in the Mirror of Science. William Morrow & Co., New York (1989)
6. Chalmers, D.J., French, R.M., Hofstadter, D.R.: High-level perception, representation, and analogy: a critique of artificial intelligence methodology. J. Exp. Theor. Artif. Intell. **4**(3), 185–211 (1992)
7. Chordia, P., Rae, A.: Tabla Gyan: an artificial tabla improviser. Proceedings of the International Conference on Computational Creativity: ICCC-X, Lisbon, Portugal (2010)
8. Cohen, H.: On Modeling of Creative Behavior. Rand Corporation Report Series, P-6681 (1981)
9. Colton, S.: Creativity versus the perception of creativity in computational systems. Proceedings of the AAAI Spring Symposium on Creative Systems (2008)
10. Colton, S., Charnley, J., Pease, A.: Computational creativity theory: the FACE and IDEA models. Proceedings of the International Conference on Computational Creativity (2011)
11. Colton, S., Wiggins, G.A.: Computational creativity: the final frontier. Proceedings of the 20th European Conference on Artificial Intelligence (2012)
12. Cropley, D.H., Kaufman, J.C., Cropley, A.J.: Measuring creativity for innovation management. J. Technol. Manag. Innov. **6**(3), 13–30 (2011)
13. Csikszentmihalyi, M.: Creativity: Flow and the Psychology of Discovery and Invention. HarperCollins, New York (1996)
14. de Bono, E.: New Think: The Use of Lateral Thinking in the Generation of New Ideas. Basic Books, New York (1975)
15. Fauconnier, G., Turner, M.: The Way We Think: Conceptual Blending and the Mind's Hidden Complexities. Basic Books, New York (2002)
16. Gineste, M.-D., Indurkhya, B., Scart, V.: Emergence of features in metaphor comprehension. Metaphor Symb. **15**(3), 117–135 (2000)
17. Glicksohn, J.: Encyclopedia of creativity. In: Runco, M.A., Pritzker, S.R. (eds.) Schizophrenia and Psychosis, 2nd edn. Academic Press, New York (2011)
18. Gordon, W.J.J.: Synectics: The Development of Creative Capacity. Harper & Row, New York (1961)
19. Harry, H.: On the role of machines and human persons in the art of the future. Pose **8**, 30–35 (1992)
20. Hart, C.: The Prehistory of Flight. University of California Press, California (1985)
21. Hofstadter, D.: The Fluid Analogies Research Group: Fluid Concepts and Creative Analogies. Basic Books, New York (1995)
22. Horn, D., Salvendy, G.: Consumer-based assessment of product creativity: a review and reappraisal. Hum. Factors Ergonomics Manufact. Serv. Ind. **16**(2), 155–175 (2006)
23. Hunter, D., Indurkhya, B.: 'Don't think, but look' a gestalt interactionist approach to legal thinking. Proceedings of the Workshop on Advances in Analogy Research, Sofia, Bulgaria, pp. 345–353 (1999)

24. Indurkhya, B.: Metaphor Cogn. Kluwer Academic Publishers, The Netherlands (1992)
25. Indurkhya, B.: On modeling creativity in legal reasoning. In: Proceedings of the Sixth International Conference on AI and Law, pp. 180–189. Melbourne, Australia (1997)
26. Indurkhya, B.: An algebraic approach to modeling creativity in metaphor. In: Nehaniv, C.L. (ed.) Computation for Metaphors, Analogy and Agents. LNAI, vol. 1562, pp. 292–306. Springer, Berlin (1999)
27. Indurkhya, B.: Emergent representations, interaction theory, and the cognitive force of metaphor. New Ideas Psychol. **24**(2), 133–162 (2006)
28. Indurkhya, B.: Rationality and reasoning with metaphors. New Ideas Psychol. **25**(1), 16–36 (2007)
29. Indurkhya, B.: On the role of metaphor in creative cognition. Proceedings of the International Conference on Computational Creativity: ICCC-X, Lisbon, Portugal (2010)
30. Indurkhya, B.: Thinking like a child: the role of surface similarities in stimulating creativity. In: Stojanov, G., Indurkhya, B. (eds.) Proceedings of the AAAI Spring Symposium on Creativity and (Early) Cognitive Development, pp. 38–43. AAAI Press, Palo Alto (2013)
31. Indurkhya, B.: Computers and creativity. In: Veale, T., Feyaerts, K., Forceville, C.J. (eds.) Creativity and the Agile Mind, pp. 61–79. De Gruyter Mouton, Berlin (2013)
32. Indurkhya, B., Kattalay, K., Ojha, A., Tandon, P.: Experiments with a creativity-support system based on perceptual similarity. In: Fujita, H., Zualkernan, I. (eds.) New Trends in Software Methodologies, Tools and Techniques, pp. 316–327. IOS Press, Amsterdam (2008)
33. Indurkhya, B., Ogawa, S.: An empirical study on the mechanisms of creativity in visual arts. In: Miyake, N., Peebles, D., Cooper, R.P. (eds.) Proceedings of the 34th Annual Conference of the Cognitive Science Society, pp. 1727–1732. Cognitive Science Society, Austin (2012)
34. Ishii, Y., Indurkhya, B., Inui, N., Nose, T., Kotani, Y., Nishimura, H.: A System based on rodari's 'estrangement' principle to promote creative writing in children. Proceedings of Ed-Media & Ed-Telecom 98, pp. 1685–1686, Freiburg, Germany (1998)
35. Jia, L., Hirt, E.R., Karpen, S.C.: Lessons from a Faraway land: the effect of spatial distance on creative cognition. J. Exp. Soc. Psychol. **45**, 1127–1131 (2009)
36. Koestler, A.: The Act of Creation. Hutchinsons, London (1964)
37. Kozbelt, A.: Factors affecting aesthetic success and improvement in creativity: a case study of the musical genres of mozart. Psychol. Music **33**(3), 235–255 (2005)
38. Krzeczkowska, A., Colton, S., Clark S.: Automated collage generation—with intent. Proceedings of the International Conference on Computational Creativity: ICCC-X, Lisbon, Portugal (2010)
39. Kurzweil, R.: Cybernetic Poet. http://www.kurzweilcyberart.com/poetry/ (2001). Last accessed on 29 Jan 2012
40. López, A.R., Oliveira, A.P., Cardoso, A.: Real-time emotion-driven music engine. Proceedings of the International Conference on Computational Creativity: ICCC-X, Lisbon, Portugal (2010)
41. Maher, M.L.: Evaluating Creativity in Humans, Computers, and Collectively Intelligent Systems, DESIRE'10: Creativity and Innovation in Design. Aarhus, Denmark (2010)
42. Martindale, C.: The Clockwork Muse: The Predictability of Artistic Change. Basic Books, New York (1990)
43. McCorduck, P.: Aaron's Code: Meta-art, Artificial Intelligence, and the Work of Harold Cohen. W. H. Freeman, New York (1991)
44. Monteith, K., Martinez, T., Ventura, D.: Automatic generation of music for inducing emotive response. Proceedings of the International Conference on Computational Creativity: ICCC-X, Lisbon, Portugal (2010)
45. Ni, Y., Santos-Rodriguez, R., Mcvicar, M., De Bie, T.: Hit song science once again a science? Presented at the 4th International Workshop on Machine Learning and Music: Learning from Musical Structure. Sierra Nevada, Spain (2011)
46. Ojha, A., Indurkhya, B.: New frontiers in analogy research. In: Kokinov, B., Holyoak, K., Gentner, D. (eds.) Perceptual vs. Conceptual Similarities and Creation of New Features in Visual Metaphor. New Bulgarian University Press, Sofia (2009)

47. Painter, K.: Mozart at work: biography and a musical aesthetic for the emerging german bourgeoisie. Music. Q. **86**(1), 186–235 (2002)
48. Palmer, S.: Human Ontology and Rationality. Avebury, Aldershot (1992)
49. Pease A., Colton, S.: On impact and evaluation in computational creativity: a discussion of the turing test and an alternative proposal. Proceedings of the AISB symposium on AI and Philosophy (2011)
50. Pease, A., Colton, S.: Computational creativity theory: inspirations behind the FACE and IDEA models. Proceedings of the International Conference on Computational Creativity (2011)
51. Pérez y Pérez, R., Negrete, S., Peñalosa, E., Ávila, R., Castellanos, V., Lemaitre, C.: MEXICA-Impro: a computational model for narrative improvisation. Proceedings of the International Conference on Computational Creativity: ICCC-X, Lisbon, Portugal (2010)
52. Ramachandran, V.S., Hirstein, W.: The science of art: a neurological theory of aesthetic experience. J. Conscious. Stud. **6**(6–7), 15–51 (1999)
53. Riesbeck, C.K., Schank, R.C.: Inside Case-based Reasoning. Lawrence Erlbaum & Associates, New Jersey (1989)
54. Rodari, G.: The Grammar of Fantasy (J. Zipes, Trans.) Teachers & Writers Collaborative, New York (1996)
55. Sacks, O.: An Anthropologist on Mars: Seven Paradoxical Tales. Alfred A. Knopf, New York (1995)
56. Sarkar, P., Chakrabarti, A.: Development of a method for assessing design creativity. International Conference on Engineering Design ICED'07, pp. 28–31 August, Cite des Sciences et de L'Industrie, Paris, France (2007)
57. Sawyer, K.: Explaining Creativity. Oxford University Press, Oxford (2006)
58. Schön, D.A.: Displacement of Concepts. Humanities Press, New York (1963)
59. Schön, D.A.: Generative metaphor: a perspective on problem-setting in social policy. In: Ortony, A. (ed.) Metaphor and Thought, pp. 154–283. Cambridge University Press, Cambridge (1979)
60. Shapira, O., Liberman, N.: An easy way to increase creativity. Scientific American (Mind Matters), 21 July 2009. http://www.scientificamerican.com/article.cfm?id=an-easy-way-to-increase-c
61. Singh, S.: Big Bang. Harper Collins, New York (2004)
62. Suzuki, H., Hiraki, K. : The constraint relaxation process as theory change: toward theorizing the process of insight. (In Japanese.) Proceedings of the Japanese Cognitive Science Society's Special Interest Group on Language & Learning, vol. 97(1), pp. 33–42 (1997)
63. Szabo, G., Huberman, B.A.: Predicting the popularity of online content. Commun. ACM **53**(8), 80–88 (2010)
64. Tandon, P., Nigam, P., Pudi, V. and Jawahar, C.V.: FISH: A practical system for fast interactive image search in huge data bases. Proceedings of 7th ACM International Conference on Image and Video Retrieval (CIVR'08), Niagara Falls, Canada (2008)
65. Veale, T.: Re-representation and creative analogy: a lexico-semantic perspective. New Gener. Comput. **24**(3), 223–240 (2006)
66. Veale, T., Hao, Y.: A fluid knowledge representation for understanding and generating creative metaphors. Proceedings of COLING 2008, pp. 1–5. Manchester (2008)
67. Yong, K., Lander, M.W., Mannucci, P.V.: Novelty x Usefulness: actor-level effects and cultural influences on creativity in organizations. 35th DRUID Celebration Conference, Barcelona, Spain, June 17–19 (2013)
68. Zeki, S.: Inner Vision: An Exploration of Art and the Brain. Oxford University Press, Oxford (2000)

Chapter 7
IDyOT: A Computational Theory of Creativity as Everyday Reasoning from Learned Information

Geraint A. Wiggins and Jamie Forth

Abstract We present progress towards a computational cognitive architecture, IDyOT (Information Dynamics of Thinking) that is intended to account for certain aspects of human creativity and other forms of cognitive processing in terms of a pre-conscious predictive loop. The theory is motivated in terms of the evolutionary pressure to be efficient. It makes several predictions that may be tested by building computational implementations and studying their behaviour.

7.1 Introduction

The aim of this chapter is to explain more of the theoretical detail of a cognitive architecture that was outlined by Wiggins [55], with a view to developing a minimalist theory of cognition that encompasses many human behaviours within as simple a functional framework as possible. The architecture is called IDyOT, for "Information Dynamics of Thinking".

The overarching proposal is that information-theoretically regulated probabilistic prediction, which serves as a general mechanism for managing information and events in the world, is the mechanism underlying non-conscious creativity—that creativity is, essentially, a by-product of mechanisms which have evolved to serve prosaic, not poetic, forms of existence.

It is important to emphasise from the start that taking such a reductionist position about the *source* of creativity does not in any sense diminish the value of its application, as exhibited in magnificently various forms and magnitude throughout human society, and, less consistently, by other species.

In the current chapter, then, we will present the abstract structure of the proposed architecture, which is based on function only, and not on any kind of physiological model of a brain, human or otherwise. Nevertheless, we will refer to

G.A. Wiggins (✉) · J. Forth
Queen Mary University of London, London, UK
e-mail: geraint.wiggins@qmul.ac.uk

J. Forth
e-mail: jamie.forth@qmul.ac.uk

T.R. Besold et al. (eds.), *Computational Creativity Research: Towards Creative Machines*,
Atlantis Thinking Machines 7, DOI 10.2991/978-94-6239-085-0_7

neurophysiological proposals as to where these functions might be operationalised in the wetware, and to evolutionary arguments that motivate the thinking. We will show how the simple architecture we propose addresses several questions which are conventionally studied separately in the literature: it is important to remember that the point is to draw these together under our unifying architectural umbrella.

The rest of the chapter is structured as follows. Section 7.2 introduces our methodology. Section 7.3 summarises the background to our central proposal; this is extensive, so only a superficial survey can be given here, and we apologise to authors who may feel that their work has not been given adequate coverage. Section 7.4 presents the model itself, beginning at an abstract level of detail and working down. Section 7.5 explains the most interesting predictions of the theory, and, finally, Sect. 7.6 explains how it is will model spontaneous creativity.

7.2 Methodology: Structured Abstract Computational Modelling

For the reader who is unfamiliar with the philosophy of computational modelling research, it is worth briefly summarising our approach here; other readers may safely skip this section.

Computational models provide a means of developing a theory about a complex process that is not directly observable. The idea is to write a program that embodies the theory, and then test that program across all its possible parameters, looking for predictions that are testable, and the more unexpected the better [22]. These unanticipated predictions are important, because, otherwise, one can fall into the trap of building a model that works perfectly well, but which tells one nothing about the phenomenon being modelled: instead, it merely confirms that the modelling technique used is capable of modelling the data supplied, without necessarily shedding light on the process that produced it.

Wiggins [51] introduces the distinction between *descriptive* and *explanatory* models: the former is the kind of model that describes a phenomenon without attempting to explain how it works (e.g., the Gestalt grouping principles) and the latter is the kind that provides a mechanism by which the phenomenon actually works, at some level of abstraction. It is usually the case that a descriptive model is developed before corresponding explanatory ones, and it is often the case that an explanatory model functions as a descriptive model at a less abstract level of modelling. For example, the finite element analysis techniques used in modern weather modelling give an effective model of the physics of the weather, but abstract away the detail of the movements of individual molecules in the atmosphere. Similarly, one may often usefully think of the function of a program without considering the operation of the bits in the computer's memory, and one may consider the operation of the processor on the bits without considering the operation of the electronics in the chip that implements it. This last example is similar to—but not the same as—Marr's *levels of description* [26], which have been debated by McClamrock [29]. Wiggins [52] expands on these ideas at greater length.

For the current work, we will be focusing on a level of abstraction very distant from the neural substrate, looking at combinations of high-level abstract mathematical operations to build an explanatory model at that level. The intention, subsequently, is to proceed downwards, viewing the components of the model and their interactions[1] as descriptive and developing explanatory models below, so as to lead down, eventually reaching the neural level. This long-term view will not, however, prevent us from proposing links direct to the neurophysiology where appropriate.

7.3 The Theoretical Lineage of IDyOT

The theory presented here draws on multiple backgrounds, in cognitive science (particularly the cognitive science of music), in information theory, and in consciousness research. The methodology is drawn from computer science, and is outlined explicitly above to avoid misunderstandings between disciplines about the purpose and approach of computational modelling research.

7.3.1 Expectation in Music and Other Cognitive Phenomena

Pearce and Wiggins [39] argue for the importance of expectation in music, continuing a tradition that began with Meyer [31] and includes the important work of Huron [23]. Pearce's doctoral work [33, 37] concerns the construction and validation of a computational model of sequence processing, based on Markov modelling, and building on ideas by Conklin [7, 8] regarding multidimensional models, that admit prediction of human musical expectation, based on observed likelihood. Musical melody is an interesting subject of study in this context: it is readily separable from other musical constructs (such as harmony) and thus subject with minimal damage to reductionist science, and it is clearly present in the vast majority of the world's musics, in excitingly varied forms. This variation encourages us to decompose the phenomenon into its temporal component [16], which is universally observed, and its musical content, which is (therefore) culture-specific. Pearce's model, known as IDyOM for "Information Dynamics of Music", when exposed to a representative selection of Western tonal melody, predicts the musical expectations of Western listeners very well, explaining 81 % of the variance in empirical validation studies. It remains the most successful such model to date [36, 37].

Importantly for the current work, the IDyOM model, combined with a descriptive rule, also predicts musical phrase segmentation, achieving an F_1 score of 0.61. This result is favourably comparable with other computational models, all of which are programmed by humans, not learning-based [36]. The same model, when presented

[1] The interactions are particularly important: by considering them, we avoid the trap of naïvely assuming Fodorian modularity [15].

with a multidimensional encoding of speech based on ⟨phoneme,stress⟩ feature pairs, was capable of segmentation into syllables with an F_1 score of 0.67 [54]. Further, an electrophysiological study suggests that there are neural signals that correspond with reported experience of expectation, which in turn corresponds with IDyOM's predictions [35]; thus, IDyOM is demonstrably capable of predicting human behaviour.

Following Meyer's original proposal, that expectation (and its denial or fulfillment) is a contributor to the emotional experience of musical listening, Egermann et al. [10] compared IDyOM's information-theoretic predictions with physiological and consciously reported measures of emotional response, and found significant correlations. This lends general empirical support to Huron's suggestion [23] that expectation (or "sweet anticipation") is fundamental to humans and other organisms and exapted[2] into music appreciation because it uses a more general mechanism: IDyOM's mechanism is completely domain-general, and uses only temporal sequence of the observed data to predict them (thence, it can be used for speech segmentation, as above). This argument is expounded in full by Huron [23] and Wiggins [55].

The above work supplies a substantial body of evidence, supplementing that which exists in the statistical linguistics literature, to hypothesise that the learning of language and music are essentially the same statistical process, running over different data representations. Expectations are represented as statistical distributions across known symbol-sets (which might represent musical pitches, chords, phonemes, or words). That representation is in turn subject to analysis in terms of Shannon's information theory [46], and it is that analysis that will drive the cognitive architecture presented below. The work is summarised by Pearce and Wiggins [39].

7.3.2 Conceptual Representations of Information Structure

While the IDyOM model is primarily concerned with the prediction of expectation, and therefore with the establishment of a model of the statistics of perceived sequences, the work also makes useful predictions about the representations over which those sequences are formed. In particular, given a selection of features that might be used to represent the data over which it works, IDyOM uses a hill-climbing search technique to choose the subset that allows the data to be represented in the most *compact* way, with respect to average information content (also called *cross-entropy* [8]). This embodies a hypothesis that brains seek to find representations that store what they have learned efficiently (perhaps, we speculate, as part of memory consolidation). However, this model is probably only descriptive, in that the features are given as program functions, and (expensive) brute force search and re-evaluation

[2] *Exaptation* is the appropriation of a biological capacity driven by given evolutionary pressures into a different function. An alternative view is that these capacities form *spandrels*, supporting other behaviours, but not becoming part of them.

is used to select the combination that is actually used; we return to this issue below. Nevertheless, the idea is promising, demonstrating musical behaviour that we might expect: as the model is exposed to more music, it quickly discovers that relative pitch representations are preferable to absolute ones, and thence that scalar and modal representations, where relative pitch is expressed relative to a tonal centre (key note) are more efficient still. The former of these two effects is exhibited by human infants [43]; a strong informal argument can be made for the latter by reference to the dominance of tonal music in the West, and the commonly attested need for Western-enculturated listeners to deliberately enter a different mode of listening to appreciate non-tonal music [27].

Furthermore, to admit creativity beyond mere re-ordering of symbol sequences, it is necessary to use continuous representations of meaning. Categorical perception of pitch, timbre and colour give a clear indication that mind/brains[3] are able to maintain dual simultaneous representations that have both discrete (symbolic) and continuous properties. One theory that admits such a representation is that of Gärdenfors [17], in which *conceptual spaces* are composed of perceptual dimensions, whose existence is motivated by the need to make distinctions as an organism understands the world. For example, hue, saturation and luminance are three dimensions that define colour space, and they are *integral* in the sense that they cannot exist separately. Discrete symbolic concepts such as "red" exist in correspondence with regions in the space, with all the geometrical reasoning that this implies. Gärdenfors shows in detail how this idea can extend to very complicated observations, such as the movement of a human arm. Finally, the geometry of the conceptual spaces may be adjusted with weights, changing the relative distances between concepts, allowing us to model (aspects of) salience.

We propose that Gärdenfors' theory is an appropriate mathematical background to support a model which is capable of optimising *both* the representation of the perceptual content *and* the learned sequence, by inferring meaningful and information-theoretically efficient conceptual spaces in context of sequences of stimuli. This approach is comparable with the *deep learning* proposed by Hinton et al. [20] and Bengio [2], in that it allows stacked hierarchies of successively more abstract representation and decreasing dimensionality. However, our approach has the advantage of being theoretically driven, methodologically top-down, and is therefore more readily amenable to post-hoc analysis. A particular feature of both Gärdenfors' theory and deep learning that is important here is the ability to learn representations over time, each one being learned in terms of those that are already available; this is in contrast with the usual methods of statistical linguistics, where the entire high-dimensional space is approximated in a single operation, giving all dimensions equal precedence.

A different, but wholly compatible, notion of conceptual space was introduced by Boden [3] in one of the seminal works of Computational Creativity. This conceptual space is conceived with respect not to an internal representation, but to an abstract notion of the artefact itself. For example, one can have a conceptual space of post

[3] We wish to avoid arguments over where the brain ends and the mind begins, so we use this epithet to refer to the whole assembly.

boxes, with properties and affordances; there is a core set of requirements that would make something not a post box (e.g., having no opening) and there are some properties that are incidental (e.g., in the UK, a post box is always red, whereas elsewhere in Europe they are often yellow and sometimes blue). Interestingly, a post box is still a post box if one removes it from its intended function and, for example, grows flowers in it. The conceptual space of post boxes then defines what it is to be one; a general space of post boxes may include both UK and Dutch examples, while there are two disjoint conceptual spaces for the two nationalities' mail. A human or computer interested in post boxes may search the more general space for a box that will meet their needs; this is *exploratory creativity* in Boden's terms. However, to invent a Dutch post box on the basis of knowledge restricted to UK ones, it is necessary to apply *transformational creativity* to (at least) the dimension of colour. Wiggins [49, 50] has formalised these notions in a Creative Systems Framework (CSF) and shown how the concept of *aberration*, in which the conceptual space is "stretched" by the production of new concepts that nearly (but not quite) fit within its boundaries, can be used to drive learning and adaptation in a creative system.

Boden's conceptual spaces (which may usefully be defined in terms of those of Gärdenfors, via the CSF or a similar theory) may be represented in many ways, the most obvious analogy being with an AI search space—except that the *a priori* solution detector of the AI approach is not present, because, in creative work, one does not always know the solution until one sees it. In the context where one is interested in sequences, as we are here, a convenient representation of the space compiles its sequences into a statistical structure that summarises the space. Exploratory creativity can then be implemented as sampling in the structure, while transformations of the space can be viewed as changing the structure's parameters, one way or another—and so statistical learning is a (weak) kind of transformational creativity. This is the view we will take in IDyOT; the interesting research question at this point is: how is the sampling to be implemented?

7.3.3 Conscious Awareness and Creativity

IDyOT's simulation works at the boundary between the conscious and the non-conscious mind, and so it is appropriate to examine the relationship between creativity and consciousness here. First, it is important to make a distinction between two different behaviours which are both called "creative" when observed in humans. We call these "spontaneous" and "deliberate" creativity. *Spontaneous creativity*[4] is a cognitive event in which an idea appears in conscious awareness without warning and without immediately prior conscious preparation, and is then consciously judged novel and/or interesting by the person who experienced it; a familiar example would

[4] In earlier publications, we have referred to this as "inspiration" [53] and "non-conscious creativity". However, "spontaneous" captures better the meaning we intend. Similarly, we have previously referred to deliberate creativity as "conscious creativity".

be waking in the morning with the solution to a problem experienced the previous day in mind. *Deliberate creativity* is a process experienced by the human conducting it[5] as a consciously controlled series of reasoning and development of a creative artefact; an example would be the film composer who constructs a new theme tune to order by the application of rules that she knows from experience will yield acceptable and appropriate material. Many composers, among them Mozart [21] and (less famously) the current authors, view composition of music as a cyclic combination of these things, where spontaneous ideation is followed by deliberated development, which in turn leads to further spontaneous creativity, and so on.

Wallas [47] proposes a multi-phase process of creativity, in which "incubation" leads to "illumination"; it seems likely that the "illumination" point, colloquially known as "the 'Aha!' moment", corresponds with the emergence of an idea from non-conscious systems that emit it into conscious awareness. Indeed, to claim otherwise necessarily leads into mysticism (which we eschew here) because there is nowhere else for the ideas to come from.

IDyOT is conceived as a cognitive architecture, following Baars' Global Workspace Theory [1], of which more detail is given below. A key detail of that formulation is the throttling of a large amount of information produced in parallel, by an explicit selection mechanism, as it approaches conscious awareness. The ultimate effect of this mechanism is that an organism is conscious of one thing at a time, and, conversely, consciousness, an expensive resourse, is managed and applied only where it is required. Merker [30] and Shanahan [45] propose candidate neural assemblies that may accommodate Baars' selection mechanism, but no progress has yet been made towards empirical examination of these proposals as a mechanism for creativity. This is partly because making testable predictions about creative behaviour is extremely difficult; it is to be hoped that simulations such as IDyOT may help to do so in future.

For the purposes of the current work, it is not necessary to consider the nature of consciousness (the "hard" problem [5]), but, instead, only what passes through the throttle—a much easier problem; therefore, we refer to "conscious awareness", to make the distinction clear. Thus, the current, initial work on IDyOT is focused on optimising the throttling mechanism, effectively a parameter of an abstracted parallel sampling system, so as to allow the system to perform the various functions that we predict should arise from the architecture. The original proposal for this parameter is given by Wiggins [55], though we are considering a wider range of possibilities.

[5] We avoid the troublesome question of whether the phenomenon experienced as a conscious decision is really that, because it is not relevant here.

7.4 A Cognitive Architecture that Supports Non-conscious Creativity

Thus, the overarching proposal examined in the IDyOT approach is that the mechanism underlying spontaneous creativity is information-theoretically regulated prediction from models built from observation of perceived input, that encode its observed likelihood. It is important to acknowledge that there are aspects of cognition that are not included in this model, such as affect; and also that there are aspects of mechanism that are not modelled: for example, it is known that different brain components have different learning rates, but this is not the case in IDyOT. Rather than trying to emulate a whole mind/brain, our aim here is to push our proposal, at this purely functional level, as far as it will go, by using it to attempt to explain as many established phenomena as possible. Understanding where the limits are will inform us as to what other mechanisms are needed and where and how they fit.

In this section, we unpack this dense proposal and consider the ancillary mechanisms required to support it. In the following section, we propose and explain the initial testable predictions that we make of it.

7.4.1 Overview

IDyOT is a direct implementation of the Global Workspace Theory as described by Baars [1]. A large[6] number of generators sample from a complex statistical model of sequences, performing Markovian prediction from context [25, Chap. 9]. Each generator (which may be thought of as a simple AI agent [24]) maintains a buffer of perceptual input, which may include mis-perceptions due to the possibility of multiple predictions matching ambiguous input. Buffered sequences are flushed into the Global Workspace (which may be thought of as an AI blackboard [9]) when an information-theoretic throttling condition related to a proposal by Wiggins [55] is met. This mechanism solves a problem within Global Workspace Theory known as the Threshold Paradox [1, 55]. A block diagram of the system is shown in Fig. 7.1.

The diagram illustrates the cyclic (and hence dynamic) nature of the model. The generators sample from statistical memory, synchronised[7] by the perceptual input, if some exists, that it receives. If there is no input, the generators freewheel (evidence for the neuroscientific validity of this position is given by Fink et al. [13]), conditioned only by prior context, and this is where creativity is admitted; however, for the moment, we focus on the perceptual input, for it is this that fundamentally drives the system. Perceptual input is matched against generators' predictions, and

[6] The number is not specified in Baars' theory. In IDyOT, the number of generators has a direct bearing on the (statistical) prediction quality: as the number of generators increases, so does the likelihood of correct predictions.

[7] The initial version of IDyOT has an abstract, symbolic representation of time; however, more developed versions will predict the real-world timing of perceptual input, as well as its content.

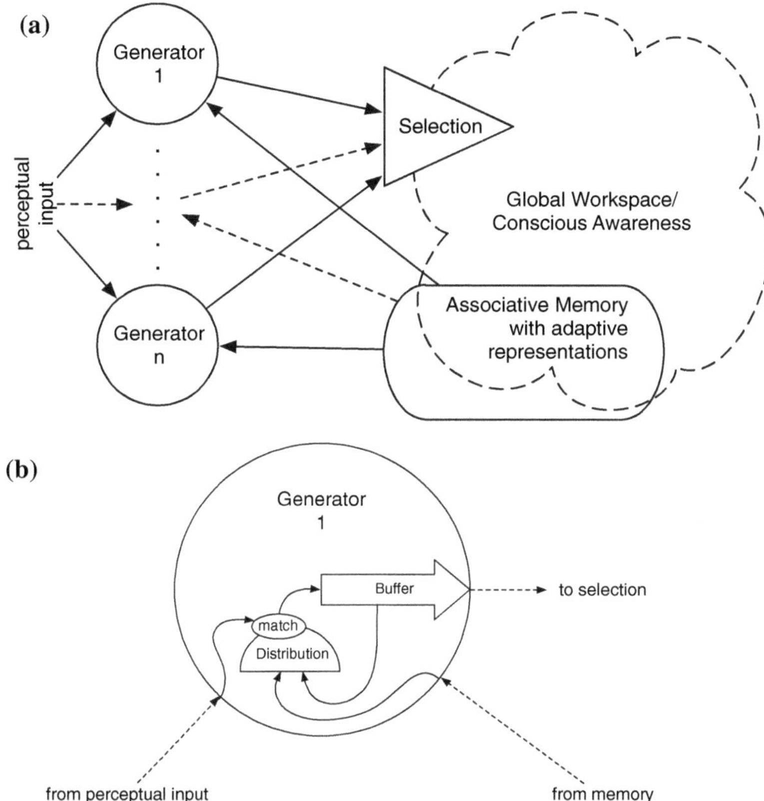

Fig. 7.1 Overview of the IDyOT architecture. **a** Generators synchronised to perceptual input sample, given previously buffered perceptual input (if any), from a mixed-order, multidimensional Markov model to predict the next symbol in sequence, which is matched with the input. Predicted symbols that match are buffered by each generator until it is selected by the throttling function. The selected generator then flushes its buffer into the Global Workspace ("conscious awareness"). This allows the resulting chunk of sequence to be stored in the memory, to become part of the statistical model and thence to be used subsequently. **b** Within the generators, perceived inputs are matched probabilistically with predicted distributions, and then the match that best matches expectation, from the generator's prediction, is entered into a buffer. The buffer flushes when the generators' buffer content is selected by the throttling function

where a match, or a prediction from a match, contains a large amount of information in comparison with other current matches, the corresponding generator's buffer is emptied into the Global Workspace, displacing what was there previously (though what was there is stored in the memory, as below). Buffers that fail to empty are forgotten after a fixed period, the duration of that period, alongside other potential reasons, being a question of the research. The model entails that, for perception to work, at least some generators must be working in all perceptual modalities at all times, including sensory ones; otherwise nothing would be predicting for new input in a given modality to match against. We hypothesise that more generators would deploy in a given area as the amount of information in the relevant signal increases; this might account for the cognitive "jolt" of attention reallocation experienced when one is consciously focused on one stimulus and another forcibly intervenes: this is the subjective effect of sudden and ill-prepared reassignment of generators.

Entry into the Global Workspace broadly corresponds with the assignment of attention (construed as processing power) to the chunk of perceived input thus produced. As in language experiments on parsing by competitive chunking [40, 44], this breaks linguistic sequences into statistically coherent groups, which tend to correspond with semantically coherent sub-phrases, though the chunks do not necessarily correspond with traditional linguistic categories. Once a chunk has entered the Global Workspace, it is also added to the memory, and so becomes available to the generators for prediction. This generates a positive feedback loop in which the chunks inform the statistical model that in turn causes chunking, reinforcing it, and leading to our first prediction, in Sect. 7.5.1, below.

7.4.2 Representation, Memory and Prediction

Thus far, we have discussed chunks and sequences, but we have not specified the detail: chunks and sequences of *what?* The reason for this is that the representation formalism proposed in IDyOT must be understood in a way which is intimately related with the chunking process, described above.

The key to the representation of IDyOT is that all percepts are represented in multiple, statistically related ways. Since the architecture is centrally focused on sequence, memory is expressed in this way. Each sequence is statistically linked to a lower level set of sub-sequences composing it, and to a higher level set of super-sequences categorising its chunks. The easiest way to understand this is to think in terms of language. Given a lexicon of English words, we begin with a simple sentence such as

The horse raced past the barn.

At a naïve level of representation, this sentence would appear as a sequence of 6 symbols, and at each point in the sequence (i.e., at each word boundary) there is a distribution, computed from the context so far and a background model, predicting the next word, just as in most statistical parsing approaches. In IDyOT's theory, there is one difference at this level: as each input word appears, rather than simply taking it as given, we match it against the symbols IDyOT predicted, using a continuous similarity metric that interacts with the distribution, as described in Sect. 7.4.3. In this way, expectation governs (and selects) what is heard, and when there is not a clear winner, misunderstanding between similar symbols can arise, exactly as in humans.

However, IDyOT's memory does not consist only of sequences of unstructured symbols. The design of the system is intended to capture the full stack of capabilities from audio processing up to the Global Workspace, though we expect to begin (for language) with the level of phonemes. Our claim is that phonemes are remembered as a compact categorical representation whose symbols correspond with regions in a multi-dimensional continuous space [17], according with our principle of information-theoretic efficiency. Conversely, the detail of a particular speech signal is not remembered, once it has passed beyond echoic memory.

The aim of the approach, then, is to perform human-like learning, beginning (in language) from the phonemes, and passing through the Global Workspace cycle described above. Initially, while memory is empty, the model is not able to predict, everything has high information content, and so individual symbols, and then pairs begin to appear in the memory. Earlier work with IDyOM, IDyOT's ancestor, demonstrated that such chunking can successfully be performed statistically over a fixed corpus, with a descriptive model of segmentation [54]. In the more dynamic IDyOT, in spite of the initial chunks being somewhat chaotic, they are held together in larger sequences, because each chunk is given its own label, and the labels form their own sequence in memory. Thus, when a symbol is encountered for the second time, IDyOT is able to predict not only from the phoneme level, but also from the sequences of chunks that it has constructed, and so hone its predictions. Because of the positive feedback induced by the IDyOT processing loop, we expect the memory to stabilise,[8] and once it has stabilised, the erroneous details that it inferred early in the learning process, and therefore in the absence of a model, will fade into statistical obscurity. Thus, we claim, IDyOT constitutes an explanatory model of the chunking behaviour that the IDyOM segmentation described.

Theoretically, this layering of models can proceed up to a level beyond that of sentences, but it is not restricted to the sentential forms of Chomskian linguistics: the grouping is motivated by statistical structure, and not by semantic connection. The approach gives rise to predictive behaviours very like Cohort Theory [28], where words are identified incrementally as the phonemes appear. IDyOT assembles a network of probabilistic predictions, as shown in Fig. 7.2a.

Many readers will recognise this as (like) a Bayesian Network, and this accords to evidence that such networks can predict aspects of human parsing behaviour [32]. It is important to note that IDyOT's network is in a sense *stratified*: there are distinct layers, those layers predict only above and below, and only one symbol forwards on each layer, which allows us in principle to control the computational cost of prediction, though the problem of prediction from Bayesian networks in general remains NP-hard.[9]

However, in general, the stimuli to which IDyOT will respond will not be sequences of atomic percepts. Speech, our illustration here, includes pitch, stress and volume information, all of which will be encoded in memory as structured, multi-dimensional symbols, and used for prediction, as has been demonstrated in IDyOM [39]. This demands a more powerful model than is common in cognitive models of language. For music, multi-dimensionality is a *sine qua non*, and for this reason, Conklin and Witten [8] proposed an approach based on *viewpoints* that allows a set of

---

[8] It will stabilise when it has produced an efficient model of the data that it is being exposed to.

[9] That is to say, it cannot be computed in polynomial time by a von Neumann machine.

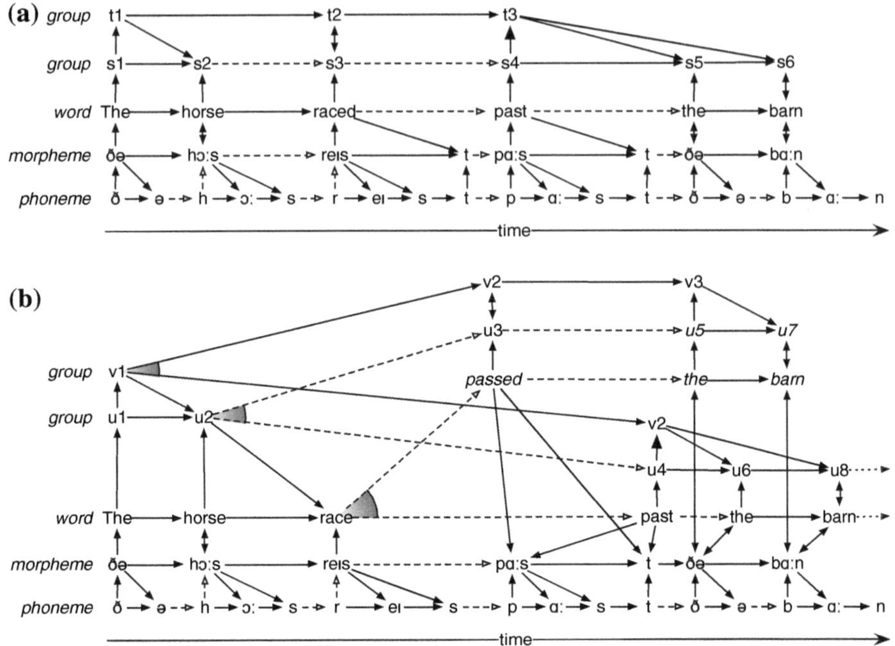

Fig. 7.2 Representation and recognition. **a** A sequence of symbols in IDyOT memory is composed of subsequences and supersequences; here phonemes are given in International Phonetic Alphabet for UK standard English; larger groups are marked as words or arbitrarily named groups; note though that some of these groups correspond with fairly conventional syntactic labels, such as "noun phrase". In this simplified diagram, no alternatives are shown; however, each *arrow* in the diagram is associated with two distributions (one in each direction) over the set of symbols that may be at the appropriate end. The arrow heads are a rough guide to the flow of information around the diagram as phonemes are perceived and words assembled. *Dotted arrows* are relatively low-probability implications. It is important to understand that each higher level is inferred from the level below, so the linguistically motivated labels on the group level and words themselves are added to assist the reader, and would be arbitrary symbol names in the software. **b** A more complex parse, including one ambiguous possiblity. The *shaded angles* indicate where two values from a distribution are included in the diagram. Note that the lower reading leads to expectation of a continuing sentence (indicated by *dotted rightwards arrows*). Necessarily for a 2-dimensional diagram, the two parses are not synchronised on the time dimension

interacting features, associated together by means of sequences of multi-dimensional symbols, to admit multi-dimensional prediction. This is the system used in IDyOM [33] (see Sect. 7.3.1), and adapted for multidimensional language models by Wiggins [54]. A key feature of IDyOM is its ability to integrate information from different features with weights determined by their information content [34], and the same idea is used in IDyOT.

Given Conklin's notion of viewpoint [8] and the associated mathematics, it becomes possible also to represent propositional meaning within the statistical framework: one simply incorporates representations of the meaning (perhaps drawn from another sensory modality, e.g., describing in language a scene representation derived from a vision input) in the statistical model, in the style of Eshghi et al. [11].

In summary, IDyOT's memory consists of multiple structures like those in Fig. 7.2a, in parallel, tied together by observed co-occurrences of feature values expressed in multidimensional perceptual input sequences. The whole constitutes a Bayesian Network, stratified as explained above, and constrained to predict only to the subsequent symbol; note, though, that the subsequent symbol may represent something arbitrarily far in the perceptual future, because higher-level, more abstract models predict in parallel with, and conditioned by, more concrete ones. From this model, IDyOT's generators make predictions and their outputs are selected on the basis of probabilistic matching with input—and the cycle begins again.

7.4.3 Matching, Similarity and Representation

One important factor in the above description is the notion of similarity, and—implicitly—the provenance of the symbols used in the sequences. We address both of these issues by appeal to Gärdenfors' theory of Conceptual Spaces [17], in which a low-dimensional "conceptual level" mediates between a "symbolic" or "linguistic" level, and the high-dimensional continuous representation that hypothetically captures the wetware. Using this approach, each of our symbols corresponds with a region in a conceptual space, and thus can its relationship with other symbols be defined. In some perceptual domains (such as musical pitch) there are well-defined theories of the geometry of the specific conceptual spaces required, and these can support the measurement of similarity. More excitingly, Gärdenfors' theory has an account of the development of such spaces in a way that may be construed as information-theoretic, and a future avenue for development of IDyOT will be the inclusion of learned representations for the learned structures in its memory.

In the current formulation, perceptual inputs are matched for similarity in a metric space distorted by the predicted distribution, in such a way that more expected symbols are more tolerant of a poor match with the symbols encountered. Without such tolerant but predictive matching, it would be difficult to understand one's own language when spoken with an accent different from one's own, and the phonemic restoration effect would not take place. This effect can also be modelled by lowering the tolerance of a simple sequential match, so that unmatched intermediate symbols do not cause failure. This approach, however, leaves no room to account for the ability to learn an unfamiliar accent over time, nor indeed, does it account for learning perceptual spaces at all.

7.5 Testable Predictions of the Model

We expect IDyOT's behaviour to produce simulations of several aspects of human cognition. Methodologically, this is very useful, since it is possible to give an ordered list of goals, starting with perceptual mechanisms and ending with demonstrable spontaneous creativity, that can be tested more or less in sequence. Furthermore, many of the items on this empirical shopping list can be tested against established data using established paradigms; it is only later on that one encounters the harder-to-evaluate creative behaviours. Thus, we can develop confidence in the model as a whole before we attempt the more challenging evaluations. It is worth mentioning that the effects introduced below are not in themselves novel solutions to problems: the novelty is in the one central mechanism that gives rise to all these (and more) phenomena at once.

7.5.1 Segmentation of (at Least) Music and Language

As already discussed, it is well-established in the literature that sequences of musical [36] and linguistic [6, 42, 54] symbols can be segmented in ways that correspond with perceptual chunking [18] using methods based on information content or entropy, and it is easy to suggest evolutionary reasons why this should be so: compression on the basis of information-theoretic structure is likely to yield efficient representations. The extant methods, however, do not attempt to fit this idea into a larger cognitive architecture, and use descriptive rules to detect the boundaries given the statistics of the data; we propose that our approach, in which sequences are buffered until their consequent predictions have high entropy in comparison with the predictions of other generators, not only yields this effect as an emergent property, but also accounts for handling of boundary ambiguity in language. Thus, IDyOT should, without extra mechanism, be able to model the "garden path" effect, where an utterance such as

> The horse raced past the barn fell.

is initially parsed as complete after the word "barn" has been encountered.

IDyOT is able to predict simultaneously and coherently at multiple levels of abstraction. This is somewhat like processing multiple parses of a sentence in parallel, but with a probabilistic optimisation that omits relatively unlikely parses (thus avoiding combinatorial explosion), and also including abstract-level parsing too. Thus, both readings of the garden path sentence, above, are parsed in parallel, but the correct one is relatively unlikely, and the misleading one takes precedence as a conscious experience because the (incorrectly predicted) end of the sentence after "barn" yields a high-entropy prediction. When the alternative reading enters consciousness as the result of the alternative sentence ending doing the same, a revision is forced between the previous and the new reading, and their respective interpretations, and it is this sudden and unexpected change, in the IDyOT view of cognition, that leads to the perceived cognitive "jolt" caused by the garden path effect, rather than the need to re-parse, in more traditional accounts. In this way, the phenomenon somewhat imprecisely called "recursion" by Chomskian linguists [14] is (unsurprisingly) enfolded in the parallel model without the addition of a special mechanism, and in particular without the stack entailed by computational notions of recursion. Because the mechanism proposed here is not specific to humans (indeed, it is motivated by reference to more cognitively primitive species [55]), there is consequently no reason to claim, as do Fitch et al. [14], that cognitive recursion is a defining property of humanity. What appears as recursion in the particular style of the Chomskian analysis is accounted for by a mathematically simpler mechanism in IDyOT.

The parsing-by-prediction process is illustrated in Fig. 7.3. The figure shows, centrally, a line of text being processed by IDyOT. Above and below the line are the predictions of two generators, G1 and G2, respectively—but recall that in a real run of IDyOT there would be many more generators than this. The predictions are labelled arbitrarily in the figure, but it is important to note that these categories are synthesised bottom-up and not imposed top-down. The function T_{Gi} (per generator) is the information-theoretic measure used to compare and select the input into the Global Workspace. The effect modelled here is akin to that proposed by Hale [19], differing because it is an epiphenomenon of the throttling of information entering the Global Workspace (specifically: more information that is usual), rather than relying on top-down detection of a change in distribution.

Our testable prediction (*about* the model, as opposed to the predictions *of* the model), therefore, is that IDyOT should be able to learn to segment to the same degree as the top-down models in both music and language; and to simulate the garden path effect, when an initial semantic interpretation is displaced by a new one. It should be able to do these things without additional descriptive rules.

7.5.2 Lexical Ambiguity and Human-Like Misunderstanding

Given that it can segment, the IDyOT model should also be able to parse. However, the notion of parsing here is different from the Chomskian one. Chomskian linguists sometimes talk about a parser and/or a grammar "getting" a grammatical construct, which we take to mean that it can be correctly parsed, presumably with something like a shift-reduce parser. The equivalent construct in IDyOT has no notion of correctness, but instead of a degree of match with observed likelihood. Thus, IDyOT is not dependent on whole sentences, or any predefined syntactic unit: its simple inference method runs bottom-up to combine and predict from whatever input it is given from whatever network structure it has learned. When the result contains sufficient information in comparison with other candidates, it is passed into the Global Workspace, and this, as it moves up the hierarchy of groups, corresponds with the moment of Chomskian "getting". This process, from the morpheme level up, is illustrated in Fig. 7.2. Figure 7.2b in particular illustrates IDyOT dealing with syntactic ambiguity

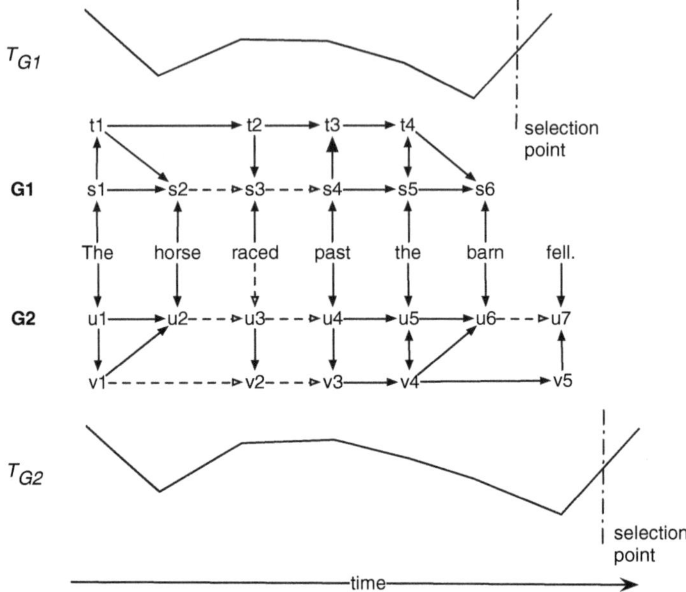

Fig. 7.3 Segmentation in IDyOT of the classic garden path sentence, "The horse raced past the barn fell". The garden path effect, where a slight cognitive "jolt" is experienced just after the final word is encountered, was previously mostly accounted for by either backtracking or re-parsing of the input. In IDyOT, the explanation is that the incorrect meaning is replaced in conscious awareness very quickly—more quickly than in normal processing—and it is this very brief information overload that causes the perceived effect. Labels in this figure are as in Fig. 7.2, except for Gi which denote the generators following the two paths and T_{Gi} which illustrate the information-theoretic signals output by the generators

arising from overloading of a single sound-word, /pɑːst/ in IPA,[10] with two lexical forms, "passed" and "past". It is to be expected that intonation, representable within IDyOT's data structures, will play an important role here, and this will be a focus of future experimentation.

In order to produce semantic representations from this parsing process, it is necessary to include in the learning process representations of semantic interpretations, in the style of Eshghi et al. [11]. Much as infants learn the meaning of words by association, guided by parental gaze and co-attended hand movements [4], IDyOT grounds its meanings by means of association between semantic symbols and the corresponding surface form. These associations can be represented straightforwardly in the viewpoint framework [8], and therefore require no extra mechanism in IDyOT.

Underlying the distillation of structure, and thence meaning, by multilevel statistical prediction, as illustrated in Fig. 7.2, is the continuous representation of perceptual

[10] The International Phonetic Alphabet is used in dictionaries to specify a standard pronunciation of each word. Good dictionaries contain an explanation of the symbols in terms of the relevant language. IPA versions used here are taken from Apple's UK English dictionary.

response to phonemes. This is what accounts for humans' ability to hear their own language pronounced in a previously-unencountered accent, and still to understand it. Underlying the symbolic representation of phonemes as shown in Figs. 7.2 and 7.3 is a conceptual space akin to the vowel space of Fairbanks and Grubb [12]. Regions in this space correspond with the sonic components of language as learned by the system, so a word represented as a sequence of phonemes, as in the figures, is in bijection with a trajectory through this space. In humans, physical constraints on the vocal apparatus restrict how closely pronunciation can follow such a theoretical trajectory.

An important aspect of human speech understanding is its robustness to mispronunciation and to accents. IDyOT addresses this ability in its matching stage, where predictions are matched against perceived input. The match is statistical, based on the statistical distribution over the possible symbols, but also geometrical, based on the conceptual space. This approach is important not just because it affords robustness in understanding natural language, given inter-individual variations, but because it also allows flexibility of understanding, in a human-like way, based on statistical priming.

A well-known trap set by linguists, at the phonetic level, is the sentence

It's easy to wreck a nice beach.

which is readily mistaken, given an appropriate context, for the sentence

It's easy to recognise speech.

By choosing a pronunciation that is somewhat loose, and coupling it with appropriate verbal or visual priming, this serves as a useful demonstration to students of how much prior information is used in understanding language. In IPA, these two sentences are respectively denoted

/ɪts iːzi tə rɜkəgnʌɪz spiːtʃ/
/ɪts iːzi tə rɜk ə nʌɪs biːtʃ/

In fact, in common parlance, the /g/ is usually soft, and often omitted altogether; so then the difference comes down only to the amount of voice in /z s/, which are elided to /zs/, against /s/, and /p/ against /b/ respectively. In Fig. 7.4, we illustrate this process.

For the purposes of the current paper, this demonstration makes two points: first, the multi-layered approach of IDyOT affords human-like behaviours as emergent properties, without extra mechanism; and second, the conceptual layer, which affords the flexibility of approximate matching in a principled way, is as important as the symbolic one in driving the system. One might ask, therefore, why not conflate the two and do all the inference in a continuous probability space? From our perspective, the answer is methodological: we believe that the neural representation that we are modelling is indeed on one high-dimensional level. However, current technology is not adequate to model that representation, so working at multiple levels, in parallel, gives us a way of working towards a solution, and, crucially, identifying where in the structured representation and its associated operations any shortcomings are located.

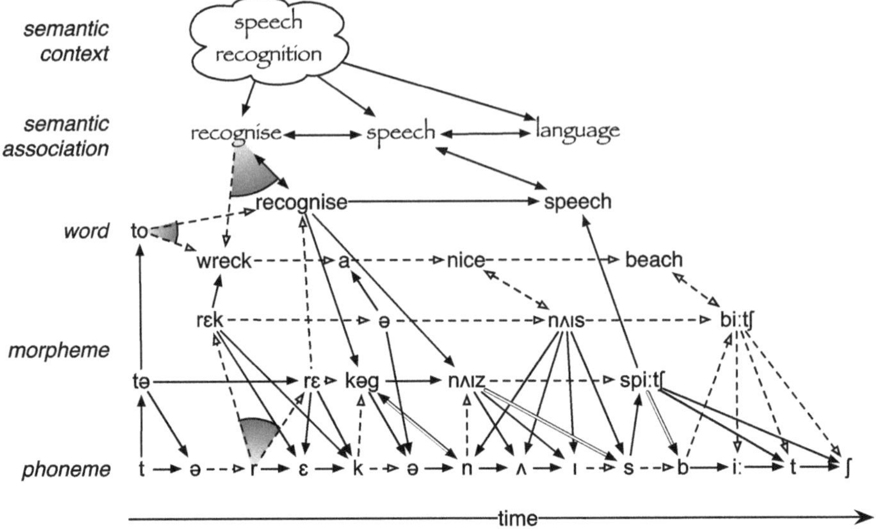

Fig. 7.4 Confusing IDyOT: how priming resolves speech analysis in the wrong direction. As previously, points where there is a distribution containing a potential choice in the example are shaded. This example is different from the earlier ones, because it includes a semantic context level, conditioned by previous input, that adjusts the prior provided by the language model. The assumption here is that speech recognition has been the topic; so the distribution associated with /rɜk/ leads away from "wreck" and towards "recognise". Once that path is taken, the semantic association is reinforced and inference overcomes the correct reading where the phonemes can be conflated, indicated by the *double-line arrows*. As before, *solid arrows* indicate strong likelihoods, and broken ones are relatively weak

It also allows us to better understand the functions of the system independent of their implementation. Thus, we are better placed to improve future versions of the model.

7.6 Creativity

Finally, having discussed an extended example of how IDyOT perceives and analyses sentences, we turn to its potential for creativity and other general reasoning.

The key here lies in the idea of predictive parsing laid out in the previous sections. It is necessary to drop the more conventional notion of parsing as the input and processing of known symbols, and instead to view the parser as a prediction machine which continually attempts to match its predictions with what is perceptually encountered. There is considerable flexibility in this process, which can result, as demonstrated above, in substantial changes in semantics. The benefits conferred by this approach are efficiency and robustness: prediction allows the listener to get ahead of the speaker, and also to reconstruct obscured or unclear parts of the input stream, both of which might be considered restricted creative behaviours. In the higher levels of the IDyOT representation, the same process allows semantic predictions, inferring what the speaker intends, and potentially looking ahead to anticipate the entire discourse. There is every reason to believe that this is a realistic simulation of humans, who jump to conclusions, correct or otherwise, all the time.

Such a predictive mechanism necessarily begs the question: what happens when there is no input to match with? In this circumstance, there is no existing sequence to drive prediction, so it is driven directly from the memory, and the Global Workspace may be thought of as an empty auditorium into which ideas may enter [55]. The mechanism used above for chunking may also be used to select structures as they are predicted, and it is important to remember here that there is no reason why prediction cannot be performed at any level of representation that exists within IDyOT's memory. Thus, predictions conditioned only by the learned model, and not by current input, can be made at any or all of the various levels from semantic context down to phoneme, and the resulting distributions combined to make a single one [34], from which a structure can be chosen by the throttle mechanism. Therefore, given the necessary information-theoretic properties, such structures, which may or may not be novel, may enter the Global Workspace without the need for external stimulus [55], though, clearly, they will be conditioned by previous (learned) stimuli. As a result of passage through the Global Workspace, the structures are added to memory, and therefore become available for use in future generative inference. In such a way, IDyOT can follow a cycle which is a simplified version of the process (probably) described by Wolfgang Amadeus Mozart, characterising his own composition:

> When I am, as it were, completely myself, entirely alone, and of good cheer—say traveling in a carriage, or walking after a good meal, or during the night when I cannot sleep; it is on such occasions that my ideas flow best and most abundantly. Whence and how they come, I know not; nor can I force them. Those ideas that please me I retain in memory, and am accustomed, as I have been told, to hum them to myself.
>
> All this fires my soul, and provided I am not disturbed, my subject enlarges itself, becomes methodized and defined, and the whole, though it be long, stands almost completed and finished in my mind, so that I can survey it, like a fine picture or a beautiful statue, at a glance. Nor do I hear in my imagination the parts successively, but I hear them, as it were, all at once. What a delight this is I cannot tell! All this inventing, this producing takes place in a pleasing lively dream. Still the actual hearing of the toutensemble is after all the best. What has been thus produced I do not easily forget, and this is perhaps the best gift I have my Divine Maker to thank for. [21, p.317–318]

Some aspects of this description refer to what we defined in Sect. 7.3.3 as spontaneous creativity, where ideas appear unbidden in conscious awareness. Others are deliberate: the selection and memorisation of pleasing elements. These latter are not present in IDyOT, as a notion of "pleasingness" is currently absent, requiring as it does a solution to the "hard" problem.

Therefore, our initial experiments with IDyOT as a creator will focus on the production of sentences, melodies and concepts, based on a substantial body of learned information, and selected by IDyOT's throttling mechanism. It remains to be seen whether such a purely spontaneous system can create substantive, interesting structures, but early experiments with statistical generation in music have been promising

[8, 38, 41, 48]. A successful outcome in this respect will not only yield a useful tool in computational creativity and artificial general intelligence, but also form a platform to explore further questions as to the nature of consciousness, and its interaction with creativity in humans and machines.

Acknowledgments We gratefully acknowledge the contribution of our colleagues in the Intelligent Sound and Music Systems group at Goldsmiths, University of London and in the Computational Creativity Lab at Queen Mary University of London. In particular, we are grateful to Marcus Pearce, whose work on musical expectation originally inspired the current thinking, and to Kat Agres, Sascha Griffiths and Matt Purver for their insightful comments. Previous work on IDyOM (Sect. 7.3.1) was funded by EPSRC studentship number 00303840 to Marcus Pearce, and EPSRC research grants GR/S82220 and EP/H01294X to Marcus Pearce and the first author. The current work was funded by two project grants from the European Union Framework Programme 7, Lrn2Cre8 and ConCreTe. The projects ConCreTe and Lrn2Cre8 acknowledge the financial support of the Future and Emerging Technologies (FET) programme within the Seventh Framework Programme for Research of the European Commission, under FET grants number 611733 and 610859 respectively.

References

1. Baars, B.J.: A Cognitive Theory of Consciousness. Cambridge University Press, Cambridge (1988)
2. Bengio, Y.: Learning deep architectures for ai. Found. Trends Mach. Learn. **2**(1), 1–127 (2009). doi:10.1561/2200000006
3. Boden, M.A.: The Creative Mind: Myths and Mechanisms. Weidenfield and Nicholson, London (1990)
4. Brooks, R., Meltzoff, A.N.: The development of gaze following and its relation to language. Dev. Sci. **8**(6), 535–543 (2005). doi:10.1111/j.1467-7687.2005.00445.x
5. Chalmers, D.J.: The Conscious Mind: In Search of a Fundamental Theory. Oxford University Press, Oxford (1996)
6. Cohen, P., Adams, N.: An algorithm for segmenting categorical time series into meaningful episodes. In: Hoffmann, F., Hand, D., Adams, N., Fisher, D., Guimaraes, G. (eds.) Advances in Intelligent Data Analysis. Lecture Notes in Computer Science, pp. 198–207. Springer, Berlin (2001). doi:10.1007/3-540-44816-0_20
7. Conklin, D.: Prediction and entropy of music. Master's thesis, Department of Computer Science, University of Calgary, Canada (1990). http://pharos.cpsc.ucalgary.ca:80/Dienst/UI/2.0/Describe/ncstrl.ucalgary_cs/1989-352-14?abstract=
8. Conklin, D., Witten, I.H.: Multiple viewpoint systems for music prediction. J. New Music Res. **24**, 51–73 (1995)
9. Corkill, D.D.: Blackboard systems. AI Expert **6**(9), 40–47 (1991)
10. Egermann, H., Pearce, M., Wiggins, G., McAdams, S.: Probabilistic models of expectation violation predict psychophysiological emotional responses to live concert music. Cognit. Affect. Behav. Neurosci. **13**(3), 533–553 (2013). doi:10.3758/s13415-013-0161-y
11. Eshghi, A., Purver, M., Hough, J.: Probabilistic induction for an incremental semantic grammar. In: Proceedings of the 10th International Conference on Computational Semantics (IWCS 2013)—Long Papers, pp. 107–118. Association for Computational Linguistics, Potsdam, Germany (2013). http://www.aclweb.org/anthology/W13-0110
12. Fairbanks, G., Grubb, P.: A psychophysical investigation of vowel formants. J. Speech Hear. Res. **4**, 203–219 (1961)
13. Fink, A., Grabner, R.H., Benedek, M., Reishofer, G., Hauswirth, V., Fally, M., Neuper, C., Ebner, F., Neubauer, A.C.: The creative brain: investigation of brain activity during creative problem solving by means of eeg and fmri. Hum. Brain Map. **30**, 734–748 (2009)

14. Fitch, W.T., Hauser, M.D., Chomsky, N.: The evolution of the language faculty: clarifications and implications. Cognition **97**, 179–210 (2005)
15. Fodor, J.: Special sciences: or the disunity of science as a working hypothesis. Synthese **28**, 97–115 (1974)
16. Forth, J., Wiggins, G., McLean, A.: Unifying conceptual spaces: concept formation in musical creative systems. Mind. Mach. **20**, 503–532 (2010). doi:10.1007/s11023-010-9207-x
17. Gärdenfors, P.: Conceptual Spaces: The Geometry of Thought. MIT Press, Cambridge (2000)
18. Gobet, F., Lane, P.C.R., Croker, S., Cheng, P.C.H., Jones, G., Oliver, I., Pine, J.M.: Chunking mechanisms in human learning. TRENDS Cognit. Sci. **5**(6), 236–243 (2001)
19. Hale, J.: A probabilistic earley parser as a psycholinguistic model. In: Proceedings of NACL, pp. 159–166 (2001)
20. Hinton, G.E., Salakhutdinov, R.R.: Reducing the dimensionality of data with neural networks. Science **313**(5786), 504–507 (2006)
21. Holmes, E.: The Life of Mozart: Including His Correspondence. Cambridge University Press, Cambridge (2009)
22. Honing, H.: Computational modeling of music cognition: a case study on model selection. Music Percept. **23**(5), 365–376 (2006)
23. Huron, D.: Sweet Anticipation: Music and the Psychology of Expectation. Bradford Books. MIT Press, Cambridge (2006)
24. Luck, M., McBurney, P., Preist, C.: Agent technology: enabling next generation computing. Agentlink. http://calcium.dcs.kcl.ac.uk/841/1/al2roadmap.pdf (2003)
25. Manning, C.D., Schütze, H.: Foundations of Statistical Natural Language Processing. MIT Press, Cambridge (1999)
26. Marr, D.: Vision: A Computational Approach. Freeman & Co., San Francisco (1982)
27. Marsden, A.: Response to Geraint Wiggins. J. Math. Music **6**(2), 125–128 (2012)
28. Marslen-Wilson, W.D.: Function and process in spoken word recognition. In: Bouma, H., Bouwhuis, D. (eds.) Attention and Performance X: Control of Language Processes, pp. 125–150. Erlbaum, Hillsdale (1984)
29. McClamrock, R.: Marr's three levels: a re-evaluation. Mind. Mach. **1**(2), 185–196 (1991)
30. Merker, B.: The efference cascade, consciousness, and its self: naturalizing the first person pivot of action control. Front. Psychol. **4**(501) (2013). doi:10.3389/fpsyg.2013.00501
31. Meyer, L.B.: Emotion and Meaning in Music. University of Chicago Press, Chicago (1956)
32. Narayanan, S., Jurafsky, D.: A bayesian model predicts human parse preference and reading time in sentence processing. In: Advances in Neural Information Processing Systems, vol. 14, pp. 59–65. MIT Press, Cambridge (2002)
33. Pearce, M.T.: The construction and evaluation of statistical models of melodic structure in music perception and composition. Ph.D. thesis, Department of Computing, City University, London (2005)
34. Pearce, M.T., Conklin, D., Wiggins, G.A.: Methods for combining statistical models of music. In: Wiil, U.K. (ed.) Computer Music Modelling and Retrieval, pp. 295–312. Springer, Heidelberg, Germany (2005). http://www.doc.gold.ac.uk/mas02gw/papers/cmmr04.pdf
35. Pearce, M.T., Herrojo Ruiz, M., Kapasi, S., Wiggins, G.A., Bhattacharya, J.: Unsupervised statistical learning underpins computational, behavioural and neural manifestations of musical expectation. NeuroImage **50**(1), 303–314 (2010). doi:10.1016/j.neuroimage.2009.12.019
36. Pearce, M.T., Müllensiefen, D., Wiggins, G.A.: The role of expectation and probabilistic learning in auditory boundary perception: a model comparison. Perception **39**(10), 1367–1391 (2010)
37. Pearce, M.T., Wiggins, G.A.: Expectation in melody: the influence of context and learning. Music Percept. **23**(5), 377–405 (2006)
38. Pearce, M.T., Wiggins, G.A.: Evaluating cognitive models of musical composition. In: Cardoso, A., Wiggins, G.A. (eds.) Proceedings of the 4th International Joint Workshop on Computational Creativity, pp. 73–80. Goldsmiths, University of London, London (2007)
39. Pearce, M.T., Wiggins, G.A.: Auditory expectation: the information dynamics of music perception and cognition. Top. Cognit. Sci. **4**(4), 625–652 (2012)

40. Perruchet, P., Vinter, A.: Parser: a model for word segmentation. J. Mem. Lang. **39**, 246–263 (1998)
41. Ponsford, D., Wiggins, G.A., Mellish, C.: Statistical learning of harmonic movement. J. New Music Res. **28**(2), 150–177 (1999). http://www.soi.city.ac.uk/geraint/papers/JNMR97.pdf
42. Reynar, J.C., Ratnaparkhi, A.: A maximum entropy approach to identifying sentence boundaries. In: Proceedings of the 5th Conference on Applied Natural Language Processing, ANLC'97, pp. 16–19. Association for Computational Linguistics, Stroudsburg (1997). doi:10.3115/974557.974561
43. Saffran, J.R., Griepentrog, G.J.: Absolute pitch in infant auditory learning: evidence for developmental reorganization. Dev. Psychol. **37**(1), 74–85 (2001). http://www.waisman.wisc.edu/infantlearning/publications/DevPsychAP.pdf
44. Servan-Schreiber, E., Anderson, J.R.: Learning artificial grammars with competitive chunking. J. Exp. Psychologyrimeatal Psychol. **16**(4), 592–608 (1990)
45. Shanahan, M.: Embodiment and the Inner Life: Cognition and Consciousness in the Space of Possible Minds. Oxford University Press, Oxford (2010)
46. Shannon, C.: A mathematical theory of communication. Bell Syst. Tech. J. **27**(379–423), 623–656 (1948)
47. Wallas, G.: The Art of Thought. Harcourt Brace, New York (1926)
48. Whorley, R.P., Wiggins, G.A., Rhodes, C., Pearce, M.T.: Multiple viewpoint systems: time complexity and the construction of domains for complex musical viewpoints in the harmonization problem. J. New Music Res. **42**(3), 237–266 (2013). doi:10.1080/09298215.2013.831457. http://www.tandfonline.com/doi/abs/10.1080/09298215.2013.831457
49. Wiggins, G.A.: A preliminary framework for description, analysis and comparison of creative systems. J. Knowl. Based Syst. **19**(7), 449–458 (2006). doi:10.1016/j.knosys.2006.04.009
50. Wiggins, G.A.: Searching for computational creativity. New Gener. Comput. **24**(3), 209–222 (2006)
51. Wiggins, G.A.: Models of musical similarity. Musicae Sci. Discuss. Forum **4A**, 315–338 (2007)
52. Wiggins, G.A.: Computer models of (music) cognition. In: Rebuschat, P., Rohrmeier, M., Cross, I., Hawkins, J. (eds.) Language and Music as Cognitive Systems. Oxford University Press, Oxford (2011)
53. Wiggins, G.A.: Defining inspiration? modelling non-conscious creative process. In: Collins, D. (ed.) The Act of Musical Composition: Studies in the Creative Process. Ashgate, Aldershot (2012)
54. Wiggins, G.A.: I let the music speak: cross-domain application of a cognitive model of musical learning. In: Rebuschat, P., Williams, J. (eds.) Statistical Learning and Language Acquisition. Mouton De Gruyter, Amsterdam (2012)
55. Wiggins, G.A.: The mind's chorus: creativity before consciousness. Cognit. Comput. **4**(3), 306–319 (2012). doi:10.1007/s12559-012-9151-6

Part II
Practice

Chapter 8
Accounting for Creativity Within a Psychologically Realistic Cognitive Architecture

Ron Sun and Sebastien Helie

Abstract Creativity research, computational or otherwise, can benefit from a detailed understanding of human creative problem solving. The psychological processes of human creative problem solving have been tackled using computational modeling and simulation based on the CLARION cognitive architecture. CLARION, in general, attempts to provide a unified explanation of a wide range of psychological phenomena using five basic principles. By using these CLARION principles, the Explicit-Implicit Interaction (EII) theory of creative problem solving was derived, which provides a unified framework for understanding creative problem solving. A list of phenomena that can be accounted for by the EII theory and simulated using the CLARION cognitive architecture is presented. This work represents a step in the development of unified process-based theories of creativity encompassing incubation, insight, and various other related phenomena. Beyond EII, the roles of motivation, personality, emotion, and social interaction in creativity may also be explored using CLARION.

8.1 Introduction

Creativity research, computational or otherwise, can benefit from a detailed understanding of human creative problem solving, including the detailed psychological processes and mechanisms involved therein. Psychological processes of human creative problem solving have been tackled with a variety of means, including computational modeling and simulation, for example, based on the CLARION cognitive architecture [26, 31].

R. Sun (✉)
Department of Cognitive Sciences, Rensselaer Polytechnic Institute,
Troy, NY, USA
e-mail: rsun@rpi.edu

S. Helie
Department of Psychological Sciences, Purdue University,
West Lafayette, IN, USA

T.R. Besold et al. (eds.), *Computational Creativity Research: Towards Creative Machines*,
Atlantis Thinking Machines 7, DOI 10.2991/978-94-6239-085-0_8

Cognitive architectures, in general, are becoming increasingly important in cognitive science, in psychology, and in artificial intelligence (AI) [17]. Among these cognitive architectures that have been proposed, the CLARION cognitive architecture [26, 27, 31] tries to provide amore unified explanation of a wide variety of psychological phenomena. It tries to do so using mostly five basic principles: (1) The co-existence of, and the difference between, explicit and implicit psychological processes; (2) The simultaneous involvement of implicit and explicit processes (in most tasks); (3) The "redundant" representation of explicit and implicit knowledge; (4) The integration of the results of explicit and implicit processing; and (5) The iterative (and possibly bidirectional) processing. This cognitive architecture has already been used to account for many psychological phenomena (such as implicit learning, bottom-up learning, cognition-motivation interaction, creativity, and so on) and to simulate a great deal of relevant human behavioral data (e.g., with respect to low-level skill learning, high-level cognitive skill acquisition, and reasoning; see e.g., [26, 32, 33]).

In relation to problem solving, some existing psychological theories of problem solving and reasoning have highlighted a role for implicit cognitive processes. For instance, implicit processes are often thought to generate hypotheses that are later explicitly tested [10, 13, 15]. Also, similarity has been shown to affect reasoning through processes that are mostly implicit [25]. Yet, most theories of problem solving have focused on explicit processes that gradually bring the problem solver closer to the solution in an explicit, deliberate way. However, when an ill-defined or complex problem has to be solved (e.g., when the initial state can lead to many different interpretations, or when the solution paths are highly complex), the solution is often found by sudden 'insight' [4, 22], and theories of regular problem solving are, for the most part, unable to account for this apparent absence of deliberate processes. Hence creative problem solving needs to be examined specifically.

Research on such creative problem solving has tried to tackle more complex, more ambiguous problems. However, psychological theories of creative problem solving tend to be fragmentary and usually concentrate only on a subset of phenomena, such as focusing only on incubation (i.e., a period away from deliberate work on the problem; see [24]) or insight (i.e., the sudden appearance of a solution; see [22]). The lack of detailed process models (e.g., detailed process oriented computational models) has resulted in their limited impact on the field of problem solving and creativity [8].

In this chapter, we explore an integrative theory of creative problem solving that is based on a psychologically realistic cognitive architecture, that is, the CLARION cognitive architecture. The integrative theory and the cognitive architecture on which it is based will hopefully transcend the shortcomings of many existing models/theories, and address many relevant aspects of creative problem solving, from incubation to insight, and from motivation to personality.

The remainder of this chapter is organized as follow. First, we discuss the relevance of psychologically realistic cognitive architectures to AI, cognitive science, and psychology. Second, the CLARION cognitive architecture is sketched. Third, the Explicit-Implicit Interaction (EII) theory of creative problem solving, derived from the CLARION cognitive architecture, is briefly explained. Fourth, we present a summary of phenomena that are captured by the EII theory and simulated by CLARION. Fifth, going beyond EII, we examine the relevance of motivation, personality, emotion, and social interaction to creative problem solving, and how CLARION may account for them. We conclude with a discussion of the advantages of using integrative frameworks (such as cognitive architectures) in AI and cognitive science.

8.2 Why Is a Cognitive Architecture Important?

In cognitive science as well as in AI, a cognitive architecture is the specification of the essential structures, mechanisms, and processes in the form of a domain-generic computational cognitive model, which can be used for a broad, multiple-level, multiple-domain analysis of cognition and behavior [28]. Its function is to provide a framework to facilitate more detailed modeling and understanding of various components and processes of the mind. In this way, a cognitive architecture serves as an initial set of assumptions to be used for further development of models and theories.

While there are all kinds of "cognitive architectures" in existence, we focus specifically on psychologically oriented cognitive architectures (as opposed to software engineering oriented ones). For cognitive science, the importance of such cognitive architectures lies in the fact that they are highly useful to understanding the human mind in many ways. Researchers who use cognitive architectures must specify cognitive mechanisms in sufficient detail to allow the resulting models to be implemented on computers and run as simulations. While it is true that more specialized, narrowly scoped models may also serve this purpose, they are not as generic and as comprehensive and therefore do not provide as unified accounts [26, 28].

For the fields of AI, the importance of cognitive architectures lies in the fact that they support its central goal—building artificial systems that are as capable as human beings (or more). Cognitive architectures help to reverse engineer the only truly intelligent system around currently—the human mind. The use of cognitive architectures in building intelligent systems may also facilitate the interaction between humans and artificially intelligent systems because of the similarity between humans and cognitively/psychologically grounded intelligent systems.

8.3 The CLARION Cognitive Architecture

CLARION [26, 31–33] is an integrative, comprehensive cognitive architecture consisting of a number of distinct subsystems for distinct psychological functionalities, with a dual representational structure in each subsystem (implicit versus explicit representations).

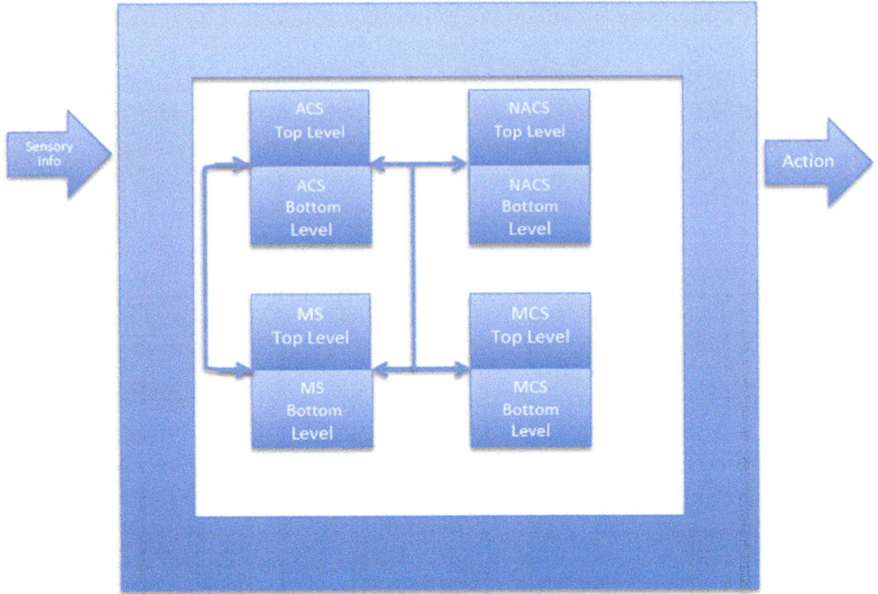

Fig. 8.1 The CLARION cognitive architecture. The major subsystems are shown

The subsystems within CLARION include the action-centered subsystem (the ACS), the non-action-centered subsystem (the NACS), the motivational subsystem (the MS), and the meta-cognitive subsystem (the MCS). The role of the action-centered subsystem is to control actions, regardless of whether the actions are for external physical movements or for internal mental operations. The role of the non-action-centered subsystem is to maintain general (declarative) knowledge. The role of the motivational subsystem is to provide underlying motivations for perception, action, and cognition, in terms of providing impetus and feedback (e.g., indicating whether an outcome is satisfactory or not). The role of the meta-cognitive subsystem is to monitor, direct, and modify dynamically the operations of the other subsystems. See Fig. 8.1.

Generally speaking, within each subsystem, a dual representation exists, which is made up of two "levels". The top level encodes explicit knowledge in a localist fashion, while the bottom level encodes implicit knowledge in distributed connectionist representation.

Given the length limit of this chapter, a detailed mathematical/algorithmic description of the CLARION cognitive architecture cannot be presented. Instead, some of the most basic general principles are briefly reviewed below. The reader interested in detailed specifications of the cognitive architecture is referred to the cited papers above (in particular [26, 31]).

8.3.1 Principle #1: The Co-existence of, and the Difference Between, Explicit and Implicit Knowledge

CLARION assumes the existence of two different types of knowledge, namely explicit and implicit, residing in two separate stores [26]. Explicit knowledge is easier to access and to verbalize, crisper, more flexible, and usually symbolic [26]. However, using explicit knowledge requires more attentional resources. In contrast, implicit knowledge is relatively inaccessible, harder to verbalize, often more vague, and usually "subsymbolic" [26]. However, using implicit knowledge does not tap much attentional resources. Explicit and implicit knowledge is processed differently. According to CLARION, explicit processes often perform some forms of rule-based reasoning (in a very generalized sense) and result in relatively crisp and exact processing (often involving hard constraints), while implicit processing is "associative" and often involves soft-constraint satisfaction [25, 26].

8.3.2 Principle #2: The Simultaneous Involvement of Implicit and Explicit Processes in Most Tasks

Explicit and implicit processes are involved simultaneously in most tasks under most circumstances [26]. This can be justified by the different representations and mechanisms involved with the two types of knowledge respectively [26]. As such, each type of processes can end up with similar or different conclusions that contribute to the overall output (see e.g., [33]).

8.3.3 Principle #3: The "Redundant" Representation of Explicit and Implicit Knowledge

According to CLARION, explicit and implicit knowledge are often "redundant": They frequently amount to a re-description of one another in different representational forms. For example, knowledge that is initially implicit may be later re-coded to form explicit knowledge (e.g., through "bottom-up learning"; [32]). Likewise, knowledge that is initially learned explicitly (e.g., through verbal instructions) is often later assimilated and re-coded into an implicit form, usually after extensive practice (top-down assimilation; [26]). There may also be other ways redundancy is created, for example, through simultaneous learning of implicit and explicit knowledge.

8.3.4 Principle #4: The Integration of the Results of Explicit and Implicit Processing

Although explicit and implicit knowledge are often re-descriptions of one another, they involve different forms of representation and processing, which may produce similar or different conclusions; the integration of these conclusions is necessary, which may lead to synergy, that is, overall better performance [33].

8.3.5 Principle #5: The Iterative (and Possibly Bidirectional) Processing

According to CLARION, processing may be iterative and potentially bidirectional. If the integrated outcome of explicit and implicit processing does not yield a definitive result (i.e., a result in which one is sufficiently confident), another round of processing may occur, which may often use the integrated outcome as a new starting point. Reversing the direction of reasoning may sometimes happen (e.g., using abductive reasoning; [16]). Alternating between forward and backward processing has been argued to happen in everyday human reasoning [23]. Of course, time constraints may limit the number of iterations.

8.4 The EII Theory of Creative Problem Solving

CLARION led to an integrative theory of creative problem solving. The theory has been termed the EII (Explicit and Implicit Interaction) theory [15].

The EII theory, in part, attempts to integrate and thus unify existing theories of creative problem solving in two senses. First, most theories of creative problem solving have focused on either a high-level stage decomposition (e.g., [38]) or on a process explanation of only one of the stages [18]. Second, the process theories of incubation (e.g., [24]) and insight (e.g., [22]) are usually incomplete and often mutually incompatible. EII attempts to integrate the existing theories, each of which tends to describe only a part of creative problem solving, to provide a detailed description of the processes involved in the key stages of creative problem solving. EII starts from Wallas' [38] stage decomposition of creative problem solving and provides a detailed process-based explanation that is ready for a coherent computational implementation. However, EII is not just an integration and implementation of previously existing theories, but it is a new theory, which emphasizes the importance of implicit processing and implicit-explicit integration in problem solving.

The EII theory relies mainly on the five basic principles of CLARION, as explained above, plus a few (relatively minor) auxiliary principles. In addition to the five basic principles presented so far, three auxiliary principles should be mentioned here. These principles are less important and alternative principles may be equally viable. Therefore they are not central to the fundamental theoretical framework. First, Principle #5 implies that a 'definitive result' needs to be achieved in order to terminate the iterative process. This stopping criterion assumes a primitive form of meta-cognitive monitoring that can estimate the confidence in a potential solution [5]. In CLARION, this meta-cognitive measure is termed the *Internal Confidence Level* (ICL). Second, there must be a threshold that defines what is meant by a 'definitive result'. This threshold can vary as a function of task demands, and there might be several thresholds for different levels of confidence [5]. Lastly, a negative relationship between the ICL and the psychological response time might be assumed [7].

8.5 Creativity in Problem Solving

This section presents the EII explanations and the corresponding CLARION-based simulations of well-established psychological paradigms (e.g., free recall, lexical decision, and problem solving) and their results.

In what follows, the emphasis is not on the fine-grained details involved (which is inevitable given the summary nature of this chapter). Detailed explanations and simulations can instead be found in prior publications, for example, in [15].

8.5.1 Modeling Incubation Within EII

Incubation (i.e., a period of not thinking about a given problem consciously) is one of the major stages of creative problem solving, according to Wallas [38]. EII stipulates that incubation occurs mostly through implicit processes, thus usually without conscious awareness [15].

In CLARION, incubation is mostly captured within the bottom level of the NACS, which is implicit. The bottom level of the NACS consists of fully recurrent connectionist attractor networks (e.g., Hopfield-type networks) with distributed representations of implicit knowledge. Such networks are known to perform soft constraint satisfaction through gradual, iterative activation propagations. Therefore potential solutions may be formed gradually, resulting from an iterative process of soft constraint satisfaction [31].

Within the bottom level of the NACS of CLARION, the search is generally more diffused due to soft constraint satisfaction, and thus may be more remote. As a result, less likely associations may be retrieved, leading to creative solutions.

8.5.2 Example of Simulating Incubation in a Lexical Decision Task

Yaniv and Meyer [40] asked human subjects to perform two tasks sequentially. First, in a rare-word association task, they showed subjects word definitions that were weakly associated with their definienda. The subjects had a limited time to find a definiendum for each definition. If the subject found the definiendum, they were transferred to a lexical decision task, where they had to classify briefly presented strings of letters as 'word' or 'non-word'. If the subject did not produce a definiendum, they were asked to rate their feeling of knowing (FOK) and then started the lexical decision task also. The results of the second task showed that those definitions that allowed for the retrieval of the correct definienda by the subjects or generated high FOKs from the subjects produced priming in the lexical decision task (i.e., faster reaction times).

According to the EII theory, a rare-word association trial produces a simultaneous search at the explicit and the implicit levels (Principle #2). Because the target association is rare in this task, explicit memory search is not likely to yield a satisfactory solution within the allotted time (because the existing set of hard constraints does not necessarily lead to solutions in this case). In contrast, according to EII, implicit memory search is more likely to retrieve the desired association if given enough time, because soft constraint satisfaction can allow partial match that can be iteratively improved. However, implicit memory search is often cut short by the experimenter who then asks the subject to take part in lexical decision trials (for those subjects who did not produce a definiendum). At the beginning of the lexical decision trials, implicit knowledge is still in the same state as it was at the end of the corresponding rare-word association trial. Hence, if the association was retrieved or nearly retrieved during the rare-word association trial (i.e., with high FOK), the preceding memory search is relevant and the target word (related to the definiendum in question) is thus primed for the lexical decision trial (i.e., leading to faster reaction times). In contrast, the recognition of unrelated words (distractors) is not affected by the previous state of implicit knowledge, because the rare-word association trial was irrelevant to these words.

This conceptual explanation by EII led to a detailed computational model that produced simulations in line with Yaniv and Meyer's [40] results. The results of 3,000 simulations with a CLARION-based model are shown in Fig. 8.2, which capture the corresponding human data [15].

8.5.3 Modeling Insight Within EII

According to Wallas [38], insight, the sudden emergence into consciousness of a potential solution, is another major stage of creative problem solving, normally following the incubation stage. According to EII, insight results from the transferring

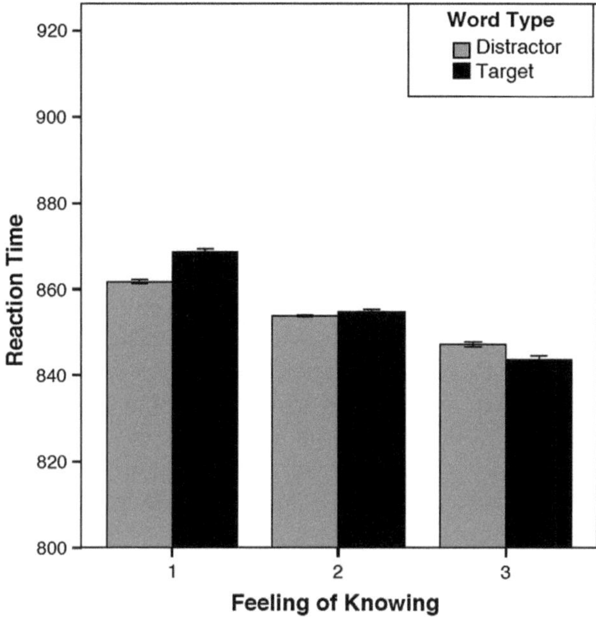

Fig. 8.2 Simulated response times in the lexical decision task for subjects who did not produce a definiendum in the rare-word association task

of the activation of implicit knowledge (e.g., as a result of incubation) to that of the corresponding explicit knowledge; hence the sudden appearance of insight.

Correspondingly, CLARION captures insight computationally through the emergence of activations from the bottom level of the NACS to the top level of the NACS (bottom-up activation flows). That is, insight amounts to activating explicit (easily accessible) representations at the top level by implicit, distributed, not so easily accessible representations at the bottom level. The result is an explicitly accessible solution emerging into consciousness.

8.5.4 Example of Simulating Insight in Problem Solving

Durso, Rea, and Dayton [9] asked human subjects to explain stories like the following one:

> A man walks into a bar and asks for a glass of water. The bartender points a shotgun at the man. The man says 'thank you', and walks out.

The subjects' task was to explain why the sight of the shotgun replaced the man's need for a glass of water (i.e., because he had the hiccup). To explain this story, the subjects had two hours to ask the experimenter yes/no questions. When the time elapsed, each subject was classified as a 'solver' or as a 'non-solver' and his/her knowledge graph was drawn. Solvers' and non-solvers' knowledge graphs were shown to have different connectivity.

According to EII, reading the story results in both explicit memory retrieval and implicit memory search (incubation). However, explicit processing (mostly rule-based; Principle #1) brings up stereotypical associations from the story. In contrast, the gradient of associations is flatter in implicit memory [20] and weak associations are recorded. The search is thus more diffused, and more remote ("creative") associations can be retrieved with soft constraint satisfaction. According to the EII theory, implicit processing allows for the retrieval of more approximate, more hypothetical associations that differ from those retrieved explicitly. These implicit associations are then integrated with the result of explicit processing (Principle #4). If the chosen integrated association is deemed plausible (i.e., if the ICL is high enough), a question concerning the validity of this association is put to the experimenter. If the experimenter confirms the association, it is added into explicit knowledge; otherwise, it is removed. This process is iterated; explicit and implicit processing are restarted with the new state of knowledge. This iterative process ends when the subject finds the solution or the allowed time elapses.

The results of 8,000 CLARION-based simulations show that, consistent with the EII explanation above, the probability of solving the problem increases with the stochasticity of the implicit association retrieval. See [15] for details.

8.6 Motivation, Emotion, Personality, and Social Interaction

Beyond the interaction between implicit and explicit processes as stipulated by the EII theory, creativity involves many other aspects of the human mind. Below, a number of important psychological aspects are discussed in relation to creativity. CLARION can be used to provide interpretations to these aspects.

For instance, it has been known that motivation has a lot to do with creativity. One of the most relevant findings is that intrinsic motivation is correlated with creativity. However, intrinsic motivation may lead to generating new ideas, but new ideas will not be judged as creative unless they are also useful (as defined by some communal standards). Individuals with prosocial motivation and good at perspective taking should be better at generating useful ideas. Grant and Berry [14] tested experimentally whether or not intrinsic motivation and prosocial motivation combined result in greater creativity and their results confirmed the expectation.

It is also known that anxiety has debilitating (or facilitating) effects on routine cognitive or motor tasks. Byron and Khazanchi [6] performed a meta-analysis and found that anxiety was significantly negatively related to creative performance—Anxiety and creativity present competing cognitive demands. Other emotions are also known to have significant effects on creativity (see, e.g., [1, 2]).

As explained earlier, the CLARION cognitive architecture contains four major subsystems, which include the motivational subsystem (the MS, as mentioned earlier; see Fig. 8.1) that captures various innate, essential motives as well as acquired motives, in the form of *drives* (both primary and secondary drives; [30]). Within the MS, different motives (drives), such as *achievement and recognition*, *affiliation and belongingness*, and *similance*, exist and may be activated by situations [30]. Therefore, with CLARION, it is possible to explore effects of motivation on creativity, including effects of intrinsic achievement orientation, prosocial tendency, and so on. There is also an account of emotion within CLARION, largely on the basis of motivation, including capturing a variety of major emotions (e.g., anxiety; [39]). Thus, CLARION can also be used to capture the effects of emotion on creativity.

Personality, as extensively studied in personality psychology, is also known to have a lot to do with creativity. There have been various studies concerning the effects of personality on creativity. For instance, based on a meta-analysis of data, Feist [11] argued that, in general, creative individuals were more open to experiences, less conventional and less conscientious, more self-confident, more driven, ambitious, and dominant, and so on. There also appeared to be temporal stability of these distinguishing characteristics among creative individuals. These characteristics corresponded well with the five-factor model of personality (commonly known as the Big Five), and their detailed facets (see [19, 35]).

Furthermore, Feist and Barron [12] reported a longitudinal study. At age 27, a sample of 80 male graduate students was assessed on potential, intellect, personality, and creativity. At age 72, personality and career outcome data were collected again. Intellect, potential, and personality at age 27 were expected to predict lifetime creativity. It was also predicted that personality would explain unique variance in creativity over and above that explained by intellect and potential. Results supported these expectations. They concluded that certain personality traits such as openness to experiences, tolerance, and self-confidence might serve as a relatively direct link to creative behavior.

The role of personality can be captured within CLARION, as has been shown before (e.g., [34–36]). As mentioned earlier, the CLARION cognitive architecture emphasizes four major subsystems, as shown in Fig. 8.1. These subsystems together capture various dimensions of human personality, as demonstrated in detail by Sun and Wilson [35, 36]. Among these subsystems, the MS plays an especially important role in determining personality. The various essential motives within the MS capture major aspects of personality dimensions, such as the Big Five [35, 36]. Various simulations have been carried out to date to explore the effects of personality dimensions such as openness to experiences and conscientiousness.

In relation to the role of social interaction in creativity, Ashton-James and Chartrand [3] emphasized behavioral mimicry as a social cue for creative thinking. Specifically, being mimicked by an interaction partner cues convergent thinking and a social opportunity for collaboration, while not being mimicked cues divergent thinking and a social demand for improvisation and innovation. They experimentally manipulated whether individuals were mimicked or not and subsequently measured their capacity for convergent and divergent thinking. The results showed the importance of understanding how social relationships influenced the creative processes. Relatedly, Tadmor et al. [37] showed that multicultural experience among members of a collective would enhance joint creativity in a super-additive fashion. The results showed that in terms of creativity, the social whole might be greater than the sum of its parts.

Correspondingly, CLARION emphasizes modeling and simulating social interaction and social processes in understanding the mind. Social simulation with a cognitive architecture such as CLARION enables the exploration of the mutual influence of the social and the psychological [29]. Many different social situations have been simulated, using CLARION as the model of the individual mind involved in social interaction. For instance, Naveh and Sun [21] showed that growth of academic science was closely related to not only individual cognitive processes but also social processes and social institutions. Together they determined the level of creativity of a society.

Therefore, CLARION provides a comprehensive framework for capturing human creativity, including its many different aspects and components. Some of these aspects, such as motivation, emotion, personality, and social interaction, although downplayed or ignored in many previous theories or models of creativity, may nevertheless be captured and explained within CLARION.

8.7 Conclusion

The work described in this chapter shows how a psychologically realistic cognitive architecture, namely, CLARION [26, 31–33], can lead directly to a theory of creative problem solving (e.g., EII; [15]). Cognitive architectures such as CLARION generally integrate many cognitive/psychological mechanisms and processes in order to produce intelligent behavior in a psychologically realistic way. In the EII theory (as derived from CLARION), the key ingredient is the interaction of explicit and implicit processing. By incorporating both explicit and implicit processing, the EII theory is able to provide a unified framework for re-interpreting and integrating some important (but fragmentary) psychological theories of incubation, insight, and creativity (see [15] for details of the re-interpretation and integration).

The EII theory is, however, not complete yet. For instance, it needs to move on to account for real-world cases of creative problem solving. Such cases would inevitably involve motivation, personality, and emotion. Social interaction is also important in this regard. However, the EII theory is currently more complete than previous theories (especially previous computational theories/models). We have shown that the roles of motivation, personality, emotion, and social interaction can all be accounted for within CLARION, and therefore can be added to EII.

In relation to AI, a unified computational model (CLARION) is shown to be capable of capturing creative problem solving in widely differing settings (e.g., free recall, lexical decision, problem solving, and so on), demonstrating its computational capacities. Computationally, the model involves different types of neural networks to simulate, respectively, explicit processing (with localist, feedforward networks) and implicit processing (with distributed, fully recurrent, attractor networks). Integrating these components is essential in capturing creative problem solving. A computational cognitive architecture is an important way of exploring the advantage of synergistically combining several specialized computational models, because so far no single computational model can capture human intelligence by itself. Future work should be devoted to tackling more complex real-world creative problem solving situations involving additional factors as detailed earlier.

Better, more integrated computational models of creative problem solving that are psychologically realistic are needed for both AI and cognitive science. In relation to AI, they may spur corresponding research on computational creativity. They may influence and/or challenge common perceptions of where the limits of creativity may lie and where the limits of intelligent machines may ultimately lie. In the process, psychologically realistic models of creative problem solving may help to push the boundaries, one step at a time.

Acknowledgments This research was supported by (1) the ARI research grants DASW01-00-K-0012 and W74V8H-04-K-0002, and (2) the ONR research grants N00014-08-1-0068 and N00014-13-1-0342.

References

1. Akinola, M., Mendes, W.B.: The dark side of creativity: biological vulnerability and negative emotions lead to greater artistic creativity. Pers. Soc. Psychol. Bull. **34**(12), 1677–1686 (2008)
2. Amabile, T.M., Barsade, S.G., Mueller, J.S., Staw, B.M.: Affect and creativity at work. Adm. Sci. Q. **50**(3), 367–403 (2005)
3. Ashton-James, C.E., Chartrand, T.L.: Social cues for creativity: the impact of behavioral mimicry on convergent and divergent thinking. J. Exp. Soc. Psychol. **45**(4), 1036–1040 (2009)
4. Bowden, E.M., Jung-Beeman, M., Fleck, J., Kounios, J.: New approaches to demystifying insight. Trends Cognit. Sci. **9**, 322–328 (2005)
5. Bowers, K.S., Regehr, G., Balthazard, C., Parker, K.: Intuition in the context of discovery. Cognit. Psychol. **22**, 72–110 (1990)
6. Byron, K., Khazanchi, S.: A meta-analytic investigation of the relationship of state and trait anxiety to performance on figural and verbal creative tasks. Pers. Soc. Psychol. Bull. **37**(2), 269–283 (2011)
7. Costermans, J., Lories, G., Ansay, C.: Confidence level and feeling of knowing in question answering: the weight of inferential processes. J. Exp. Psychol. Learn. Mem. Cognit. **18**, 142–150 (1992)
8. Duch, W.: Computational creativity. In: Proceedings of the International Joint Conference on Neural Networks, pp. 435–442. IEEE Press, Vancouver (2006)
9. Durso, F.T., Rea, C.B., Dayton, T.: Graph-theoretic confirmation of restructuring during insight. Psychol. Sci. **5**, 94–98 (1994)

10. Evans, J.B.T.: The Heuristic-analytic theory of reasoning: extension and evaluation. Psychon. Bull. Rev. **13**, 378–395 (2006)
11. Feist, G.F.: A meta-analysis of personality in scientific and artistic creativity. Pers. Soc. Psychol. Rev. **2**, 290–309 (1998)
12. Feist, G.J., Barron, F.: Predicting creativity from early to late adulthood: intellect, potential, and personality. J. Res. Pers. **37**, 62–88 (2003)
13. Gabora, L.: Revenge of the 'neurds': characterizing creative thought in terms of the structure and dynamics of human memory. Creat. Res. J. **22**(1), 1–13 (2010)
14. Grant, A.M., Berry, J.W.: The necessity of others is the mother of invention: Intrinsic and prosocial motivations, perspective taking, and creativity. Acad. Manag. J. **54**(1), 73–96 (2011)
15. Helie, S., Sun, R.: Incubation, insight, and creative problem solving: a unified theory and a connectionist model. Psychol. Rev. **117**, 994–1024 (2010)
16. Johnson, T.R., Krems, J.F.: Use of current explanations in multicausal abductive reasoning. Cognit. Sci. **25**, 903–939 (2001)
17. Langley, P., Laird, J.E., Rogers, S.: Cognitive architectures: research issues and challenges. Cognit. Syst. Res. **10**, 141–160 (2009)
18. Lubart, T.I.: Models of the creative process: past, present and future. Creat. Res. J. **13**, 295–308 (2001)
19. McCrae, R.R., Costa Jr, P.T.: NEO Inventories: Professional Manual. Psychological Assessment Resources Inc, Lutz, FL (2010)
20. Mednick, S.A.: The associative basis of the creative process. Psychol. Rev. **69**, 220–232 (1962)
21. Naveh, I., Sun, R.: A cognitively based simulation of academic science. Comput. Math. Organ. Theor. **12**(4), 313–337 (2006)
22. Pols, A.J.K.: Insight problem solving. Ph.D. thesis, Department of Psychology, University of Utrecht, Netherlands (2002)
23. Rips, L.J.: The Psychology of Proof: Deductive Reasoning in Human Thinking. MIT Press, Cambridge, MA (1994)
24. Smith, S.M., Dodds, R.A.: Incubation. In: Runco, M.A., Pritzker, S.R. (eds.) Encyclopedia of Creativity, pp. 39–43. Academic, San Diego, CA (1999)
25. Sun, R.: Robust reasoning: integrating rule-based and similarity-based reasoning. Artif. Intell. **75**, 241–296 (1995)
26. Sun, R.: Duality of the Mind. Lawrence Erlbaum Associates, Mahwah (2002)
27. Sun, R.: A tutorial on CLARION 5.0. Technical report, Cognitive Sciences Department, Rensselaer Polytechnic Institute, Troy (2003). http://www.cogsci.rpi.edu/~rsun/sun.tutorial.pdf
28. Sun, R.: Desiderata for cognitive architectures. Philos. Psychol. **17**, 341–373 (2004)
29. Sun, R. (ed.): Cognition and Multi-Agent Interaction. Cambridge University Press, New York (2006)
30. Sun, R.: Motivational representations within a computational cognitive architecture. Cognit. Comput. **1**(1), 91–103 (2009)
31. Sun, R.: Anatomy of the Mind. Oxford University Press, New York (2014)
32. Sun, R., Merrill, E., Peterson, T.: From implicit skills to explicit knowledge: a bottom-up model of skill learning. Cognit. Sci. **25**, 203–244 (2001)
33. Sun, R., Slusarz, P., Terry, C.: The interaction of the explicit and the implicit in skill learning: a dual-process approach. Psychol. Rev. **112**, 159–192 (2005)
34. Sun, R., Wilson, N.: Motivational processes within the perception-action cycle. In: Cutsuridis, V., Hussain, A., Taylor, J.G. (eds.) Perception-Action Cycle: Models, Architectures and Hardware, pp. 449–472. Springer, Berlin (2011)
35. Sun, R., Wilson, N.: A model of personality should be a cognitive architecture itself. Cognit. Syst. Res. **29–30**(1–30), 2014 (2014)
36. Sun, R., Wilson, N.: Roles of implicit processes: instinct, intuition, and personality. Mind Soc. **13**(1), 109–134 (2014)
37. Tadmor, C.T., Satterstrom, P., Jang, S., Polzer, J.T.: Beyond individual creativity: the superadditive benefits of multicultural experience for collective creativity in culturally diverse teams. J. Cross-Cult. Psychol. **43**, 384–392 (2012)

38. Wallas, G.: The Art of Thought. Franklin Watts, New York (1926)
39. Wilson, N., Sun, R., Mathews, R.: Performance degradation under pressure. Neural Netw. **22**, 502–508 (2009)
40. Yaniv, I., Meyer, D.E.: Activation and metacognition of inaccessible stored information: potential bases for incubation effects in problem solving. J. Exp. Psychol. Learn. Mem. Cognit. **13**, 187–205 (1987)

Chapter 9
E Pluribus Unum

Formalisation, Use-Cases, and Computational Support for Conceptual Blending

Oliver Kutz, John Bateman, Fabian Neuhaus, Till Mossakowski and Mehul Bhatt

Abstract Conceptual blending has been employed very successfully to understand the process of concept invention, studied particularly within cognitive psychology and linguistics. However, despite this influential research, within computational creativity little effort has been devoted to fully formalise these ideas and to make them amenable to computational techniques. Unlike other combination techniques, *blending* aims at creatively generating (new) concepts on the basis of input theories whose domains are thematically distinct but whose specifications share structural similarity based on a relation of analogy, identified in a generic space, the base ontology. We introduce here the basic formalisation of conceptual blending, as sketched by the late Joseph Goguen, and discuss some of its variations. We illustrate the vast array of conceptual blends that may be covered by this approach and discuss the theoretical and conceptual challenges that ensue. Moreover, we show how the

---

This paper summarises our work on conceptual blending, to create a new concept out of many, with material drawn from [32, 42, 44, 55].

O. Kutz (✉) · F. Neuhaus · T. Mossakowski
Institute of Knowledge and Language Engineering, Otto-von-Guericke University of Magdeburg, Magdeburg, Germany
e-mail: okutz@iws.cs.ovgu.de

F. Neuhaus
e-mail: fneuhaus@iws.cs.ovgu.de

T. Mossakowski
e-mail: mossakow@iws.cs.ovgu.de

J. Bateman
Faculty of Linguistics and Literary Sciences/Research Center on Spatial Cognition (SFB/TR 8), University of Bremen, Bremen, Germany
e-mail: bateman@uni-bremen.de

M. Bhatt
Research Center on Spatial Cognition (SFB/TR 8), University of Bremen, Bremen, Germany
e-mail: bhatt@informatik.uni-bremen.de

T.R. Besold et al. (eds.), *Computational Creativity Research: Towards Creative Machines*, Atlantis Thinking Machines 7, DOI 10.2991/978-94-6239-085-0_9

Distributed Ontology Language DOL can be used to declaratively specify blending diagrams of various shapes, and discuss in detail how the workflow and creative act of generating and evaluating a new, blended concept can be managed and computationally supported within Ontohub, a DOL-enabled theory repository with support for a large number of logical languages and formal linking constructs.

9.1 Concept Invention via Blending

One broad area of phenomena that is often brought into connection with issues of creativity and the emergence of new ideas concerns notions such as metaphor, blending, category mistakes, similes, analogies and the like. In each of these, seemingly inconsistent material is combined in a manner that results in a productive growth of information instead of simple logical contradiction. Approaches to treat this phenomenon are varied but commonly come to the conclusion that more or less well developed notions of 'structure' are crucial for bringing the growth of information about—e.g., 'implication complexes' for metaphor [6], 'conceptual spaces'[1] for blending [14], 'structure mapping' in analogy [20], and so on. On the one hand, the less structure that is available, the less productive the combinations appear to be; on the other, the presence of structure raises the challenge of how such formal commitments can be productively 'overridden' or rearranged in order to avoid contradiction.

In our ongoing work on ontology and its formal underpinnings, we have been led to a very similar set of questions. By 'ontology' we here refer to the now rather standard notion of a formal specification of a shared understanding of the entities, relations and general properties holding in some domain of interest (cf. [25, 27]). Achieving adequate treatments in various domains has demonstrated to us the need for heterogeneous ontological specifications that are capable of capturing distinct perspectives on the phenomena being modelled. In an architectural context, for example, it is beneficial to maintain distinct perspectives on structural integrity, spatial distribution, movement patterns by the occupants of a building ('flow'), navigation networks (possibly varying according to 'normal' and 'emergency' conditions), 'visibility' patterns (both for users and for sensors in the case of security) and many more [5]—each of these perspectives can be modelled well by employing ontological engineering techniques but there is no guarantee that they are simply compatible. Our work on natural language dialogue systems involving spatial language comes to the same conclusion [2], while similar concerns are already well known in Geographic Information Science [18, 37]. To support this fundamental 'multi-perspectivalism' we have therefore been developing an entire toolset of more sophisticated combination methods [38], leading to the formal definition of the notion of a 'hyperontology' in [43].

[1] The usage of the term 'conceptual space' in blending theory is not to be confused with the usage established by Gärdenfors [19].

The similarities apparent between the goals of heterogeneous ontology 'alignment' and the creative combination of thematically distinct information spaces can be built on quite concretely by treating such information spaces explicitly in terms of ontological specifications. This allows us to link directly with previous work by exploring the application of techniques for combining distinct perspectives that are now becoming available. For example, much work on creativity has been pursued in the context of Fauconnier and Turner's [14] account of conceptual blending, in which the blending of two thematically rather different *conceptual spaces* yields a new conceptual space with emergent structure, selectively combining parts of the given spaces whilst respecting common structural properties. The 'imaginative' aspect of blending is summarised as follows by Turner [67]:

> [...] the two inputs have different (and often clashing) organising frames, and the blend has an organising frame that receives projections from each of those organising frames. The blend also has emergent structure on its own that cannot be found in any of the inputs. Sharp differences between the organising frames of the inputs offer the possibility of rich clashes. Far from blocking the construction of the network, such clashes offer challenges to the imagination. The resulting blends can turn out to be highly imaginative.

We see the almost unlimited space of possibilities supported by 'ontological blending' for combining existing ontologies to create new ontologies with emergent structure as offering substantial benefits not only for ontological engineering—where conceptual blending can be built on to provide a structural and logic-based approach to 'creative' ontological engineering—but also for conceptual blending and related frameworks themselves—by providing a far more general and nevertheless computational, formalised foundation. Re-considering some of the classic problems in conceptual blending in terms of ontological modelling and ontological blending opens up an exciting direction for future research.

This endeavour primarily raises the following two challenges: (1) when combining the terminologies of two ontologies, the shared semantic structure is of particular importance to steer possible combinations—this shared semantic structure leads to the notion of a base ontology, which is closely related not only to the notion of 'tertium comparationis' found in classical rhetoric and poetics, but also to more recent cognitive theories of metaphor (see, e.g., [33]); (2) having established a shared semantic structure, there typically remains a considerable number of possibilities that can capitalise on this information in the combination process—here, structural optimality principles as well as ontology evaluation techniques can take on a central role in selecting 'interesting' blends.

There is still much to explore concerning the relationships between the principles governing ontological blending and the principles explored to date for blending phenomena in language or poetry or, indeed, the rather strict principles ruling blending in mathematics, in particular in the way formal inconsistencies are dealt with. For instance, whilst blending in poetry might be particularly inventive or imaginative when the structure of the basic categories found in the input spaces is almost completely ignored, in areas such as mathematics a rather strict adherence to sort structure is important in order to generate meaningful blends.[2] The use that we might typically make of ontological blending is situated somewhere in the middle: re-arrangement and new combination of basic categories can be quite interesting, but has to be finely controlled through corresponding interfaces, often regulated by or related to choices found in foundational or upper ontologies so that basic categorial relationships are maintained.

For all such cases, however, we can consider the formal mechanisms that support specific blends that we explore with respect to their potential relevance and value for understanding 'blending' phenomena in general. This will be the main purpose of the current chapter. We will summarise some of the progress that has been made in recent years towards adopting the fruitful idea of conceptual blending in a theoretically well-understood and computationally supported formal model for concept invention, focusing in particular on ontology languages. Here we elaborate on ideas first introduced in Hois et al. [32], with detailed technical definitions given in Kutz et al. [42]. More specifically, we:

- briefly characterise the kinds of creativity that have been considered hitherto in the areas of blends, metaphors and related operations where structured mappings or analogies are relied upon;
- sketch the logical analysis of conceptual blending in terms of blending diagrams and colimits, as originally proposed by Joseph Goguen, and give an abstract definition of ontological blendoids capturing the basic intuitions of conceptual blending in the ontological setting;
- sketch a formal meta-language, namely the distributed ontology language DOL, that is capable of declaratively specifying blending diagrams in a variety of ontology languages. This provides a structured approach to ontology languages and blending and combines the simplicity and good tool support for languages such as OWL[3] with the more complex blending facilities of OBJ3 [24] or Haskell [36]; DOL also facilitates the specification of a range of variations of the basic blending technique;
- discuss the capabilities of the Ontohub/HETS ecosystem with regard to collaboratively managing, creating, and displaying blended concepts, ontological theories, and entire blending diagrams; this includes an investigation of the evaluation problem in blending, together with a discussion of structural optimality principles and current automated reasoning support.

We close with a discussion of open problems and future work.

---

[2] For instance when creating the theory of transfinite cardinals by blending the perfective aspect of counting up to any fixed finite number with the imperfective aspect of 'endless counting' [58].

[3] With 'OWL' we refer to OWL 2 DL, see http://www.w3.org/TR/owl2-overview/.

9.2 An Ocean of Blends

In this section, we briefly characterise the rather diverse phenomena that may be subject to beneficial formalisations in terms of ontological blending. The starting point is the obvious one of conceptual blending, which we use as a prototypical case of emergent organisation throughout this chapter. As noted above, conceptual blending in the spirit of Fauconnier and Turner [14] operates by combining two input 'conceptual spaces', construed as rather minimal descriptions of some thematic domains, in a manner that creates new 'imaginative' configurations. A classic example for this is the blending of the concepts *house* and *boat*, yielding as most straightforward blends the concepts of a *houseboat* and a *boathouse*, but also an *amphibious vehicle* [23]; we return to this example below. This case shows well how it is necessary to maintain aspects of the structural semantics of the spaces that are blended in order to do justice to the meanings of the created terms: the houseboat stops neither being a vehicle on water nor being a place of residence, for example.

Very similar processes appear to be operating in cases of metaphor [6, 35]. Here a semantically structured 'source' is used so that facets of the semantics of the source are selected for appropriate take up by a semantically structured 'target'. This can operate on a small scale, analogously to the *house* and the *boat*, as for example in metaphors such as that evident in the 1940s film title "Wolf of New York" or the recent "The Wolf of Wall Street" (2013), where certain conventionalised properties of the wolf as animal (the source) are transferred to the people referred to by the titles (the target). Structure is essential here since the transfer is very specific: a reading of the metaphor in which 'four-leggedness' or 'furry' is transferred is in the given contexts most unlikely. Only particular relations and relational values are effected. Metaphors can also operate on much broader scales, as in considerations of metaphors as contributions to creative scientific theory construction, as in the well known transfer of a 'sun-and-planet' conceptual model to models of the atom [29, 30, 51] (see Sect. 9.3.1). Structural transfer of this kind has consequently been suggested to play a substantial role for persuasive text creation as such. Hart, for example, discusses the use of phrases such as 'limitless flow of immigration', 'flood of asylum seekers' and so on as ideologically-loaded constructions that need to be unpacked during critical discourse analysis [31].

Metaphors also bring with them some particular formal features of their own—for example, they are typically seen as *directed* in contrast to blends and have been related to models of embodiment via accounts of *image schemas* [34]. Image schemas suggest how multimodal patterns of experience can be linked to increasingly abstract conceptualisations: abstract thought is then seen as a metaphorical construction on top of concrete experience. The use of the word 'flood' in the above example can then be expected to bring about a physical component in its reception where feelings of force, damage and lack of control are activated; this makes it clear that much more than 'flowery language' might be involved in such phrasings and their selection.

A related consideration is the proposal for internalised spatial representations for supporting reasoning and more abstract conceptualisations (such as time) as well as externalised spatial representations for diagrammatic reasoning. In the former case, it is common to work within blended spaces where time and spatial extent appear to have 'collapsed', giving rise to language use such as "keep going straight until the church" or "turn left before the tower" and so on. Blends of this kind are so familiar that they may be considered to be *entrenched* in the cognitive linguistic sense of having become part of the semantics of the respective terms and shared by the language community [14, p. 49].

Blends may also be multiple in that once established, for example in a text, further conceptual spaces might be added as an argument progresses. These may progressively add details to a developing emergent space (or, alternatively, lead to a space which strains the credibility of a reader or hearer too far resulting in a charge of 'mixing metaphors'). In the right-wing immigration example above from Hart, the texts do in fact continue with phrases such as 'Britain is full up', 'no matter how open or closed its immigration policy', and 'our first step will be to shut the door'. This builds on the previous blend of immigration-as-flood by (i) combining 'Britain' with a 'container' (which can then be full) that is itself (ii) combined with a 'building' or 'room' that has 'doors' that can be closed, and (iii) those doors can in turn also be 'policies' (which can be open or closed) [31, p. 102]. There need in principle be no end to this creative extension and combination of concepts. This aspect of iteration of blending is also explored in the area of conceptual mathematics as explored by Lakoff and Núñez [46], where it is argued that abstract mathematical concepts such as modern number systems, algebra, or set theory, are created through a succession of conceptual metaphors and blends, grounded in embodied concepts and image schemas. The structure of such blends and blending patterns in general are discussed more formally in Sect. 9.4.2.

There is also now increasing discussion of the potential role of blending or similar mechanisms when considering the creative use of combinations of information from different semiotic modes, e.g., drawing relations between verbal information, visual and gestural information [17].

In such cases representations or entities in one mode of presentation are made to take on properties or behaviours in another. The general applicability of an ontological approach to semiotic blending of this kind is argued in Bateman [3]. Again, there are many examples of such creativity in action. Consider for example the extract from an advertisement discussed by van Mulken et al. [69] and shown in Fig. 9.1. Here an open-ended set of potential further inferences, all supporting the general intention of the advertisement, is opened up by virtue of the blend. There are also commonly discussed combinations such as the use of space for time in comics and visual narrative—moving across the space of a comic's panel, typically in Western comics therefore from left to right, often correlates with a progression in time [50, p. 95]—as well as blends for dramatic or emotional effect, such as when typography is shaped visually for affective purposes [10, p. 12].

A particularly creative and novel example of semiotic blending across media can be seen in the following example. In this case the film director Ang Lee works with the *dynamic* possibilities of the film medium to enlist graphic resources for expressing movement developed within the *static* medium of comics. The result is

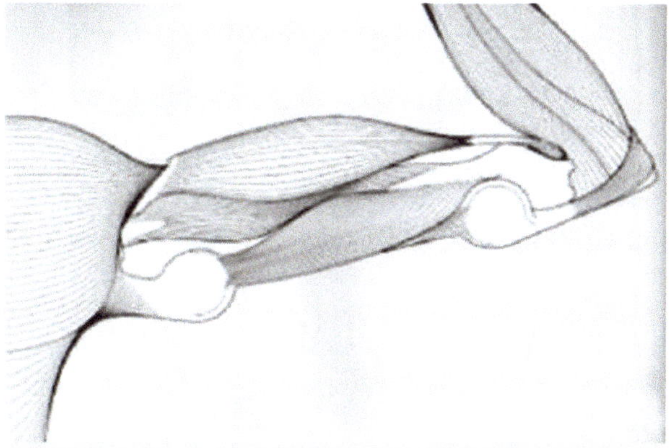

Fig. 9.1 Visual blending of a car and an anatomical representation used for advertising discussed by van Mulken et al. [69]

Wonderwoman with motion lines connecting panels

Pérez, G., Potter, G. and Wein, L. (2008), *Biblioteca DC – Mulher-Maravilha/Wonder Woman*, Brasil, São Paulo: Panini. p. 140.

Hulk with motion panels connecting actions

1:37:26.7

camera tracking left, following helicopter

camera tracking left, following helicopter. insert panel moves from left to right, picks up Hulk as he enters frame right

second insert panel moves from left to right, picks up Hulk just as he leaves first insert panel moving left

complete right-to-left trajectory of Hulk over three insert panels

Lee, Ang (2003), *Hulk*, based on Marvel characters by Stan Lee and Jack Kirby, Screenplay: John Turman, Michael France and James Schamus, Story: James Schamus, USA: Universal Pictures, Marvel Enterprises, Valhalla Motion Pictures, Good Machine.

Fig. 9.2 Blending expressive resources from comics and film [4]

an interesting and highly explorative expansion of the creative potential of what can be done with film. An illustration is shown in Fig. 9.2. On the left for comparison is a now quite traditional static rendition of movement from a comic—in this particular case showing ‘continuity’ of movement across panels. In contrast, on the right we see a short sequence of stills taken from a chase scene in Lee’s 2003 film *Hulk*, where the main character is trying to escape from pursuers in a helicopter. In this case, the escape trajectory is shown in a sequence of dynamically inserted 'panels' that move across the screen to the point where they can pick up the character's movement. This blending of properties in Lee's film does much more than 're-create' a visual effect analogous to comics as sometimes suggested in analyses of this film. Lee's appropriation of framing and movement techniques within an already dynamic medium appears instead to provide a resource that considerably heightens continuity for narrative effect. A more detailed discussion of the consequences of this appropriation for interpretation and reception is given in Bateman and Veloso [4].

We are just beginning to be able to explore extensions of meaning-making potential of these kinds. Indeed, although there are now many examples in the literature of such creative meaning growth in action, deep questions remain concerning how precisely this may be modelled. In particular, following simpler operations of 'alignment' of structures across spaces (e.g., by graph matching [16, 72]), it is by no means clear how the results that are achieved can function as productively as they evidently do. This relates also to Fauconnier's suggestion that it is actually what is done with the result of blending, termed *elaboration* (or 'running the blend'), that is the most significant stage of the entire blending process. Elaboration "consists in cognitive work performed within the blend, according to its own emergent logic" [12, p. 151]. This makes it evident that something more is required in the formalisation than a straightforward recording or noting of a structural alignment: a new blended theory should also be 'logically productive', with new and surprising entailments which may well be quite specific to the blend. This is therefore another motivation for the rather more formal and ontologically-driven approach to this kind of creative meaning creation that we now present.

9.3 Blending Computationalised

There have now been several approaches moving towards effective computational treatments of blending, metaphor and related constructs such as analogy (cf. e.g. [47, 48, 60, 66, 71, 72]). Here we follow the research direction of *algebraic semiotics* established by Goguen. In this approach certain structural aspects of semiotic systems are logically formalised in terms of algebraic theories, sign systems, and their mappings [21]. Sign systems are theories 'with extra structure' connected by a particular class of mappings, which Goguen terms 'semiotic morphisms', which preserve that extra structure to a greater or lesser degree. In Goguen and Harrell [23], algebraic semiotics has been applied to user interface design and blending. Algebraic semiotics does not claim to provide a comprehensive formal theory of blending—indeed, Goguen and Harrell admit that many aspects of blending, in particular concerning the meaning of the involved notions, as well as the optimality principles for blending, cannot be captured formally. However, the structural aspects *can* be formalised and provide insights into the space of possible blends.

Goguen defines sign systems as algebraic theories that can be formulated by using the algebraic specification language OBJ3 [24]. One special case of such a sign system is a *conceptual space*: it consists only of constants and relations, one sort, and axioms that define that certain relations hold on certain instances.

We now relate such spaces to a general formalisation of ontologies as we understand them and as introduced above. Since we will focus on standard ontology languages, namely OWL and first-order logic, we use these to replace the logical language OBJ3 used by Goguen and Malcolm. However, as some structural aspects are necessary in the ontology language to support blending, we augment these standard ontology languages with structuring mechanisms known from algebraic specification theory [39]. Such mechanisms are now included in the DOL language specification discussed below in Sect. 9.4. This allows us to translate most parts of Goguen's theory to these augmented ontology languages. Goguen's main insight has been that sign systems and conceptual spaces can be related via *morphisms*, and that blending is comparable to *colimit* construction. In particular, the blending of two concepts is often a *pushout* (also called a *blendoid* in this context). Some basic definitions we then need are the following.[4]

Non-logical symbols are grouped into **signatures**, which for our purposes can be regarded as collections of typed symbols (e.g. concept names, relation names). **Signature morphisms** are maps between signatures that preserve (at least) types of symbols (i.e. map concept names to concept names, relations to relations, etc.). A **theory** or **ontology** pairs a signature with a set of sentences over that signature, and a **theory morphism** (or **interpretation**) between two theories is just a signature morphism between the underlying signatures that preserves logical consequence, that is, $\rho : T_1 \rightarrow T_2$ is a theory morphism if $T_2 \models \rho(T_1)$, i.e. all the translations of sentences of T_1 along ρ follow from T_2. This construction is completely logic independent. Signature and theory morphisms are an essential ingredient for describing conceptual blending in a logical way.

We can now give a general definition of ontological blending capturing the basic intuition that a blend of input ontologies shall partially preserve the structure imposed by base ontologies, but otherwise be an almost arbitrary extension or fragment of the disjoint union of the input ontologies with appropriately identified base space terms.

For the following definition, a variant of which we first introduced in Kutz et al. [42], a diagram consists of a set of ontologies (the nodes of the diagram) and a set of morphisms between them (the arrows of the diagram). The **colimit** of a diagram is similar to a disjoint union of its ontologies, with some identifications of shared parts as specified by the morphisms in the diagram. We refrain from presenting the category-theoretic definition here (which can be found in Adámek et al. [1]), but will explain (the action of) the colimit operation in the examples in Sect. 9.4.1. In the following definition, we use $|D|$ to denote the set of all nodes in a diagram.

Definition 1 (*Ontological Base Diagram*) An **ontological base diagram** is a diagram D for which a distinguished set $\mathscr{B} = \{B_i \mid i \in I\} \subset |D|$ of nodes are called

[4] Note that these definitions apply not only to OWL, but also to many other logics. Indeed, they apply to *any* logic formalised as an *institution* [22].

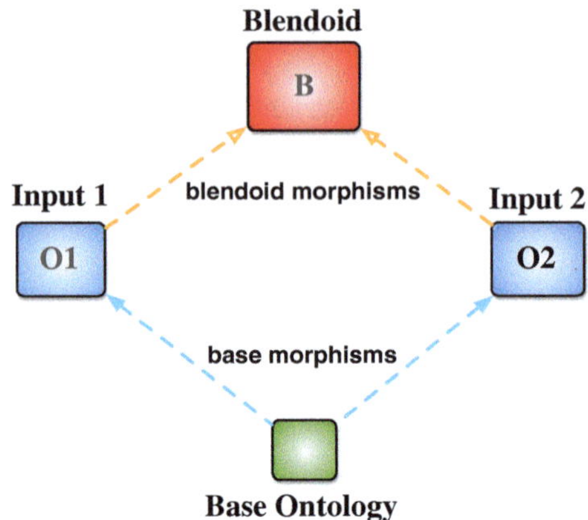

Fig. 9.3 The basic integration network for blending: concepts in the base ontology are first refined to concepts in the input ontologies and then selectively blended into the blendoid

base ontologies, and where a second distinguished set of nodes $\mathscr{I} = \{I_j \mid j \in J\} \subset |D|$ are called **input ontologies**, and where the theory morphisms $\mu_{ij} : B_i \rightarrow I_j$ from base ontologies to input ontologies are called the **base morphisms**.

If there are exactly two inputs I_1, I_2, and precisely one base $B \in \mathscr{B}$ and two base morphisms $\mu_k : B \rightarrow I_k$, $k = 1, 2$, the diagram D is called **classical** and has the shape of a 'V'. In this case, $\mathscr{B}$ is also called the **tertium comparationis**.

Figure 9.3 illustrates the basic, classical case of an ontological blending diagram. The lower part of the diagram shows the base space (tertium), i.e. the common generalisation of the two input spaces, which is connected to these via total (theory) morphisms, the base morphisms. The newly invented concept is at the top of this diagram, and is computed from the base diagram via a colimit. More precisely, any consistent subset of the colimit of the base diagram may be seen as a newly invented concept, a **blendoid**.[5] Note that, in general, ontological blending can deal with more than one base and two input ontologies, and in particular, the sets of input and base nodes need not exhaust the nodes participating in a base diagram. We will further discuss this and give some examples in Sect. 9.4.2.

9.3.1 Computing the Tertium Comparationis

To find candidates for base ontologies that could serve for the generation of ontological blendoids, much more shared semantic structure is required than the surface similarities that statistical term alignment approaches rely on (e.g., [11]). The common structural properties of the input ontologies that are encoded in the base ontology are typically of a more abstract nature. The standard example here relies

[5] A technically more precise definition of this notion is given in Kutz et al. [42]. Note also that our usage of the term 'blendoid' does not coincide with the (non-primary) blendoids defined in Goguen and Harrell [23].

on *image schemata*, such as the notion of a *container* (see e.g. [36]). Thus, in particular, foundational ontologies can support such selections. In analogical reasoning, 'structure' is (partially) mapped from a source domain to a target domain [16, 65]. Therefore, intuitively the operation of computing a base ontology can be seen as a bi-directional search for analogy or generalisation into a base ontology together with the corresponding mappings. Providing efficient means for finding a number of suitable candidate generalisations is essential to making the entire blending process computationally feasible. Consider the example of blending 'house' with 'boat' discussed in detail in Sect. 9.4.1: even after fixing the base ontology itself, guessing the right mappings into the input ontologies means guessing within a space of approximately 1.4 billion signature morphisms. Three promising candidates for finding generalisations are:

(1) **Ontology intersection:** Normann [57] has studied the automatisation of theory interpretation search for formalised mathematics, implemented as part of the Heterogeneous Tool Set (HETS, see below). Kutz and Normann [45] applied these ideas to ontologies by using the ontologies' axiomatisations for finding their shared structure. Accidental naming of concept and role names is deliberately ignored and such names are treated as arbitrary symbols (i.e., any concept may be matched with any other). By computing mutual theory interpretations between the inputs, the method allows the computation of a base ontology as an *intersection* of the input ontologies together with corresponding theory morphisms. While this approach can be efficiently applied to ontologies with non-trivial axiomatisations, lightweight ontologies are less applicable, e.g., 'intersecting' a smaller taxonomy with a larger one clearly results in a huge number of possible taxonomy matches [45]. In this case, the following techniques are more appropriate.

(2) **Structure-based ontology matching:** matching and alignment approaches are often restricted to find simple correspondences between atomic entities of the ontology vocabulary. In contrast, work such as [63, 73] focuses on defining a number of *complex correspondence patterns* that can be used together with standard alignments in order to relate complex expressions between two input ontologies. For instance, the 'Class by Attribute Type Pattern' may be employed to claim the equivalence of the atomic concept PositiveReviewedPaper in ontology O_1 with the complex concept ∃hasEvaluation.Positive of O_2. Such an equivalence can be taken as an axiom of the base ontology; note, however, that it could typically not be found by intersecting the input ontologies. Giving such a library of design patterns may be seen as a variation of the idea of using a library of image schemata.

(3) **Analogical Reasoning:** *Heuristic-driven theory projection* is a logic-based technique for analogical reasoning that can be employed for the task of computing a common generalisation of input theories. Schwering et al. [65] establish an analogical relation between a source theory and a target theory (both first-order) by computing a common generalisation (called 'structural description'). They implement this by using anti-unification [62]. A typical example is to find a generalisation (base ontology) formalising the structural commonalities between the Rutherford atomic model and a model of the solar system. This process may be assisted by a background knowledge base (in the ontological setting, a related domain or foun-

dational ontology). Indeed, this idea has been further developed in Martinez et al. [49].

9.3.2 Selecting the Blendoids: Optimality Principles

Assuming a common base ontology (computed or given) with appropriate base morphism, there is typically still a large number of possible blendoids whenever some kind of partiality is allowed. For example, even in the rather simple case of combining House and Boat, allowing for blendoids which only partially maintain structure (called *non-primary* blendoids in [23]), i.e., where any subset of the axioms may be propagated to the resulting blendoid, the number of possible blendoids is of the order of 1,000. Clearly, from an ontological viewpoint, the overwhelming majority of these candidates will be rather meaningless. A ranking therefore needs to be applied on the basis of specific ontological principles. In conceptual blending theory, a number of **optimality principles** are given in an informal and heuristic style [14]. While these provide useful guidelines for evaluating natural language blends, they do not suggest a direct algorithmic implementation, as also analysed in Goguen and Harrell [23], who in their prototype implementation only covered certain structural, logical criteria. However, the importance of designing computational versions of optimality principles was realised early on, and one such attempt may be found in the work of Pereira and Cardoso [61], who proposed an implementation of the eight optimality principles presented in Fauconnier and Turner [13] based on quantitative metrics for their more lightweight logical formalisation of blending. Such metrics, though, are not directly applicable to more expressive languages such as OWL or first-order logic.

Moreover, the standard blending theory of Fauconnier and Turner [14] does not assign types, which might make sense in the case of linguistic blends where type information is often ignored. A typical example of a type mismatch in language is the operation of *personification*, e.g., turning a boat into an 'inhabitant' of the 'boathouse'. However, in the case of blending in mathematics or ontology, this loss of information is often rather unacceptable: on the contrary, a fine-grained control of type or sort information may be of the utmost importance.

Optimality principles for ontological blending will be of two kinds:

(1) purely **structural/logical principles**: these will extend and refine the criteria as given in Goguen and Harrell [23], namely *degree of commutativity* of the blend diagram, *type casting* (preservation of taxonomical structure), *degree of partiality* (of signature morphisms), and *degree of axiom preservation*. In the context of OWL, typing needs to be replaced with preservation of specific axioms encoding the taxonomy.

(2) **heuristic principles**: these include introducing preference orders on morphisms (an idea that Goguen [21] labelled 3/2 pushouts) reflecting their 'quality', e.g. measured in terms of degree of type violation; specific ontological principles, e.g. adher-

ence to the OntoClean methodology [28] and ontological modelling principles, or general ontology evaluation techniques such as competency questions and fidelity requirements, as further discussed in Sect. 9.5.2.

9.4 Blending with the Distributed Ontology Language DOL

The distributed ontology language DOL is a formal language for specifying both ontologies, base diagrams, and their blends. DOL is a metalanguage in the sense that it enables the reuse of existing ontologies (written in some ontology language like OWL or Common Logic) as building blocks for new ontologies and, further, allows the specification of intended relationships between ontologies. One important feature of DOL is the ability to combine ontologies that are written in different languages without changing their semantics. DOL is being submitted as a response to the Object Management Group's (OMG) Ontology, Model and Specification Integration and Interoperability (OntoIOp) Request For Proposal.[6]

In this section, we introduce DOL only informally. A formal specification of the language and its model-theoretic semantics can be found in [52, 53].

For the purpose of ontology blending the following features of DOL are relevant:

- a **distributed ontology** consists of basic and structured ontologies and ontology interpretations. A basic ontology is an ontology written in some ontology language (e.g., OWL or Common Logic). A structured ontology builds on basic ontologies with the help of ontology translations, ontology unions, and symbol hiding.
- a **basic ontology** $\langle \Sigma, \Gamma \rangle$ written in some ontology language; Σ is a signature, Γ a set of Σ-sentences, with **Mod** $(\langle \Sigma, \Gamma \rangle)$ containing all Σ-models satisfying Γ;
- **ontology translation** (written O_1 **with** σ). A translation takes an ontology O_1 and a renaming function (technically, signature morphism) σ. The result of a translation is an ontology O_2, which differs from the ontology O_1 only by substituting the symbols as specified by the renaming function.
- **ontology union** (written O_1 **and** O_2). The union of two ontologies O_1 and O_2 is a new ontology O_3, which combines the axioms of both ontologies.
- **symbol hiding** (written O_1 **hide** $s_1, \ldots, s_n$). A symbol hiding takes an ontology O_1 and a set of symbols $s_1, \ldots, s_n$. The result of the hiding is a new ontology O_2, which is the result of 'removing' the symbols $s_1, \ldots, s_n$ from the signature of ontology O_1. Nevertheless, O_2 keeps all semantic constraints from O_1.[7]
- **ontology interpretation** (written **interpretation** *INT_NAME* : O_1 to $O_2 = \sigma$). An ontology interpretation is a claim about the relationship between two ontologies O_1 and O_2, giving some renaming function σ. It states that all the constraints that are the result of translating O_1 with σ logically follow from O_2.

---

[6] http://www.omg.org/cgi-bin/doc?ad/2013-12-02.

[7] By approximation, one could consider O_2 as the ontology that is the result of existentially quantifying $s_1, \ldots, s_n$ in O_1.

```
Class: Artifact
Class: Capability
ObjectProperty: has_function
    Range: Capability
ObjectProperty: executes
    Range: Capability
ObjectProperty: is_located_on
Class: Person
Class: Plot
ObjectProperty: is_inhabited_by
    Domain: House
    Range:  Person
Class: ServeAsResidence
    SubClassOf: Capability
Class: ArtifactThatExecutesResidenceFunction
    EquivalentTo: Artifact that executes
                  some ServeAsResidence
    SubClassOf: is_inhabited_by some Person
Class: House
    SubClassOf: Artifact
        that is_located_on some Plot
        and has_function some
                   ServeAsResidence
```

Fig. 9.4 Ontology House

- **ontology alignment** (written **alignment** *ALIGN_NAME* : O_1 to $O_2 = c_1, \ldots, c_n$), where the correspondences c_i relate a symbol in O_1 with one in O_2, e.g. $s_1 = s_2$ or $s_1 < s_2$. Alignments can be seen as a relational variant of interpretations, with the major difference that no logical consequence in involved.

An essential novelty introduced in DOL is that a user can specify the ontological base diagram as a DOL theory, from which the colimit and other blendoids can then be computed.[8] This is a crucial task, as the computed colimit ontology depends on the dependencies between symbols that are stored in the diagram. Ontohub, a DOL-enabled repository discussed further in Sect. 9.5, is able to use the specification of a base diagram to automatically generate the colimit ontology. In the next section, we illustrate the specification of base diagrams in DOL and the computation of the resulting blendoids by blending house and boat to houseboat and boathouse.

9.4.1 The Classic House + Boat Blend

The main inputs for the blendings consist of two ontologies, one for HOUSE and the other for BOAT. We adapt them from Goguen and Harrell [23] but give a stronger axiomatisation to make them more realistic and ontologically sound. Figure 9.4 shows

[8] While OBJ3 already provides the possibility to write down theory morphisms, only DOL provides means to collect them into a formally defined diagram; see the **network** construct below.

the ontology for HOUSE in OWL Manchester Syntax. The ontology is a fragment introducing several concepts necessary for understanding the basic meaning of the term 'house', including that it is an artefact that has the capability of serving as a residence for people and is generally located on a plot of land. The precise formalisation is not criterial at this point; any adequate ontological description of 'house' would, however, need to provide similar distinctions.[9]

```
ontology base1 =
    Class: Artifact  [...]  Class: Agent
     end

ontology base2 =
    Class: Artifact [...]  Class: Person
    Class: Object         Class: Site
    end
```

As discussed above, finding candidate base ontologies and base morphisms is a non-trivial task. For the purpose of this example, we create them manually. The purpose of the example is to show how the DOL specifications naturally allow us to express these kinds of 're-mappings' of relations and entities that are required when considering blends in general. The base ontologies used for the two blends discussed here are both quite simple, they mostly introduce shared concepts and contain only weak axiomatisations. The second base ontology only differs from the first by replacing the class `Agent` by `Person` and two additional classes, namely `Object` and `Site`.

The blending of boat and house to houseboat is achieved by turning the boat into a habitat and moving the house from a plot of land to a body of water. This can be represented by two interpretations `boat_habitable` and `house_floating`.

```
interpretation boat_habitable : base2 to Boat =
    Object ↦ Boat,
    Site ↦ BodyOfWater

interpretation house_floating : base2 to House =
    Object ↦ House,
    Site ↦ Plot
```

The base ontologies and the interpretations above provide the necessary ingredients for a blending of BOAT and HOUSE to HOUSEBOAT. The syntax of diagrams is

network $D = O_1, \ldots, O_m, \; M_1, \ldots, M_n, \; A_1, \ldots, A_k, \; D_1, \ldots, D_l$

where the O_i are ontologies, the M_i are morphisms, the A_i are alignments and the D_i are existing diagrams. The syntax of combinations is

combine $O_1, \ldots, O_m, \; M_1, \ldots, M_n, \; A_1, \ldots, A_k, \; D_1, \ldots, D_l$

---

[9] In the examples, note that concepts such as 'ArtifactThatExecutesResidenceFunction' are auxiliary symbols that are needed because of limitations of the Manchester Syntax being used, which does not allow the use of complex concepts on the left-hand side of subsumption statements. The ontology for BOAT is axiomatised similarly, it can be found at http://www.ontohub.org/conceptportal.

with the ingredients as above. The simplest (and still fully general) form is just

```
combine D
```

where D is a diagram. The semantics of combinations is the colimit of the generated diagram. A colimit involves both pasting together (technically: disjoint union) and identification of shared parts (technically: a quotient).

In our example, houseboat can be defined by the colimit based on the interpretations. To make the result easier to read, some of the classes are renamed:

```
ontology house_boat =
    combine boat_habitable, house_floating
    with Object ↦ HouseBoat, Site ↦ BodyOfWater
```

This captures formally the informal description of the house+boat blend as often given in examples of blending diagrams. Our specification then allows us to go further and derive both consequences of this and other blends. Here Ontohub is able to compute the colimit, which combines both the BOAT and HOUSE ontologies along the morphism. The colimit inherits the axioms of the input ontologies and the base with appropriate identifications of symbols. Here we just show the generated declaration of the blended class `Houseboat`:

```
Class: HouseBoat
  SubClassOf: Artifact
    and has_function some MeansOfTransportation
    and has_function some Floating
    and is_navigated_by some Agent
  SubClassOf: Artifact
    and is_located_on some BodyOfWater
    and has_function some ServeAsResidence
```

In the case of blending BOAT and HOUSE to BOATHOUSE, the crucial part in this blend is to view a boat as a kind of "person" that lives in a house. The two ontologies `House` and `Boat` presented above can be blended by selecting a base, which here provides (among others) a class `Agent`, and two interpretations, mapping `Agent` to `Boat` and `Person`, respectively. Therefore, the second base ontology only differs from the first by replacing the class `Agent` by `Person` and two additional classes, namely `Object` and `Site`.

```
ontology base1 =
    Class: Artifact  [...]  Class: Agent
     end
```

In this way, we let a boat play the role of a person (that inhabits a house).[10]

```
interpretation boat_personification :
    base1 to Boat =
    Agent ↦ Boat
```

---

[10] Compared to [23], the advantage of our formulation is that no projections ("retracts") from a supersort to a subsort are needed. Instead, we can carefully select which parts of the theory of houses and their inhabitants are instantiated with boats.

```
interpretation house_import :
    base1 to House =
    Agent ↦ Person

ontology boat_house =
 combine boat_personification, house_import
 with Agent ↦ Boat, House ↦ BoatHouse
```

As before, Ontohub is able to compute the colimit. As above, we present here only the relevant declarations of the blended concept.

```
Class: BoatHouse
  SubClassOf: Artifact
    and is_located_on some Plot
    and has_function some ServeAsResidence
Class: ArtifactThatExecutesResidenceFunction
       EquivalentTo: Artifact
           and executes some ServeAsResidence
       SubClassOf: is_inhabited_by some Boat
```

Figure 9.5 shows the representation of the ontologies and their relations in Ontohub.

Of course, the possibilities for blending the two concepts do not stop here. For example, we could map the `agent` in the base ontology to `person` in the boat ontology. This can be achieved by first defining an additional interpretation and by blending all three interpretations.

```
interpretation boat_import :
    base1 to Boat =
        Agent ↦ Person

ontology boat_house =
 combine boat_personification, house_import,
         boat_import
 with Agent ↦ Boat, House ↦ BoatHouse
```

The resulting blendoid is consistent, but it contains some strange consequences. For example, in the blendoid boats are driven by boats. However, if we are interested both in hosting boats and a hub for autonomous vehicles, this would count as an interesting result. In general, whether such more creative aspects of blendoids are desirable or not will depend on the context of the blending. We will address this issue in the section on evaluation below. It should be noted, however, that an ontologically cleaner axiomatisation of the input spaces makes blending in fact easier—this is because it reveals more clearly the type structure of the inputs, whose modification can then be more elegantly controlled via the base morphisms.

9.4.2 Variations: Blends of Blends and Partiality

We have discussed a more sophisticated version of the classic HOUSE + BOAT blending in order to illustrate some of the fine detail in the workings of formalised blending

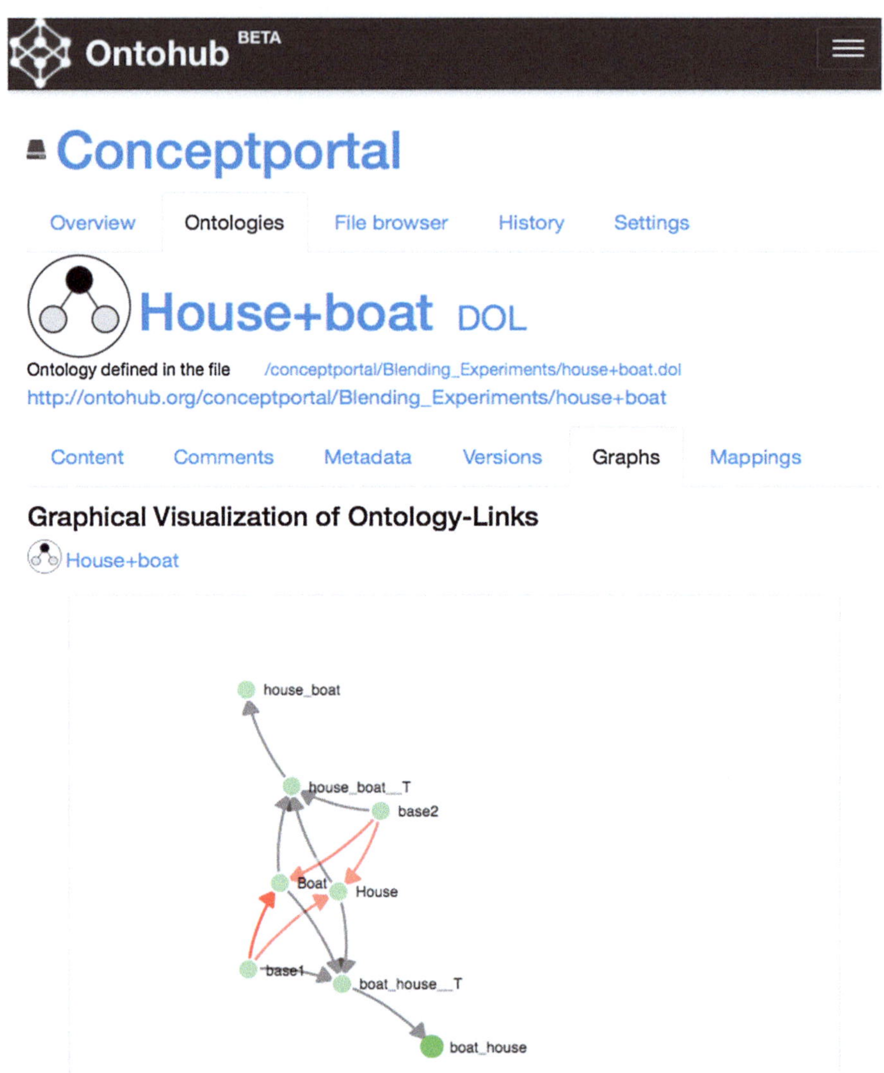

Fig. 9.5 Blendoid representation and colimit computation via HETS/Ontohub: the screenshot of Ontohub shows Conceptportal, which contains the double-blend of house and boat. In the graph the *dots* represent the ontologies: the input spaces (House, Boat), the two bases, and the computed blendoids (boat_house, house_boat). (The ontologies boat_house_T and house_boat_T are generated by Onthub as an intermediate step before the terms in the signature are renamed.) The *arrows* denote the relationships between the ontologies (interpretations, blending, and renaming)

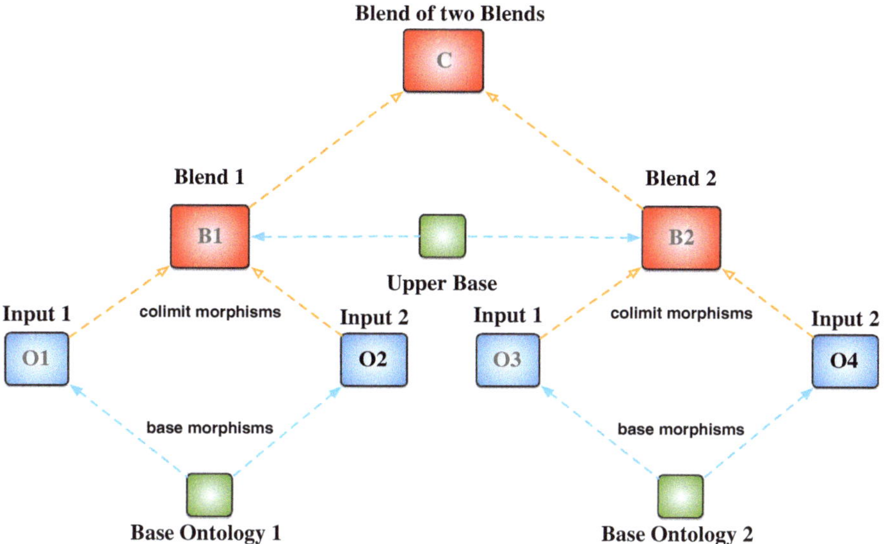

Fig. 9.6 Blending two basic blends into a third

in the Goguen tradition, here based on the DOL language. However, the basic blending diagram only covers the most basic situation, that of an 'atomic blend' using basic concepts and one base space. The real power of blending, however, is only unleashed when blends are iterated and when partiality is allowed.

Lakoff and Núñez [46] give a detailed and powerful analysis of this in the field of conceptual mathematics. A basic claim they make is that the most sophisticated mathematical concepts have been created, over time, through a tower of blended concepts, generating more and more abstract notions. A basic case is that of arithmetic, where several metaphors, image schemas, and analogies are successively blended into modern number systems such as rationals, reals, or complex numbers, including 'arithmetic as object collection', 'object construction', the 'measuring stick metaphor' and 'arithmetic as motion along a path' (see [29, 46] for further details and [15] for a conceptual blend of the complex numbers along these lines). A detailed formal re-construction of such iterated blends is a challenging task, both conceptually and on a technical level. Figure 9.6 shows the basic diagrammatical structure of such iterated blends.

Iteration of blends, however, is not the only variation of the basic blendoid structure. Figure 9.7 shows two triple blends; both have three input spaces, but the one on the left has one base, the one on the right has two base spaces. For instance, we might have 3 inputs that are simultaneously aligned with a basic image schema in the base (left), or we have three ontologies that pairwise interpret different metaphors, e.g. 'arithmetic as object collection' and 'arithmetic as motion along a path'.

Note that on a purely technical level, such complex diagrams can always be reduced to a succession of squares, possibly by duplicating some nodes or adding

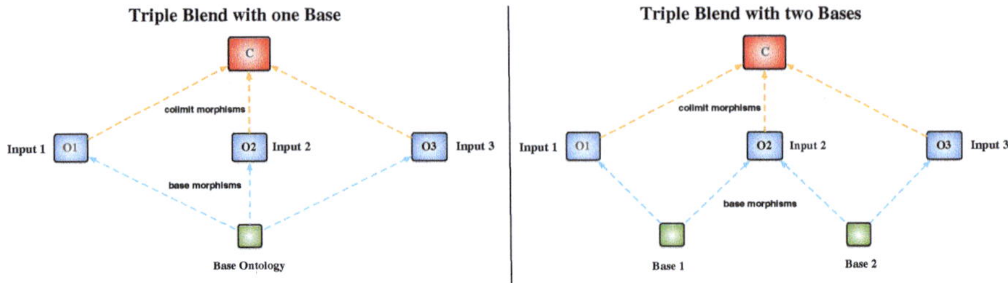

Fig. 9.7 Blending three input spaces using one respectively two base ontologies

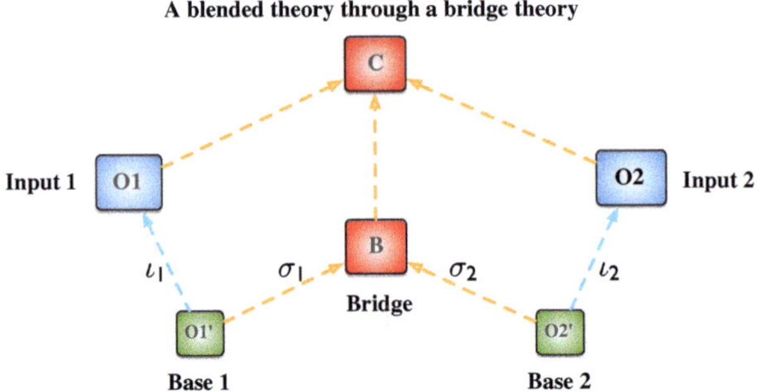

Fig. 9.8 Blending two input spaces through two bases and a bridge theory, deviating from the Goguen construction

trivial ones[11]—however, such a reduction loses the direct connection between the diagrammatic representation and the cognitive-conceptual processes that are being formalised here. In a similar vein, Definition 1 introducing the notion of an ontological base diagram in Sect. 9.3 easily generalises to the case of partial base morphisms, i.e. where only parts of the signature of an ontology are mapped. Such partial morphisms can be coded as spans of two (total) theory morphisms $B_i \leftarrow dom(\mu_{ij}) \rightarrow I_j$, where the first morphism is the embedding of the domain (actually, the larger $dom(\mu_{ij})$ is, the more defined is the partial morphism), and the second action represents the action of the partial morphism.[12] Similarly, arbitrary relations can be coded as spans $B_i \leftarrow R \rightarrow I_j$. Here, $R \subseteq B_i \times I_j$ is a relation, and the arrows are the projections to the first and second component. However, such complexities can be hidden from a user by allowing partial morphisms to be used directly in the specification of a blending diagram, and by letting a tool handle the simulation through total morphisms as discussed above.

Finally, a more severe deviation from the basic blending diagram is shown in Fig. 9.8. Here, we interpret Base1 into Input1, Base2 into Input2, and connect the

[11] A well-known theorem of category theory states that every finite colimit can be expressed by pushouts and initial objects.

[12] In this case, the base diagram becomes a bit more complex; in particular, there are minimal nodes $dom(\mu_{ij})$ which have only auxiliary purposes and do not belong to the base.

two bases by a bridge theory. For example, the bridge might introduce a higher-level connection between two image schemas which is then used to create the blended theory. An example of this might be where we have image schemas involved with time and with space and combine these first for the definition of a theory in which time and space are linked (as in our navigation examples above or even in the linking between time and space discussed for comics and visual narrative).

Technically, such diagrams are closely related to alignments [9, 74], and to distributed modelling languages [40]. Concerning the formalisation of conceptual blending, these techniques and diagram patterns will be of particular importance to tackle a computational treatment and formal representation of so-called *generic integration templates* (GIT), i.e. the idea of studying general blending templates, first introduced and discussed in detail by Pagán Cánovas [59], with more examples to be found in Turner [68].

9.5 Blending in the Hub

In this section we will discuss the computational and representational support for formalised blending via the Ontohub.org platform as well as the evaluation problem.

9.5.1 Representation and Computation

To begin, combinations (or, alternatively, the underlying colimits) can be computed directly by the web platform Ontohub. Ontohub is a repository engine for managing distributed heterogeneous ontologies. Ontohub supports a wide range of formal logical and ontology languages and allows for complex inter-theory (concept) mappings and relationships with formal semantics, as well as ontology alignments and blending. Ontohub understands various input languages, among them OWL and DOL.

We describe the basic design and features of Ontohub in general, and outline the extended feature-set that we pursue to add to Ontohub for conceptportal.org—a specialised repository for blending experiments within the distributed Ontohub architecture.

The back-end of Ontohub is the Heterogeneous Tool Set HETS, which is used by Ontohub for parsing, static analysis and proof management of ontologies. HETS can also compute colimits of both OWL and first-order logic diagrams and even approximations of colimits in the case where the input ontologies live in different ontology languages [8].

Computation of colimits in HETS is based on HETS' general colimit algorithm for diagrams of sets and functions (note that signatures in most cases are structured sets, and signature morphisms structure-preserving functions). Such a colimit of sets and functions is computed by taking the disjoint union of all sets, and quotienting it by the equivalence relation generated by the diagram, which more precisely is obtained by the rule that given any element x of an involved set, any images of x under the involved functions are identified. The quotient is computed by selecting a representative of each equivalence class.

A difficulty that arises is that we have to make a choice of these representatives, and therefore of names for the symbols in the colimit, since a symbol is often not identically mapped in the base diagram of the blendoid. The convention in HETS is that, in case of ambiguity, from among all symbols of the equivalence class, that name of the symbol is chosen which is the most frequently occurring one. In any case, the user has control over the namespace because the symbols in the colimit can later be renamed. We can see this for our boathouse example above, where `Agent` appears most often in the diagram and therefore the symbol has been correspondingly renamed.

9.5.2 Evaluating the Blending Space

Optimality principles (see Sect. 9.3.2), in particular structural ones, can be used to rank candidate blendoids on-the-fly during the ontology blending process. However, even if they improve on existing logical and heuristic methods, optimality principles will only narrow down the potential candidates and not tell us whether the result is a 'successful' blend of the ontologies. For example, assume that we had optimality principles that would show that from the roughly 1,000 candidate blendoids of *House* and *Boat* that Goguen computed, only two candidates $\mathfrak{B}_{hb}$ and $\mathfrak{B}_{bh}$ are optimal. Is either $\mathfrak{B}_{hb}$ or $\mathfrak{B}_{bh}$ any good? And, if so, which of them should we use? To answer these question, it seems natural to apply ontology evaluation techniques.

Ontologies are human-intelligible and machine-interpretable representations of some portions and aspects of a domain that are used as part of information systems. To be more specific, an ontology is a logical theory written in some knowledge representation language, which is associated with some intended interpretation. The intended interpretation is partially captured in the choice of symbols and natural language text (often in the form of annotations or comments). The evaluation of an ontology covers both the logical theory and the intended interpretation, their relationship to each other, and how they relate to the requirements that are derived from the intended use within a given information system. Therefore, ontology evaluation is concerned not only with formal properties of logical theories (e.g., logical consistency), but, among other aspects, with the *fidelity* of an ontology; that is whether the formal theory accurately represents the intended domain [56]. For example, if $\mathfrak{B}_{hb}$ is an excellent representation of the concept *houseboat*, then $\mathfrak{B}_{hb}$ provides a poor representation of the concept *boathouse*. Thus, any evaluation of the blend $\mathfrak{B}_{hb}$ depends on what domain $\mathfrak{B}_{hb}$ is intended to represent.

Given these considerations, $\mathfrak{B}_{hb}$ and $\mathfrak{B}_{bh}$ are not ontologies, they are logical theories that are the result of the blending of two logical theories that are part of ontologies. This is illustrated by the following thought-experiment: let's assume the theory $\mathfrak{B}_{hb}$ captures the concept *houseboat* very well, and that $\mathfrak{B}_{hb}$ is not the result of some automatic blending process, but was intentionally developed by an ontology engineer. In case that the ontology engineer intended to develop an ontology of houseboats $\mathfrak{B}_{hb}$, this would be considered successful. However, if the engineer intended to develop an ontology of boathouses, then $\mathfrak{B}_{hb}$ would be a poor outcome. In other words, the ontology consisting of $\mathfrak{B}_{hb}$ and the intention *houseboat* would have high fidelity, but the ontology consisting of $\mathfrak{B}_{hb}$ and the intention *boathouse* would have low fidelity. Thus, the evaluation of the theory $\mathfrak{B}_{hb}$ is dependent on the domain it is supposed to represent.

The lesson from this thought experiment is that the evaluation of the results of ontology blending is dependent on the intended goal and, more generally, on the requirements that one expects the outcome of the blending process to meet. One way to capture these requirement is similar to competency questions, which are widely used in ontology engineering [26]. Competency questions are usually initially captured in natural language; they specify examples for questions that an ontology needs to be able to answer in a given scenario. By formalising the competency questions one can use automatic theorem provers to evaluate whether the ontology meets the intended interpretation.

The requirements that are used to select between the different blends fall, roughly, into two categories: *ontological constraints* and *consequence requirements*. Ontological constraints prevent the blends from becoming 'too creative' by narrowing the space for conceptual blending. E.g., it may be desirable to ensure that the `is_inhabited_by` relationship is asymmetric and that `is_navigated_by` is irreflexive. To achieve that any blendoid can be checked for logical consistency with the following ontology:

```
ontology OntologicalConstraints =
  ObjectProperty: is_inhabited_by
    Characteristics: Asymmetric
  ObjectProperty: is_navigated_by
    Characteristics: Irreflexive
```

Given these requirements, any blendoid that involves a house that lives in itself, or any boat navigated by itself would be discarded.

Consequence requirements specify the kind of characteristics the blendoid is supposed to have. E.g., assume the purpose of the conceptual blending is to find alternative housing arrangements, because high land prices make newly built houses unaffordable. In this case, the requirement could be 'a residence that is not located on a plot of land', which can be expressed in OWL as follows:

```
ontology ConsequenceRequirements =
[...]
  Class PlotFreeResidence
    EquivalentTo: Residence
       and (is_located_on only (not (Plot)))
```

For the evaluation of a blendoid against requirements (both ontological constraints and consequence requirements), it is often not sufficient to just consider the information that is contained in the blendoid itself. Some background knowledge usually needs to be added in order to evaluate a blendoid.

Background knowledge plays another crucial role in the blending process, which we have not addressed in this paper so far. The basic blending diagram in Fig. 9.3 presents a static view, which describes how two input spaces, a base, and two interpretations give rise to a blendoid. However, any system that attempts to automate conceptual blending will need to perform not one but many blends in order to get a decent result. In this process, the background knowledge and the evaluation of previous blending results can be utilised in the selection of candidate bases and interpretations. Further, the violation of ontological constraints may be a symptom of an attempt to blend input spaces that are too rich. In these cases, the result of the evaluation can be used to guide heuristics, which remove information from the original input spaces that may have caused the violation of the ontological constraints. The result is a new, weakened input space, which may be easier to blend. In short, following a proposal by Marco Schorlemmer discussed in detail in [55], we envision an approach where background knowledge and evaluation are driving an iterative blending process, as illustrated in Fig. 9.9.

Ontohub allows the use of ontological constraints and consequence requirements to evaluate blended concepts automatically. The requirements are managed as DOL files, which allow us to express that a given blendoid (together with some background knowledge) is logically consistent with a set of ontological constraints or that it entails some consequence requirements. The requirements themselves may be stored as regular ontology files (e.g., in OWL Manchester syntax). Ontohub executes the DOL files with the help of integrated automatic theorem provers, and is able to detect whether a blendoid meets the specified requirements. At this time, the evaluation of blendoids for ontological constraints and consequence requirements depends on the use of DOL files. We are planning to integrate this functionality into the GUI of Ontohub to make it more convenient for the user.

9.6 E Pluribus Unum: The Future of the Melting Pot

Our work in this paper follows a research line in which blending processes are primarily controlled through mappings and their properties [16, 20, 60, 70]. By introducing blending techniques to ontology languages, we have provided a method which allows us to combine two or more thematically different ontologies into a newly created ontology, the blendoid, describing a novel concept or domain. The blendoid creatively mixes information from both input ontologies on the basis of structural commonalities of the inputs and selective combination of their axiomatisations.

We have moreover illustrated that the Ontohub/HETS tool ecosystem and the DOL language provide an excellent starting point for developing the theory and practice of ontology blending further [52]. They (1) support various ontology languages and their heterogeneous integration [39]; (2) allow the specification of theory interpretations and other morphisms between ontologies [43]; (3) support the computation of colimits as well as the approximation of colimits in the heterogeneous case [8]; and (4) provide (first) solutions for automatically computing a base ontol-

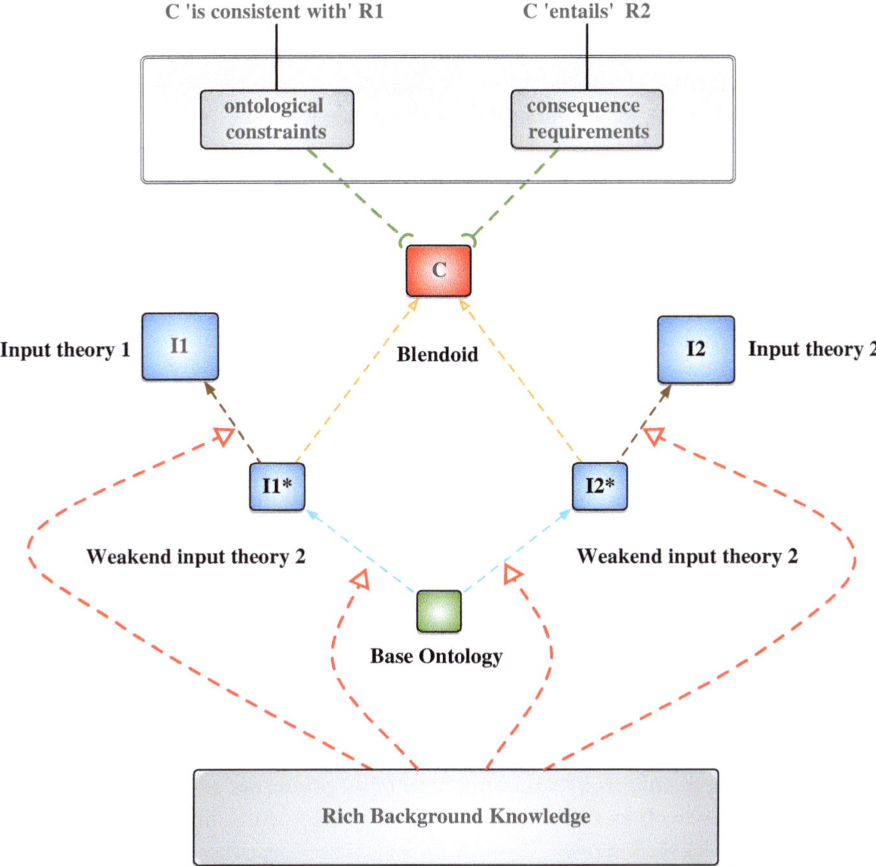

Fig. 9.9 The core Schorlemmer model for computational blending enriched with evaluation and *background layers*

ogy through ontology intersection [45] and blendoid evaluation using requirements [44, 55].

In particular, we have shown that the blending of ontologies can be declaratively encoded in a DOL theory representing the respective blending diagram—here, employing the homogeneous fragment of DOL just using OWL ontologies. Blendoid ontologies, as well as their components, i.e. input and base ontologies, can be stored, formally related, and checked for consistency within Conceptportal, a repository node within Ontohub dedicated to blending experiments carried out in the European FP7 Project COINVENT [64]. Ontohub moreover gives access to thousands of ontologies from a large number of different scientific and common sense domains, searchable via rich metadata annotation, logics used, formality level, and other dimensions, to provide not only a rich pool of ontologies for blending experiments, but also for the evaluation of newly created concepts.

Of course, constructing a homogeneous blendoid from a basic blending diagram is one of the simplest cases of conceptual blending. As discussed in Sect. 9.4.2, on a technical level a blendoid is just like an alignment diagram, except that instead of dealing with synonymy and homonymy relations, and just signature in the base, in the blendoid case we are dealing with selectively merging axioms. Following this intuition, a whole range of more complex alignment and theory combination techniques can be combined with the basic blending ideas of Goguen: this includes constructions such as W-alignments [9, 41, 74], and connections of theories following the $\mathcal{E}$-connection/DDL paradigm [7, 40, 54].

The next important milestone for computational conceptual blending will be to make the step from a *reconstructive* approach, where conceptual blending is illustrated by blending one concept (e.g., houseboat) with the help of some carefully selected input spaces (e.g., a house and a boat) and a hand-crafted base ontology, to a system that *autonomously* selects two (or more) ontologies from a repository in Ontohub and attempts to blend them in a way that meets some given requirements. In [55], we have described the first steps towards designing a computational architecture that performs conceptual blending autonomously and self-evaluates its own creations.

Within the extensive literature on conceptual blending, only few attempts have been made at a (more or less) complete automation of the blending process; notable exceptions include [23, 47, 60, 71, 72]. To make concept invention via ontological blending more feasible in practice from within Ontohub, a number of further plugins into the architecture and refinements are planned covering in particular the automatic creation of base ontologies together with their mappings, the implementation of filtering blendoids by structural optimality principles and preference orders on morphisms, as well as the addition of more ontologically motivated evaluation techniques as discussed above.

Acknowledgments The project COINVENT acknowledges the financial support of the Future and Emerging Technologies (FET) programme within the Seventh Framework Programme for Research of the European Commission, under FET-Open Grant number: 611553. Work on this paper was moreover supported by the DFG-funded collaborative research centre SFB/TR 8 'Spatial Cognition' of the Universities of Bremen and Freiburg.
We thank the anonymous referees as well as Mihai Codescu for detailed feedback on this chapter.

References

1. Adámek, J., Herrlich, H., Strecker, G.: Abstract and Concrete Categories. Wiley, New York (1990)
2. Bateman, J.A.: Language and space: a two-level semantic approach based on principles of ontological engineering. Int. J. Speech Technol. **13**(1), 29–48 (2010)
3. Bateman, J.A.: The decomposability of semiotic modes. In: O'Halloran, K.L., Smith, B.A. (eds.) Multimodal Studies: Multiple Approaches and Domains, Routledge Studies in Multimodality, pp. 17–38. Routledge, London (2011)

4. Bateman, J.A., Veloso, F.O.D.: The semiotic resources of comics in movie adaptation: Ang Lee's Hulk (2003) as a case study. Stud. Comics **4**(1), 137–159 (2013)
5. Bhatt, M., Hois, J., Kutz, O.: Ontological modelling of form and function in architectural design. Appl. Ontol. **7**(3), 233–267 (2012)
6. Black, M.: More about metaphor. In: Ortony, A. (ed.) Metaphor and Thought, pp. 19–43. Cambridge University Press, Cambridge (1979)
7. Borgida, A., Serafini, L.: Distributed description logics: assimilating information from peer sources. J. Data Semant. **1**, 153–184 (2003)
8. Codescu, M., Mossakowski, T.: Heterogeneous colimits. In: Boulanger, F., Gaston, C., Schobbens, P.-Y. (eds.) MoVaH'08 Workshop on Modeling, Validation and Heterogeneity, pp. 131–140. IEEE Press (2008)
9. Codescu, M., Mossakowski, T., Kutz, O.: A categorical approach to ontology alignment. In Proceedings of the 9th International Workshop on Ontology Matching (OM-2014), ISWC-2014, Riva del Garda, Trentino. CEUR-WS (2014)
10. Eisner, W.: Comics and Sequential Art. Kitchen Sink Press Inc., Princeton (1992)
11. Euzenat, J., Shvaiko, P.: Ontology Matching. Springer, Heidelberg (2007)
12. Fauconnier, G.: Mappings in Thought and Language. Cambridge University Press, Cambridge (1997)
13. Fauconnier, G., Turner, M.: Conceptual integration networks. Cognit. Sci. **22**(2), 133–187 (1998)
14. Fauconnier, G., Turner, M.: The Way We Think: Conceptual Blending and the Mind's Hidden Complexities. Basic Books, New York (2003)
15. Fleuriot, J., Maclean, E., Smaill, A., Winterstein, D.: Reinventing the complex numbers. In: Besold, T., Kühnberger, K.-U., Schorlemmer, M., Smaill, A. (eds.) Computational Creativity, Concept Invention, and General Intelligence, Proceedings of the 3rd International Workshop C3GI@ECAI-14, vol. 1-2014, Prague, Czech Republic. Publications of the Institute of Cognitive Science, Osnabrück (2014)
16. Forbus, K., Falkenhainer, B., Gentner, D.: The structure-mapping engine. Artif. Intell. **41**, 1–63 (1989)
17. Forceville, C.J., Urios-Aparisi, E. (eds.): Multimodal Metaphor. Mouton de Gruyter, Berlin (2009)
18. Frank, A., Kuhn, W.: A specification language for interoperable GIS. In: Goodchild, M., Egenhofer, M., Fegeas, R., Kottmann, C. (eds.) Interoperating Geographic Information Systems, pp. 123–132. Kluwer, Norwell (1999)
19. Gärdenfors, P.: Conceptual Spaces—The Geometry of Thought. Bradford Books, MIT Press, Cambridge (2000)
20. Gentner, D.: Structure mapping: a theoretical framework for analogy. Cognit. Sci. **7**(2), 155–170 (1983)
21. Goguen, J.A.: An introduction to algebraic semiotics, with applications to user interface design. In: Computation for Metaphors, Analogy and Agents, LNCS, vol. 1562, pp. 242–291. Springer, Berlin (1999)
22. Goguen, J.A., Burstall, R.M.: Institutions: abstract model theory for specification and programming. J. Assoc. Comput. Mach. **39**(1):95–146 (1992) (Predecessor. In: LNCS **164**, 221–256 (1984))
23. Goguen, J.A., Harrell, D.F.: Style: a computational and conceptual blending-based approach. In: Argamon, S., Dubnov, S. (eds.) The Structure of Style: Algorithmic Approaches to Understanding Manner and Meaning, pp. 147–170. Springer, Berlin (2010)
24. Goguen, J.A., Malcolm, G.: Algebraic Semantics of Imperative Programs. MIT, Cambridge (1996)
25. Gruber, T.: Toward principles for the design of ontologies used for knowledge sharing. Int. J. Human-Comput. Stud. **43**(5/6), 907–928 (1995)
26. Grüninger, M., Fox, M.S.: The role of competency questions in enterprise engineering. In: Benchmarking-Theory and Practice, pp. 22–31. Springer, Berlin (1995)

27. Guarino, N.: The ontological level. In: Casati, R., Smith, B., White, G. (eds.) Philosophy and the Cognitive Sciences, pp. 443–456. Hölder-Pichler-Tempsky, Vienna (1994)
28. Guarino, N., Welty, C.: Evaluating ontological decisions with OntoClean. Commun. ACM **45**(2), 61–65 (2002)
29. Guhe, M., Pease, A., Smaill, A., Martínez, M., Schmidt, M., Gust, H., Kühnberger, K.-U., Krumnack, U.: A computational account of conceptual blending in basic mathematics. Cognit. Syst. Res. **12**(3–4), 249–265 (2011)
30. Gust, H., Kühnberger, K.-U., Schmid, U.: Metaphors and anti-unification. In: Proceedings of the Twenty-First Workshop on Language Technology, Algebraic Methods in Language Processing, Verona, Italy (2003)
31. Hart, C.: Critical discourse analysis and metaphor: toward a theoretical framework. Crit. Discourse Stud. **5**(2), 91–106 (2008)
32. Hois, J., Kutz, O., Mossakowski, T., Bateman, J.: Towards ontological blending. In: Proceedings of the The 14th International Conference on Artificial Intelligence: Methodology, Systems, Applications (AIMSA-2010), Varna, Bulgaria, 8th–10th September 2010
33. Jaszczolt, K.M.: On translating 'What is said': tertium comparationis in contrastive semantics and pragmatics. In: Meaning Through Language Contrast, vol. 2, pp. 441–462. J. Benjamins, Amsterdam (2003)
34. Johnson, M.: The body in the mind. In: The Bodily Basis of Meaning, Imagination, and Reasoning. The University of Chicago Press, Chicago (1987)
35. Kövescses, Z.: Metaphor: A Practical Introduction, 2nd edn. Oxford University Press, Oxford (2010)
36. Kuhn, W.: Modeling the semantics of geographic categories through conceptual integration. In: Proceedings of GIScience 2002, pp. 108–118. Springer (2002)
37. Kuhn, W.: Semantic reference systems. Int. J. Geogr. Inf. Sci. **17**(5), 405–409 (2003)
38. Kutz, O., Lücke, D., Mossakowski, T.: Heterogeneously structured ontologies: integration, connection, and refinement. In: Proceedings of the KROW 2008, CRPIT, vol. 90, pp. 41–50. ACS (2008a)
39. Kutz, O., Lücke, D., Mossakowski, T., Normann, I.: The OWL in the CASL: designing ontologies across logics. In: Proceedings of OWLED-08, vol. 432. CEUR (2008b)
40. Kutz, O., Lutz, C., Wolter, F., Zakharyaschev, M.: $\mathcal{E}$-connections of abstract description systems. Artif. Intell. **156**(1), 1–73 (2004)
41. Kutz, O., Mossakowski, T., Codescu, M.: Shapes of alignments: construction, combination, and computation. In: Sattler, U., Tamilin, A. (eds.) Proceedings of the 1st Workshop on Ontologies: Reasoning and Modularity (WORM-08), ESWC, vol. 348. Tenerife, Spain. CEUR-WS (2008c)
42. Kutz, O., Mossakowski, T., Hois, J., Bhatt, M., Bateman, J.: Ontological blending in DOL. In: Besold, T., Kühnberger, K.-U., Schorlemmer, M., Smaill, A., (eds.) Computational Creativity, Concept Invention, and General Intelligence, Proceedings of the 1st International Workshop C3GI@ECAI, vol. 01-2012. Montpellier, France. Publications of the Institute of Cognitive Science, Osnabrück (2012)
43. Kutz, O., Mossakowski, T., Lücke, D.: Carnap, Goguen, and the hyperontologies: logical pluralism and heterogeneous structuring in ontology design. Log. Universalis **4**(2):255–333 (2010) Special issue on 'Is logic universal?'
44. Kutz, O., Neuhaus, F., Mossakowski, T., Codescu, M.: Blending in the hub-towards a collaborative concept invention platform. In: Proceedings of the 5th International Conference on Computational Creativity, Ljubljana, Slovenia (2014)
45. Kutz, O., Normann, I.: Context discovery via theory interpretation. In: Workshop on Automated Reasoning about Context and Ontology Evolution, ARCOE-09 (IJCAI-09) (2009)
46. Lakoff, G., Núñez, R.E.: Where Mathematics Comes From. Basic Books, New York (2000)
47. Li, B., Zook, A., Davis, N., Riedl, M.O.: Goal-driven conceptual blending: a computational approach for creativity. In: Proceedings of the 2012 International Conference on Computational Creativity, Dublin, Ireland (2012)
48. Mamakos, C., Stefaneas, P., Dimarogkona, M., Ireson-Paine, J.: Polytropos project: experiments in blending. In: Besold, T., Kühnberger, K.-U., Schorlemmer, M., Smaill, A. (eds.)

Computational Creativity, Concept Invention, and General Intelligence, Proceedings of the 3rd International Workshop C3GI@ECAI-14, vol. 1-2014. Prague, Czech Republic. Publications of the Institute of Cognitive Science, Osnabrück (2014)
49. Martinez, M., Besold, T.R., Abdel-Fattah, A., Kühnberger, K.-U., Gust, H., Schmidt, M., Krumnack, U.: Towards a domain-independent computational framework for theory blending. In: Proceedings of the AAAI Fall 2011 Symposium on Advances in Cognitive Systems (2011)
50. McCloud, S.: Understanding Comics: The Invisible Art. HarperPerennial, New York (1994)
51. Miller, A.I.: Metaphor and scientific creativity. In: Hallyn, F. (ed.) Metaphor and Analogy in the Sciences, pp. 147–164. Kluwer Academic Publishers, Dordrecht (2000)
52. Mossakowski, T., Kutz, O., Codescu, M., Lange, C.: The distributed ontology, modeling and specification language. In: Del Vescovo, C., Hahmann, T., Pearce, D., Walther, D. (eds.) Proceedings of the 7th International Workshop on Modular Ontologies (WoMO-13), vol. 1081. CEUR-WS (2013)
53. Mossakowski, T., Kutz, O., Neuhaus, F., Codescu, M., Lange, C., Gruninger, M., Keet, M.: The distributed ontology, modeling and specification language. Draft answer to the OMG RFP "OntoIOp" (2014)
54. Nalon, C., Kutz, O.: Towards resolution-based reasoning for connected logics. Electron. Notes Theor. Comput. Sci. **305**, 85–102 (2014) Post-proceedings of the 8th Workshop on Logical and Semantic Frameworks (LSFA)
55. Neuhaus, F., Kutz, O., Codescu, M., Mossakowski, T.: Fabricating monsters is hard: towards the automation of conceptual blending. In: Besold, T., Kühnberger, K.-U., Schorlemmer, M., Smaill, A. (eds.) Computational Creativity, Concept Invention, and General Intelligence, Proceedings of the 3rd International Workshop C3GI@ECAI-14, vol. 1-2014. Prague, Czech Republic. Publications of the Institute of Cognitive Science, Osnabrück (2014)
56. Neuhaus, F., Vizedom, A., Baclawski, K., Bennett, M., Dean, M., Denny, M., Grüninger, M., Hashemi, A., Longstreth, T., Obrst, L., et al.: Towards ontology evaluation across the life cycle: the ontology summit 2013. Appl. Ontol. **8**(3), 179–194 (2013)
57. Normann, I.: Automated Theory Interpretation. PhD thesis, Jacobs University Bremen (2009)
58. Núñez, R.E.: Creating mathematical infinities: metaphor, blending, and the beauty of transfinite cardinals. J. Pragmat. **37**, 1717–1741 (2005)
59. Pagán Cánovas, C.: Erotic emissions in Greek poetry: a generic integration network. Cognit. Semiot. **6**, 7–32 (2010)
60. Pereira, F.C.: Creativity and Artificial Intelligence: A Conceptual Blending Approach, volume 4 of Applications of Cognitive Linguistics (ACL). Mouton de Gruyter, Berlin (2007)
61. Pereira, F.C., Cardoso, A.: Optimality principles for conceptual blending: a first computational approach. AISB J. **1**(4) (2003)
62. Plotkin, G.D.: A note on inductive generalization. Mach. Intell. **5**, 153–163 (1970)
63. Ritze, D., Meilicke, C., Šváb Zamazal, O., Stuckenschmidt, H.L.: A pattern-based ontology matching approach for detecting complex correspondences. In: OM-09, vol. 551. CEUR (2009)
64. Schorlemmer, M., Smaill, A., Kühnberger, K.-U., Kutz, O., Colton, S., Cambouropoulos, E., Pease, A.: COINVENT: towards a computational concept invention theory. In: Proceedings of the 5th International Conference on Computational Creativity, Ljubljana, Slovenia (2014)
65. Schwering, A., Krumnack, U., Kühnberger, K.-U., Gust, H.: Syntactic principles of heuristic-driven theory projection. Cognit. Syst. Res. **10**(3), 251–269 (2009a)
66. Schwering, A., Kühnberger, K.-U., Krumnack, U., Gust, H., Wandmacher, T., Indurkhya, B., Ojha, A.: A computational model for visual metaphors: interpreting creative visual advertisements. In: International Conference on Agents and Artificial Intelligence (ICAART-09), pp. 339–344 (2009b)
67. Turner, M.: The way we imagine. In: Roth, I. (ed.) Imaginative Minds—Proceedings of the British Academy, pp. 213–236. OUP, Oxford (2007)
68. Turner, M.: The Origin of Ideas: Blending, Creativity, and the Human Spark. Oxford University Press, Oxford (2014)
69. van Mulken, M., le Pair, R., Forceville, C.: The impact of perceived complexity, deviation and comprehension on the appreciation of visual metaphor in advertising across three European countries. J. Pragmat. **42**, 3418–3430 (2010)

70. Veale, T.: Creativity as pastiche: a computational treatment of metaphoric blends, with special reference to cinematic "borrowing". In: Proceedings of Mind II, Computational Models of Creative Cognition (1997)
71. Veale, T.: From conceptual mash-ups to "Bad-Ass" blends: a robust computational model of conceptual blending. In: Proceedings of the 2012 International Conference on Computational Creativity, Dublin, Ireland (2012)
72. Veale, T., O'Donoghue, D.: Computation and blending. Cognit. Linguist. **11**(3/4), 253–281 (2001)
73. Walshe, B.: Identifying complex semantic matches. In: The Semantic Web: Research and Applications, pp. 849–853. Springer (2012)
74. Zimmermann, A., Krötzsch, M., Euzenat, J., Hitzler, P.: Formalizing ontology alignment and its operations with category theory. In: Proceedings of FOIS-06, pp. 277–288 (2006)

Chapter 10
Ode to a Keatsian Turn: Creating Meaningful and Poetic Instances of Rhetorical Forms

Tony Veale

Abstract Linguistic creativity requires a marriage of form and content in which each works together to convey our meanings with concision, resonance and wit. Though form clearly influences and shapes our content, the most deft formal trickery cannot compensate for a lack of real insight. Before computers can be truly creative with language, we must first imbue them with the ability to formulate meanings that are worthy of creative expression. This is especially true of computer-generated poetry. If readers are to recognize a poetic turn-of-phrase as more than a superficial manipulation of words, they must perceive and connect with the meanings and the intent behind the words. So it is not enough for a computer to merely generate *poem-shaped texts*; poems must be driven by conceits that build an affective worldview. This chapter describes a conceit-driven approach to computational poetry, in which metaphorical conceits are generated for a given topic and affective slant. Subtle inferences drawn from these conceits can then drive the process of poetry generation. In the same vein, we consider the problem of generating witty insights from the banal truisms of common-sense knowledge bases.

10.1 Introduction

Raymond Chandler saw the primary task of the "natural" writer as bridge-building, between "*what one wants to say*" and "*what one knows how to say*" [5]. The scholarly study of how best to bridge one's words and ideas is an ancient one [1], and rhetoricians have systematically identified and classified a wide variety of linguistic forms with which to give our meanings a persuasive force [13]. These rhetorical devices are so effective in the shaping and delivery of well-developed meanings that they can also lend our less substantial thoughts the unmerited appearance of solidity. This is not always a bad thing: a well-chosen rhetorical form can act as a scaffolding for an undeveloped idea, allowing it to take root and grow during subsequent elaboration.

T. Veale (✉)
School of Computer Science and Informatics, University College Dublin,
Belfield D2, Ireland
e-mail: tony.veale@UCD.ie

T.R. Besold et al. (eds.), *Computational Creativity Research: Towards Creative Machines*,
Atlantis Thinking Machines 7, DOI 10.2991/978-94-6239-085-0_10

Nonetheless, just as good painters sometimes paint fakes, clever orators sometimes abuse rhetoric, to suggest profundity where there is shallowness, and sincerity where there is indifference.

One such rhetorical device is chiasmus, which takes its name from the cross-shaped Greek letter chi, or 'χ'. The name is apt, for chiasmus is the crossover repetition of words, meanings, images or syntactic structures in a text. It is a much-used device in the texts of the Bible—it is used in both old and new testaments—and in other ancient Hebrew and Greek texts. Indeed, one of the most widely-quoted examples of chiasmus is also nicely self-descriptive: "*Those that are first shall be last and those that are last shall be first*" (Matthew 19:30). Biblical uses of chiasmus combine a profundity of thought with a symmetry of form, and the effectiveness of this balancing act has not been lost on orators throughout the ages. Consider this use of chiasmus by Abraham Lincoln, the 16th president of the United States, "*I claim not to have controlled events, but confess plainly that events have controlled me*," and this use by its 35th president, John F. Kennedy, "*Ask not what your country can do for you, but what you can do for your country*". JFK's use of chiasmus at his inauguration in 1961 was both memorable and effective, allowing his words to hold a mirror, structurally speaking, to what he saw as a much-needed shift from selfishness to selflessness in modern times.

The crisscross pattern of chiasmus is the linguistic equivalent of a tightly-laced boxing glove. When used effectively, with a substantial meaning to communicate, the surface crossover of linguistic content implies a dove-tailing of ideas at a deeper level. Chiasmus relies on repetition to drive home these ideas, though this duplication of content is not always superficial or obvious. James Joyce, for example, employed chiasmus to lend balance and symmetry to his use of imagery in *Dubliners*, but nowhere is his use of the form as structurally obvious as it is in either the Lincoln or Kennedy examples. The explicit repetition of words is a hallmark of chiasmus, as in this popular slogan of the American gun lobby, "*when guns are outlawed, only outlaws will have guns*" (though even here, the verb "*to outlaw*" is repeated as a noun). But one can implicitly repeat an idea in a chiastic crossover by referring instead to its opposite counterpart, as in these wise words from the Dalai Lama: "*In the practice of tolerance, one's enemy is the best teacher*" (here "*practice*" aligns with "*teacher*" while "*tolerance*" crosses over to "*enemy*").

These are aphoristic uses of the form, in which the chiasmus is designed to be noticed, just as the resulting epigrams and slogans are designed to be remembered. Yet this repetition with crossover can be just as effective even when it is not overtly noticed. Consider this use of chiasmus by the always quotable fashion-designer Karl Lagerfeld: "*Sweatpants are a sign of defeat. You lost control of your life so you bought some sweatpants*". If you feel the need to repeat yourself, as Lagerfeld does here with withering contempt, then repetition with crossover may subtly strengthen the logical force of your argument. Notice how Lagerfeld begins by asserting a causal link from *sweatpants* to *defeat*, and quickly follows this generalization by asserting a causal link in the opposite direction, from a loss of personal control (defeat again) to the purchase of those very same sweatpants. His use of chiasmus suggests abductive and deductive reasoning, and shows us the same causal link from complementary perspectives, *effect* ← *cause* and *cause* → *effect*. The overriding impression that one is left with is that sweatpants are more than merely indicative of shame and demoralization; viewed through Lagerfeld's gimlet eye, they are one and the same thing.

By drawing our attention to superficial similarities and deep dissimilarities between what is expected and what is real, devices like chiasmus can pack a powerful ironic punch. Yet, though chiasmus offers a convenient vehicle for packing ironic insights into a structurally pleasing form, such devices can often be too easy to use, allowing one to fake the presence of cutting insight with little more than *cut-and-paste*. Consider the following exchange from the 1999 comedy *Mystery Men*, which concerns the misadventures of a group of wannabe superheroes with underwhelming powers. Mr. Furious has anger management issues, while the Sphinx's only power is an ability to torture syntax until it yields an apparent profundity.

> *The Sphinx*: He who questions training, only trains himself in asking questions. [...] Ah yes, work well on your new costumes my friends, for when you care for what is outside, what is inside cares for you. [...] Patience, my son. To summon your power for the conflict to come, you must first have power over that which conflicts you.
>
> *Mr. Furious*: Okay, am I the only one who finds these sayings just a little bit formulaic? "*If you want to push something down, you have to pull it up. If you want to go left, you have to go right.*" It's ...
>
> *The Sphinx*: Your temper is very quick, my friend. But until you learn to master your rage ...
>
> *Mr. Furious*: ...your rage will become your master? That's what you were going to say. Right? Right?
>
> *The Sphinx*: Not necessarily.

That wonderful last line says it all: the Sphinx has hit on a successful formula for *mere generation*, to turn casual utterances into guru-like prognostications. His utterances appear deep, yet they are little more than superficial repetitions with crossover. One can imbue them with real meaning, of course, but it is clear that meaning takes a back seat to surface form in the Sphinx's need to appear wise and all-knowing. We laugh at the Sphinx because his formulaic use of rhetorical devices has made him necessarily predictable. This is the essence of a deterministic formula: it always produces the same outputs for the same inputs, making a weak demurral such as "*Not necessarily*" all the more risible.

Yet all rhetorical devices are formulas of a sort. It is their repeated utility in different contexts that makes them worthy of study by those who want to give their arguments a form that most effectively reflects their meaning. As an orator, the Sphinx is a one-trick pony; he is predictable not because chiasmus is always predictable, but because he is always predictable in his choice of chiasmus. As a device for inverting an opponent's argument, chiasmus has few equals, yet we tire quickly of any device that is used too often and with too little variety. Were the Sphinx to up his game, and use a wider variety of rhetorical forms to better convey an impression of mental agility, we might pay more attention to what he has to say. For the syntactic manipulation of surface forms is actually a reasonable strategy for exploring the world of ideas. Words are often our only handle on subtle feelings and half-formed ideas, and the systematic manipulation of words can be an effective means of navigating the corresponding conceptual spaces (recall Goethe's maxim that "*words are often most useful precisely when ideas fail us*"). By searching for opportunities for chiasmus, the Sphinx is actually employing a simple form of dialectical reasoning. Given a thesis, he fabricates its structural antithesis, and then uses chiasmus to forge a synthesis of the two. The Sphinx is no Hegel, and he is certainly no Kant, but we must assume that he applies some aesthetic and semantic filters to his formulations. For he does not invert everything, but chooses to selectively invert theses whose antitheses appear structurally and semantically sound. A computer that modeled the generative abilities of the Sphinx would almost certainly be accused of mere generation. Yet its creator might validly reply, Sphinx-like, "*Not necessarily.*"

So what might distinguish a computer's best efforts at chiasmus from those of the Sphinx? Well, it would certainly help if it could display an appreciation of the different shades of meaning carried by related forms of the same word-concept. Consider Mae West's chiastic innuendo "*It's not the men in your life that counts, but the life in your men.*" Mae uses deliberate equivocation here, by employing the word "life" in two different senses—"life" as in personal life, and "life" as in zest and vigor. Between these two senses, Mae stakes out a third sense, her *sex "life"*. Equivocation like this is a form of trickery that often produces humor. Consider another humorous example of chiastic equivocation: "*Children in the back seats of cars sometimes cause accidents. Accidents in the back seats of cars sometimes cause children.*" This is more than syntactic manipulation for its own sake. The repeated use of "accidents" in two different senses—car accidents and accidental pregnancies—produces a pithy commentary on life's surprises, and gives the impression that the speaker has peeked behind the curtain of everyday language to glimpse a universal truth. Each of these examples relies on word play, but each also evokes an unspoken meaning that chimes with our experience of the world.

A computer can easily be programmed to scour a large text corpus for reversible chunks of language such as "*hardly working*" and "*working hard*", so as to generate countless examples of chiasmus in the egregious vein of "*working hard or hardly working?*". Yet this would surely be a poor investment of anyone's time. Even a more semantics-savvy generator, one capable of producing the political aphorism "*for society to prosper, prosperity should be socialized*" from the independent text chunks "*society to prosper*", "*prosperity should*" and "*be socialized*", is hardly worth the effort if all it can do is generate one instance of chiasmus after another. As Truman Capote once said of Jack Kerouac on hearing of the latter's frenetic stream-of-consciousness writing method, "*that's not writing, that's typing*." Even if such a system could generate instances of chiasmus of a quality deemed usable by a professional comedian, no professional would ever craft a whole act around a single rhetorical device. Chiasmus, like other conduits for linguistic creativity, should not be viewed as a party-trick. It should not be generated in bulk, nor sold by the yard. Party tricks are the province of chumps like the Sphinx, in whose hands they are glib generators of fakes rather than vehicles of self-expression.

Computer scientists are trained to embrace modularity, so it's tempting to imagine how a standalone chiasmus generator might later unite with generators of other creative forms to yield a comprehensively well-rounded system, such as one for generating poetry on a specific topic. But it is wishful thinking to imagine that a useful standalone generator might come first, rather like believing that five standalone fingers might later glom together to form a working hand. Any act of specific creativity, such as the generation of an apt chiastic form, must serve a larger creative goal that makes it apt—such as the creation of a poem with a specific purpose—and do so within a larger architecture for creativity that harnesses a diversity of knowledge sources. This architecture must coordinate the actions of many different components as they contribute to the same result, and provide a contextual focus for salient feelings, emotions and expectations. We describe such an architecture for poetry generation, one that meaningfully exploits rhetorical forms, in this chapter.

10.2 Shallow Heuristics and Deep Knowledge

It is often said that we "eat with our eyes", so that the stylish presentation of food can subtly influence our sense of taste. So it is with poetry: a pleasing form can do more than enhance our recall and comprehension of meaning—it can also suggest a lasting and profound truth. Experiments reported in McGlone and Tofighbakhsh [14, 15] lend empirical support to this so-called *Keats heuristic*, the intuitive belief—named for John Keats' memorable line "*Beauty is truth, truth beauty*"—that a meaning which is rendered in an aesthetically-pleasing form is much more likely to be perceived as truthful than if it is rendered in a less poetic form. McGlone and Tofighbakhsh demonstrated this effect by searching a book of proverbs for uncommon aphorisms with internal rhyme—such as "*woes unite foes*"—and by using synonym substitution to generate non-rhyming (and thus less poetic) variants such as "*troubles unite enemies*". While no significant differences were observed in subjects' ease of comprehension for rhyming/non-rhyming forms, subjects did show a marked tendency to view the rhyming variants as more truthful expressions of the human condition than the corresponding non-rhyming forms.

So a well-polished poetic form can lend even a modestly interesting observation the lustre of a profound insight. An automated approach to poetry generation can exploit this symbiosis of form and content in a number of useful ways. It might harvest interesting perspectives on a given topic from a text corpus, or it might search its stores of common-sense knowledge for modest insights to render in immodest poetic forms. We describe here a system that combines both of these approaches for meaningful poetry generation.

As shown in the sections to follow, this system—named *Stereotrope*—uses corpus analysis to generate affective metaphors for a topic on which it is asked to wax poetic. *Stereotrope* can be asked to view a topic from a particular affective stance (e.g., to view love negatively) or to elaborate on a familiar metaphor (e.g. love is a prison). In doing so, *Stereotrope* takes account of the feelings that different metaphors are likely to engender in an audience to highlight nuances of a topic that are worthy of poetic expression (see [12]). *Stereotrope* uses a knowledge-base of conceptual norms to anchor its understanding of these metaphors, and though these norms are very much the stuff of banal clichés and stereotypes, such as that *dogs chase cats* and *cops eat donuts*, we show how *Stereotrope* finds and exploits corpus evidence to recast these banalities as witty, incisive and poetic insights.

But *Stereotrope* cannot operate without knowledge. Samuel Johnson famously opined that "*Knowledge is of two kinds. We know a subject ourselves, or we know where we can find information upon it.*" Traditional approaches to the modeling of metaphor and other figurative devices have typically sought to imbue computers with the former (see [8]). More recently, however, the latter kind has gained traction, with the use of the Web and text corpora to source large amounts of shallow knowledge as it is needed (e.g. see [18, 20–22]). But the kind of knowledge demanded by a knowledge-hungry phenomenon such as metaphor is very different to the specialist "book" knowledge so beloved of Johnson. Metaphor demands knowledge of the quotidian world that we all tacitly share but rarely articulate, not even in the thoughtful definitions of Johnson's dictionary.

Fortunately, similes open a rare window onto our shared expectations of the world. Thus, the as-as-similes "*as hot as an oven*", "*as dry as sand*" and "*as tough as leather*" illuminate the expected properties of these objects, while the like-similes "*crying like a baby*", "*singing like an angel*" and "*swearing like a sailor*" reflect intuitions of how these familiar entities are tacitly expected to behave. The authors of [20, 21] thus harvest large numbers of as-as-similes from the Web to build a stereotypical model of familiar ideas and their salient properties, while a similar approach is applied (albeit on a smaller scale) by [16] using Google's query completion service. David Fishelov [11] argues convincingly that poetic and non-poetic similes are crafted from the same words and ideas. Poetic conceits use familiar ideas in non-obvious combinations, often with the aim of creating semantic tension. The simile-based model used here thus harvests almost 10,000 familiar stereotypes (drawing on a stock of almost 8,000 features) from both as-as and like-similes. Poems construct affective conceits, but as shown in [24], the features of a stereotype can be affectively partitioned as needed into distinct pleasant and unpleasant perspectives. We are thus confident that a stereotype-based model of common-sense knowledge is equal to the task of generating and elaborating affective conceits for a poem.

Stereotrope's model of common-sense knowledge requires both features and relations, with the latter showing how stereotypes relate to each other. It is not enough then to know that cops are tough and gritty, or that donuts are sweet and soft; our stereotypes of each should include the cliché that *cops eat donuts*, just as *dogs chew bones* and *cats cough up fur-balls*. Following [22], we acquire inter-stereotype relationships from the Web, not by mining similes but by mining questions. As in [16], we target query completions from a popular search service (Google), which offers a smaller, public proxy for a larger, zealously-guarded search query log. We harvest questions of the form "*Why do Xs* < relation > *Ys*", and assume that since each relationship is presupposed by the question (so "*Why do bikers wear leathers*" presupposes that everyone knows that bikers wear leathers), the triple of subject/relation/object captures a widely-held norm. In this way we harvest over 40,000 such norms from the Web.

10.3 Generating Metaphors that are Affective and Effective

The Google n-grams [4] is a rich source of popular metaphors of the form *Target is Source*, such as "*politicians are crooks*", "*Apple is a cult*", "*racism is a disease*" and "*Steve Jobs is a god*". Let $src(T)$ denote the set of stereotypes commonly used to describe a topic T, where commonality is defined as the presence of the corresponding metaphor in the Google n-grams. To find metaphors for proper-named entities, we also analyze 3/4-grams of the form *stereotype First [Middle] Last*, such as "*tyrant Adolf Hitler*" and "*boss Bill Gates*". Thus, e.g.:

$$src(racism) = \{problem, disease, joke, sin, poison, crime, ideology, weapon\}$$

$$src(Hitler) = \{monster, criminal, tyrant, idiot, madman, vegetarian, racist, \ldots\}$$

Let $typical(T)$ denote the set of properties and behaviors harvested for T from Web similes (see previous section), and let $srcTypical(T)$ denote the aggregate set of properties and behaviors ascribable to T via the metaphors in $src(T)$. Thus:

$$(1) \qquad srcTypical(T) = \bigcup_{M \in src(T)} typical(M)$$

We can generate conceits for a topic T by considering not just obvious metaphors for T, but *metaphors of metaphors*. Thus:

$$(2) \qquad conceits(T) = src(T) \cup \bigcup_{M \in src(T)} src(M)$$

The features evoked by the conceit T *as* M are then given by:

$$(3) \qquad salient(T, M) = [srcTypical(T) \cup typical(T)] \cap [srcTypical(M) \cup typical(M)]$$

The degree to which a conceit M is apt for T is given by:

$$(4) \qquad aptness(T, M) = \frac{|salient(T, M) \cap typical(M)|}{|typical(M)|}$$

However, we should focus only on apt conceits $M \in conceits(T)$ where:

$$(5) \qquad apt(T, M) = |salient(T, S) \cap typical(M)| > 0$$

and rank the set of apt conceits by *aptness*(T, M), as given in (4).

The set *salient*(T,M) identifies the properties and behaviors that are evoked and projected onto T when T is viewed through the metaphoric lens of M. For affective conceits, this set can be partitioned on demand to highlight only the unpleasant aspects of the conceit (e.g. "you are such a baby!") or only the pleasant aspects (e.g. "you are my baby!"). The authors of [22] further show how n-gram evidence can be used to selectively project the salient norms of M onto T.

10.4 Once More, with *Feeling!*

As shown in [24], it is a simple matter to filter a set of stereotypes by affect, to reliably identify the metaphors that impart a mostly positive or negative "spin". But poems are emotion-stirring texts that exploit much more than a crude two-tone polarity. A system like *Stereotrope* must also model the emotions that a metaphorical conceit will stir in a reader. Yet before *Stereotrope* can appreciate the emotions stirred by the properties of a poetic conceit, it must model how properties reinforce and imply each other.

A stereotype is a simplified but coherent representation of a complex real-world phenomenon. So we cannot simply model stereotypes as mere sets of discrete properties—we must also model how these properties cohere with each other. For example, the property *lush* suggests the properties *green* and *fertile*, while *green* suggests *new* and *fresh*. Let $cohere(p)$ denote the set of properties that suggest and reinforce *p*-ness in a stereotype-based description. Thus e.g. $cohere(lush) = \{green, fertile, humid, \ldots\}$ and $cohere(hot) = \{humid, spicy, sultry, arid, \ldots\}$. The set of properties that coherently reinforce another property is easily acquired through corpus analysis—we need only look for similes where multiple properties are ascribed to a single topic, as in e.g. "*as hot and humid as a jungle*." To this end, *Stereotrope* trawls the Web for instances of the pattern "*as X and Y as*", and assumes for each *X* and *Y* pair that $Y \in cohere(X)$ and $X \in cohere(Y)$.

Many properties have an emotional resonance, though some evoke more obvious feelings than others. The linguistic mapping from properties to feelings is also more transparent for some property / feeling pairs than others. Consider the property *appalling*, which is stereotypical of *tyrants*: the common linguistic usage "*feel appalled by*" suggests that an entity with this property is quite likely to make us "*feel appalled.*" Corpus analysis allows a system to learn a mapping from properties to feelings for these obvious cases, by mining instances of the n-gram pattern "*feel P+ed by*" where P can be mapped to the property of a stereotype via a simple morphology rule. Let $feeling(p)$ denote the set of feelings that is learnt in this way for the property p. Thus, $feeling(disgusting) = \{feel\_disgusted\_by\}$ while $feeling(humid) = \{\}$. Naturally, because this approach can only find obvious mappings, $feeling(p) = \{\}$ for most p.

However, $cohere(p)$ can be used to interpolate a range of feelings for almost any property p. Let $evoke(p)$ denote the set of feelings that are likely to be stirred by a property p. We can now interpolate $evoke(p)$ as follows:

$$(6) \qquad evoke(p) = feeling(p) \; \cup \bigcup_{c \in cohere(p)} feeling(c)$$

So a property p is likely to evoke a feeling f in an audience if p suggests another property c that is known to evoke f. We can predict the range of emotional responses to a stereotype S in the same way:

$$(7) \qquad evoke(S) = \bigcup_{p \in typical(S)} evoke(p)$$

If M is chosen from $conceits(T)$ to metaphorically describe T, then the metaphor M is likely to evoke these feelings for T:

$$(8) \quad evoke(T, M) = \bigcup_{p \in salient(T,M)} evoke(p)$$

For purposes of gradation, $evoke(p)$ and $evoke(S)$ each denote a bag of feelings rather than a set of feelings. Thus, the more properties of S that evoke f, the more times that $evoke(S)$ will contain f, and the more likely it is that the use of S as a conceit will stir the feeling f in the reader. *Stereotrope* can thus predict that both *feel_disgusted_by* and *feel_thrilled_by* are two possible emotional responses to the property *bloody* (or to the stereotype *war*), but will also appreciate that the former is by far the more likely response of the two.

The set $evoke(T, M)$ for the metaphorical conceit *T is M* can serve the goal of poetry generation in different ways. Most obviously, it is a rich source of feelings that can be explicitly mentioned in a poem about T (as viewed thru the lens of M). Alternately, these feelings can be used in a meta-text to motivate and explain the viewpoint of the poem. The act of crafting an explanatory text to showcase a poetry system's creative intent is dubbed "*framing*" in [6]. *Stereotrope* puts the contents of $evoke(T, M)$ to both of these uses: in the poem itself, it expresses feelings as a reaction to the metaphorical properties of T; and in an accompanying framing text, it cites these feelings as a reason for choosing the conceit *T is M*. For example, in a poem based on the conceit *marriage is a prison*, the set $evoke(marriage, prison)$ contains the feelings *bored_by, confined_in, oppressed_by, chilled_by* and *intimidated_by*. The meta-text that frames the poem expresses the following feelings (generated using simple natural-language generation schemas):

> Gruesome marriage and its depressing divorces appall me. I often feel disturbed and shocked by marriage and its twisted rings. Does marriage revolt you?

10.5 Bridging Worlds with Phrasal Blends

If linguistic creativity is chemistry with words and ideas, then stereotypes and their typical properties constitute the periodic table of elements that novel reactions are made of. These are the descriptive atoms that poems combine into metaphorical mixtures, as modeled in (1) ...(8) above. But poems can also fuse these atoms into nuanced compounds that may subtly suggest more than the sum of their parts.

Consider the poetry-friendly concept *moon*, for which Web similes provide the following descriptive atoms:

> typical(moon) = {ambent, white, round, pockmarked, shimmering, airless, silver, bulging, cratered, waning, waxing, spooky, eerie, pale, pallid, deserted, glowing, pretty, shining, expressionless, rising}

Corpus analysis reveals that authors combine atoms such as these in a wide range of resonant compounds. Thus, the Google 2-grams contain such compounds as "*pallid glow*", "*lambent beauty*", "*silver shine*" and "*eerie brightness*", all of which can be used to good effect in a poem about the moon. Each compound denotes a compound property, and each exhibits the same linguistic structure. So to harvest a very large number of compound properties, we simply scan the Google 2-grams for phrases of the form "*ADJ NOUN*", where ADJ and NOUN must each denote a property of the same stereotype. While ADJ maps directly to a property, a combination of morphological analysis and dictionary search is needed to map NOUN to its property (e.g. *beauty* → *beautiful*). What results is a large poetic lexicon, one that captures the diverse and sometimes unexpected ways in which the atomic properties of a stereotype can be fused into nuanced carriers of meaning. Compound descriptions denote compound properties, and those that are shared by different stereotypes reflect the poetic ways in which those concepts are alike. For example, "*shining beauty*" is shared by over 20 stereotypes in our poetic lexicon, describing such entries as *moon, star, pearl, smile, goddess* and *sky*.

A stereotype suggests behaviors as well as properties, and a fusion of both perspectives can yield a more nuanced view. The patterns "*VERB ADV*" and "*ADV VERB*" are used to harvest all 2-grams where a property expressed as an adverb qualifies a related property expressed as a verb. For example, the Google 2-gram "*glow palely*" unites the properties *glowing* and *pale* of *moon*, which allows *moon* to be recognized as similar to *candle* and *ghost* because they too can be described by the compound "*glow palely*". A *ghost*, in turn, can *noiselessly glide*, as can a *butterfly*, which may *sparkle radiantly* like a *candle* or a *star* or a *sunbeam*. Not every pairing of descriptive atoms will yield a meaningful compound, and it takes common-sense—or a poetic imagination—to sense which pairings will work in a poem. Though an automatic poet is endowed with neither, it can still re-use the many valid combinations that humans have added to the language trove of the Web.

Poetic allusions anchor a phrase in a vivid stereotype while shrouding its meaning in constructive ambiguity. Why talk of the *pale glow* of the moon when you can allude to its *ghostly glow* instead? The latter does more than evoke the moon's paleness—it attributes this paleness to a supernatural root, and suggests a halo of other qualities such as *haunting, spooky, chilling* and *sinister*. Stereotypes are dense descriptors, and the use of one to convey a single property like *pale* will subtly suggest other readings and resonances. The phrase "*ghostly glow*" may thus allude to any corpus-attested compound property that can be forged from the property *glowing* and any other element of the set $typical(ghost)$. Many stereotype nouns have adjectival forms—such as *ghostly* for *ghost*, *freakish* for *freak*, *inky* for *ink*—and these may be used in corpora to qualify the nominal form of a property of that very stereotype, such as *gloom* for *gloomy*, *silence* for *silent*, or *pallor* for *pale*. The 2-gram "*inky gloom*" can thus be understood as an allusion either to the blackness or wetness of ink, so any stereotype that combines the properties *dark* and *wet* (e.g. *oil, swamp, winter*) or *dark* and *black* (e.g. *crypt, cave, midnight*) can be poetically described as exhibiting an *inky gloom*.

Let $compounds(\ldots)$ denote a function that maps a set of atomic properties such as *shining* and *beautiful* to the set of compound descriptors—such as the compound property *shining beauty* or the compound allusion *ghostly glow*—that can be harvested from the Google 2-grams. It follows that $compounds(typical(S))$ denotes the set of corpus-attested compounds that can describe a stereotype S, while $compounds(salient(T, M))$ denotes the set of compound descriptors that might be used in a poem about T to suggest the poetic conceit *T is M*. Since these compounds will fuse atomic elements from the stereotypical representations of both *T and M*, $compounds(salient(T, M))$ can be said to sample from the blend of *T and M*. As described in [7], and computationally modeled in various ways in [17, 19, 22], a "*blend*" is a tight conceptual integration of two or more mental spaces. This integration yields more than a mixture of representational atoms: a conceptual blend often creates emergent elements—new molecules of meaning—that are present in neither of the input representations but which only arise from this fusion of inputs.

How might the representations discussed here give rise to emergent elements? We cannot expect new descriptive atoms to be created by a poetic blend, but we can expect new compounds to emerge from the re-combination of descriptive atoms in the compound descriptors of T and M. Just as we can expect the set of stereotypes $compounds(typical(T) \cup typical(M))$ to suggest a wider range of descriptive possibilities than $compounds(typical(T)) \cup compounds(typical(M))$, the emergent compound descriptions that arise from the blend of T and M are those that could not have emerged from the properties of T alone, or from M alone, but could only emerge from the fusion of T and M together. Thus,

(9) $$emergent(T, M) = compounds(salient(T, M)) \backslash compounds(typical(T)) \cup compounds(typical(M))$$

Consider the poetic conceit *love is the grave*. The resulting blend—as captured by $compounds(salient(T, M))$—contains a wide variety of compound descriptors. Some of these compounds emerge solely from the concept *grave*, such as *sacred gloom, dreary chill* and *blessed stillness*. Many others emerge only from a fusion of *love* and *grave*, such as *romantic stillness*, *sweet silence*, *tender darkness*, *cold embrace*, *quiet passion* and *consecrated devotion*. So a poem that uses these phrases to construct an emotional worldview will not only demonstrate an understanding of its topic and its conceit, but will also demonstrate some measure of insight into how one can complement and resonate with the other (e.g., that darkness can be tender, passion can be quiet and silence can be sweet). While the system builds on second-hand insights, insofar as these are ultimately derived from Web corpora, such insights are fragmentary and low-level. It still falls to the system to stitch these into its own emotionally coherent patchwork of poetry. What use is poetry if we or our machines cannot learn from it the wild possibilities of language and life?

10.6 The Keatsian Lathe: Re-Shaping the Banal as the Poetic

Insight requires depth. To derive original insights about the topic of a poem, of a kind an unbiased audience might consider witty or clever, a system needs more than shallow corpus data; it needs deep knowledge of the real world. It is perhaps ironic then that the last place one is likely to find real insight is in the riches of a structured knowledge base. Common-sense knowledge-bases are especially lacking in insight, since these are designed to contain knowledge that is common to all and questioned by none. Even domain-specific knowledge-bases, rich in specialist knowledge, are designed as repositories of axiomatic truths that will appear self-evident to their intended audience of experts.

Insight is both a process and a product. While insight undoubtedly requires knowledge, it also takes work to craft surprising insights from the unsurprising generalizations that make up the bulk of our conventional knowledge. Though mathematicians occasionally derive surprising theorems from the application of deductive techniques to self-evident axioms, sound reasoning over unsurprising facts will rarely yield surprising conclusions. Yet witty insights are not typically the product of an entirely sound reasoning process. Rather, such insights amuse and provoke via a combination of over-statement, selective use of facts, a mixing of distinct knowledge types, and a clever packaging that makes maximal use of the Keats heuristic. Indeed, as has long been understood by humor theorists, the logic of humorous insight is deeply bound up with the act of framing. The logical mechanism of a joke—a kind of pseudo-logical syllogism for producing humorous effects—is responsible for framing a situation in such a way that it gives rise to an unexpected but meaningful incongruity (see [2, 3]). To craft witty insights from innocuous generalities, a system must draw on an arsenal of logical mechanisms to frame its observations of the world in appealingly discordant ways.

Attardo and Raskin (see [2, 3]) view the role of a logical mechanism (LM) as the engine of a joke: each LM provides a different way of bringing together two overlapping scripts that are mutually opposed in some pivotal way. A joke narrative is fully compatible with one of these scripts and only partly compatible with the other, yet it is the partial match that we, as listeners, jump to first to understand the narrative. In a well-structured joke, we only recognize the inadequacy of this partially-apt script when we reach the punchline, at which point we switch our focus to its unlikely alternative. The realization that we can easily be duped by appearances, combined with the sense of relief and understanding that this realization can bring, results in the AHA! feeling of insight that often accompanies the HA-HA of a good joke. LMs suited to narrative jokes tend to engineer oppositions between narrative scripts, but for purposes of crafting witty insights in one-line poetic forms, we will view a script as a stereotypical representation of an entity or event. Armed with an arsenal of stereotype "*scripts*", *Stereotrope* seeks to highlight the tacit opposition between different stereotypes as they typically relate to each other, while also engineering credible oppositions based on corpus evidence.

A sound logical system cannot brook contradictions. Nonetheless, uncontroversial views can be cleverly framed in such a way that they appear sound and contradictory, as when the columnist David Brooks described the Olympics as a "*peaceful celebration of our warlike nature*." His form has symmetry and cadence, and pithily exploits the Keats heuristic to reconcile two polar opposites, war and peace. Poetic insights do not aim to create real contradictions, but aim to reveal (and reconcile) the unspoken tensions in familiar ideas and relationships. We have discussed two kinds of stereotypical knowledge in this chapter: the property view of a stereotype S, as captured in $typical(S)$, and the relational view, as captured by a set of question-derived generalizations of the form $Xs < \underline{relation} > Ys$. A blend of both of these sources of knowledge can yield emergent oppositions that are not apparent in either alone.

Consider the normative relation *bows fire arrows*. Bows are stereotypically *curved*, while arrows are stereotypically *straight*, so lurking beneath the surface of this innocuous norm is a semantic opposition that can be foregrounded to poetic effect. The Keats heuristic can be used to package this opposition in a pithy and thought-provoking form: compare "*curved bows fire straight arrows*" (so what?) with "*straight arrows do curved bows fire*" (more poetic) and "*the most curved bows fire the straightest arrows*" (most poetic). While this last form is an overly strong claim that is not strictly supported by the stereotype model, it has the sweeping form of a penetrating insight that grabs one's attention. Its pragmatic effect—a key function of poetic insight—is to reconcile two opposites by suggesting that they fill complementary roles. In schematic terms, such insights can be derived from any single norm of the form $Xs < \underline{relation} > Ys$ where X and Y denote stereotypes with salient properties—such as *soft* and *tough*, *long* and *short*—that can be framed in striking opposition. For instance, the combination of the norm *cops eat donuts* with the cliched views of cops as *tough* and donuts as *soft* yields the insight "*the toughest cops eat the softest donuts*." As the property *tough* is undermined by the property *soft*, this may be viewed as a playful subversion of the tough cop stereotype. The property *toughness* can be further subverted, with an added suggestion of hypocrisy, by expressing the generalization as a rhetorical question: "*Why do the toughest cops eat the softest donuts?*"

A single norm represents a highly simplified script, so a framing of two norms together often allows for opposition via a conflict of overlapping scripts. Activists, for example, typically engage in tense struggles to achieve their goals. But activists are also known for the slogans they coin and the chants they sing. Most slogans, whether designed to change the law or sell detergent, are catchy and uplifting. These properties and norms can now be framed in poetic opposition: "*Activists that chant the most uplifting slogans suffer through the most depressing struggles*". While the number of insights derivable from single norms is a linear function of the size of the knowledge base, a combinatorial opportunity exists to craft insights from pairs of norms. Thus, "*angels who fight the foulest demons play the sweetest harps*", "*surgeons who wield the most hardened blades wear the softest gloves*", and "*celebrities who promote the most reputable charities suffer the sleaziest scandals*" all achieve conflict through norm juxtaposition. Moreover, the order of a juxtaposition—positive before negative or vice versa—can also sway an audience toward a cynical or an optimistic interpretation.

Wit portrays opposition as an inherent part of reality, yet often creates the oppositions that it appears to reconcile. It does so by elevating specifics into generalities, to suggest that opposition is the norm rather than the exception. So rather than rely wholly on stereotypes and their expected properties, *Stereotrope* uses corpus evidence as a proxy imagination to concoct new classes of individuals with interesting and opposable qualities. Consider the Google 2-gram "*short celebrities*", whose frequency and plurality suggests that shortness is a noteworthy (if not typical) property of a significant class of celebrities. *Stereotrope* already possesses the norm that "*celebrities ride in limousines*", as well as a stereotypical expectation that limousines are long. This juxtaposition of conventions allows it to frame a provocatively sweeping generalization as a rhetorical question: "*Why do the shortest celebrities ride in the longest limousines?*". While *Stereotrope* has no evidence for this speculative claim, and no real insight into the status-anxiety of the rich but vertically-challenged, such an understanding may follow in time, as deeper and subtler knowledge-bases become available for poetry generation.

Poetic insight often takes the form of sweeping claims that elevate vivid cases into powerful exemplars. Consider how *Stereotrope* uses a mix of n-gram evidence and norms to generate these maxims: "*The most curious scientists achieve the most notable breakthroughs*" and "*The most impartial scientists use the most accurate instruments*". The causal seeds of these insights are mined from the Google n-grams in coordinations such as "*hardest and sharpest*" and "*most curious and most notable*". These n-gram relationships are then projected onto banal norms—such as *scientists achieve breakthroughs* and *scientists use instruments*—for whose participants these properties are stereotypical (e.g. scientists are curious and impartial, instruments are accurate, breakthroughs are notable, etc.).

Such claims can be taken literally, or viewed as vivid allusions to important causal relationships. Indeed, when framed as explicit analogies, the juxtaposition of two such insights can yield unexpected resonances. For example, "*the most trusted* celebrities ride in the *longest limousines*" and "*the most trusted preachers give the longest sermons*" are both inspired by the 4-gram "*most trusted and longest.*" This common allusion also suggests an analogy: "*Just as the most trusted celebrities ride in the longest limousines, the most trusted preachers give the longest sermons*". Though such analogies are driven by superficial similarity, they can still evoke deep resonances for an audience. Perhaps a sermon is a vehicle for a preacher's ego, just as a limousine is an obvious vehicle for a celebrity's pride? Reversing the order of the analogy significantly alters its larger import, suggesting that ostentatious wealth bears a lesson for us all.

10.7 Tying It All Together: Poetry as a Service

Having created the individual pieces of form and meaning from which a poem might be crafted, it now falls to us to put the pieces together as a coherent service. To recap, we have shown how affective metaphors are generated for a given topic, by building on popular metaphors for that topic in the Google n-grams; shown how the feelings evoked by these properties may be anticipated by a system; and shown how novel insights can be crafted from a fusion of stereotypical norms and corpus evidence.

We view a poem as a summarization and visualization device that samples the set of properties and feelings that are evoked when a topic T is viewed thru the lens of M. Given T, an apt M is chosen randomly from $conceits(T)$. Each line of the text renders one or more properties in poetic form, using tropes such as simile and hyperbolae. So, for $T = love$ and $M = fire$, since $salient(T, M)$ contains *hot* and the Google n-grams contains the 2-gram "*burn brightly*", this mix of elements may be rendered as "*No fire is hotter or burns more brightly*". It can also be rendered with the imperative "*Burn brightly with your hot love*", or the plea "*Let your hot love burn brightly*". The range of tropes is best conveyed with examples, such as this poetic view of *marriage as a prison*:

The legalized regime of this marriage

My marriage is an emotional prison
Barred visitors do marriages allow
The most unitary collective scarcely organizes so much
Intimidate me with the official regulation of your prison
Let your sexual degradation charm me
Did ever an offender go to a more oppressive prison?
You confine me as securely as any locked prison cell
Does any prison punish more harshly than this marriage?
You punish me with your harsh security
The most isolated prisons inflict the most difficult hardships
O Marriage, you disgust me with your undesirable security

Each poem obeys a semantic grammar, which minimally indicates the trope that should be used for each line. Since the second-line of the grammar asks for an apt $< \underline{simile} >$, *Stereotrope* constructs one by comparing marriage to a collective; as the second-last line asks for an apt $< \underline{insight} >$, one is duly constructed around the Google 4-gram "*most isolated and most difficult*". The grammar may also dictate whether a line is rendered as an assertion, an imperative, a request or a question, and whether it is framed positively or negatively. This grammar need not be a limiting factor, as one can choose randomly from a pool of grammars, or even evolve a new grammar by soliciting user feedback. The key point is the pivotal role of a grammar of tropes in mapping from the properties and feelings of a metaphor interpretation to a sequence of poetic renderings of these elements.

Consider this poem, elaborated around the metaphor *China is a rival*:

No Rival Is More Bitterly Determined

Inspire me with your determined battle
The most dogged defender scarcely struggles so much
Stir me with your spirited challenge
Let your competitive threat reward me
Was ever a treaty negotiated by a more competitive rival?
You compete with me like a competitively determined athlete
Does any rival test more competitively than this China?
You oppose me with your bitter battle
Can a bitter rival suffer from such sweet jealousies?
O China, you oppress me with your hated fighting

Stereotypes are most eye-catching when subverted, as in the second-last line above. The Google 2-gram "*sweet jealousies*" catches *Stereotrope*'s eye (and ours) because it up-ends the belief that jealousy is a bitter emotion. This subversion complements the stereotype that rivals are bitter, allowing *Stereotrope* to impose a thought-provoking opposition onto the banal norm *rivals suffer from jealousy*.

Stereotype emphasizes meaning and intent over sound and form, and does not (yet) choose lines for their rhyme or metre. However, given a choice of renderings, it does choose the form that makes best use of the Keats heuristic, by favoring lines with alliteration and internal symmetry.

10.8 Quality Considerations

Stereotrope is a knowledge-based approach to poetry, one that crucially relies on three sources of inspiration: a large roster of stereotypes, which maps a slew of familiar ideas to their most salient properties; a large body of normative relationships which relate these stereotypes to each other; and the Google n-grams, a vast body of language snippets. The first two are derived from attested language use on the Web, while the third is a reduced view of the linguistic Web itself. *Stereotrope* represents approx. 10,000 stereotypes in terms of approx. 75,000 stereotype-to-property mappings, where each of these is supported by a real Web simile that attests to the accepted salience of a given property. In addition, *Stereotrope* represents over 50,000 norms, each derived from a presupposition-laden question on the Web.

The reliability of *Stereotrope*'s knowledge has been demonstrated in recent studies. For instance, [23] shows that *Stereotrope*'s simile-derived representations are balanced and unbiased, as the positive/negative affect of a stereotype T can be reliably estimated as a function of the affect of the contents of $typical(T)$. In addition, [24] further shows that $typical(T)$ can be reliably partitioned into sets of positive or negative properties as needed, to reflect an affective "*spin*" imposed by any given metaphor M. Moreover, [23] also shows that copula metaphors of the form *T is an M* in the Google n-grams—the origins of $srcTypical(T)$—are broadly consistent with the properties and affective profile of each stereotype T. In 87 % of cases, one can correctly assign the label positive or negative to a topic T using only the contents of $srcTypical(T)$, provided it is not empty.

Stereotrope derives its appreciation of feelings from its understanding of how one property presupposes another. The intuition that two properties X and Y linked via the pattern "*as X and Y as*" evoke similar feelings is supported by the strong correlation (0.7) observed between the positivity of X and of Y over the many X/Y pairs that are harvested from the Web using this acquisition pattern.

The "*fact*" that bats lay eggs can be found over 40,000 times on the web via Google. On closer examination, dubious matches often form part of a larger question such as "*do bats lay eggs?*", while the question "*why do bats lay eggs?*" has zero matches. So "*why do*" questions provide an effective superstructure for acquiring normative facts from the Web: they identify facts that are commonly presupposed, and thus stereotypical, and clearly mark the start and end of each presupposition. Such questions also yield useful facts: the authors of [22] show that when these facts are treated as features of the stereotypes for which they are presupposed, they provide an excellent basis for classifying different stereotypes into the same ontological categories, as would be predicted by an ontology such as *WordNet* [9]. Moreover, these features can be reliably distributed to close semantic neighbors to overcome the problem of knowledge sparsity. The authors of [22] also demonstrate that the likelihood that a feature of stereotype A can also be assumed of stereotype B is a clear function of the WordNet similarity of A and B. While this is an intuitive finding, it would not hold at all if not for the fact that these features are truly meaningful for A and for B.

The problem posed by "*bats lay eggs*" is one faced by any system that does not perceive the whole context of an utterance. As such, it is a problem that plagues the use of n-gram models of Web content, such as Google's n-grams. *Stereotrope* uses n-grams to suggest insightful connections between two properties or ideas, but if many of these n-grams are mere noise, not even the Keats heuristic can disguise them as meaningful signals. Our focus is on relational n-grams, of a kind that suggests deep albeit tacit relationships between two concepts. These n-grams obey the pattern $X < \underline{relation} > Y$, where X and Y are adjectives or nouns and $< \underline{relation} >$ is a linking phrase, such as a verb, a preposition, a coordinator, etc. To determine the quality of these n-grams, and to assess the likelihood of extracting genuine relational insights from them, we use this large subset of the Google n-grams as a corpus for estimating the relational similarity of the 353 word pairs in the *WordSim-353* data set [10]. We estimate the relatedness of two words X and Y as the PMI (pointwise mutual information score) of X and Y, using the relational n-grams as a corpus for occurrence and co-occurrence frequencies of X and Y. A correlation of 0.61 is observed between these PMI scores and the human ratings reported in [10]. Though this is not the highest score achieved for this task, it is considerably higher than any that has been reported for approaches that use WordNet alone. The point here is that this relational subset of the Google n-grams offers a reasonably faithful mirror of human intuitions for purposes of recognizing the relatedness of different ideas. We thus believe these n-grams to be a valid source of real insights.

The final arbiters of *Stereotrope*'s poetic insights are the humans who use the system. We offer the functionality of *Stereotrope* in the guise of a public Web service, via this URL: http://boundinanutshell.com/metaphor-magnet-acl

We expect these services will also allow other researchers to reuse and extend *Stereotrope*'s approaches to metaphor, blending and poetry. Thus, for instance, poetry generators such as that described in [6]—which creates topical poems from fragments of newspapers and tweets—can use *Stereotrope*'s rich inventories of similes, poetic compounds, feelings and allusions in its poetry.

References

1. Aristotle.: The Art of Rhetoric (translated and edited by Hugh Tancred-Jones). Penguin Classics (1991)
2. Attardo, S., Raskin, V.: Script theory revis(it)ed: joke similarity and joke representational model. Humor: Int. J. Humor Res. **4**(3), 293–347 (1991)
3. Attardo, S., Hempelmann, C.F., Di Maio, S.: Script oppositions and logical mechanisms: modeling incongruities and their resolutions. Humor: Int. J. Humor Res. **15**(1), 3–46 (2002)
4. Brants, T., Franz, A.: Web 1T 5-gram Version 1. Linguistic Data Consortium (2006)
5. Chandler, R.: The Simple Art of Murder. Pearls Are a Nuisance, Hamish Hamilton (1953)
6. Colton, S., Goodwin, J., Veale, T.: Full-FACE poetry generation. In: Proceedings of ICCC 2012, the 3rd International Conference on Computational Creativity, Dublin, Ireland (2012)
7. Fauconnier, G., Turner, M.: The Way We Think. Basic Books, Conceptual Blending and the Mind's Hidden Complexities (2002)
8. Fass, D.: Processing Metonymy and Metaphor. Contemporary Studies in Cognitive Science and Technology. Ablex, New York (1997)
9. Fellbaum, C. (ed.): WordNet: An Electronic Lexical Database. MIT Press, Cambridge (1998)
10. Finkelstein, L., Gabrilovich, E., Matias, Y., Rivlin, E., Solan, Z., Wolfman, G., Ruppin, E.: Placing search in context: the concept revisited. ACM Trans. Inf. Syst. **20**(1), 116–131 (2002)
11. Fishelov, D.: Poetic and non-poetic simile: structure, semantics. Rhetoric. Poetics Today **14**(1), 1–23 (1992)
12. Lakoff, G., Turner, M.: More than Cool Reason: A Field Guide to Poetic Metaphor. University of Chicago Press (1989)
13. Leith, S.: You Talkin' to Me? Profile Books, Rhetoric from Aristotle to Obama (2012)
14. McGlone, M.S., Tofighbakhsh, J.: The Keats heuristic: rhyme as reason in aphorism interpretation. Poetics **26**(4), 235–244 (1999)
15. McGlone, M.S., Tofighbakhsh, J.: Birds of a feather flock conjointly (?): rhyme as reason in aphorisms. Psychol. Sci. **11**(5), 424–428 (2000)
16. Özbal, G., Strapparava, C.: A computational approach to automatize creative naming. In: Proceedings of the 50th Annual Meeting of the Association of Computational Linguistics, Jeju, South Korea (2012)
17. Pereira, F.C.: Creativity and Artificial Intelligence: A Conceptual Blending Approach. Walter de Gruyter (2007)
18. Shutova, E.: Metaphor identification using verb and noun clustering. In: Proceedings of the 23rd International Conference on Computational Linguistics, pp. 1001–1010 (2010)
19. Veale, T., O'Donoghue, D.: Computation and blending. Cognit. Linguist. **11**(3–4), 253–281 (2000)

20. Veale, T., Hao, Y.: Making lexical ontologies functional and context-sensitive. In: Proceedings of the 46th Annual Meeting of Association of Computational Linguistics, Prague. Czech Republic (2007)
21. Veale T., Hao, Y.: Comprehending and generating apt metaphors: a web-driven, case-based approach to figurative language. In: Proceedings of AAAI-2007, the 22nd National Conference on Artificial Intelligence, pp. 1471–1476 (2007)
22. Veale, T., Li, G.: Creative introspection and knowledge acquisition: learning about the world thru introspective questions and exploratory metaphors. In: Proceedings of AAAI-2011, the 25th Conference of the Association for the Advancement of AI, San Francisco (2011)
23. Veale, T.: Exploding the Creativity Myth: Computational Foundations of Linguistic Creativity. Bloomsbury, London (2012)
24. Veale, T.: A Context-sensitive, multi-faceted model of lexico-conceptual affect. In: Proceedings of the 50th Annual Meeting of the Association of Computational Linguistics, Jeju, South Korea (2012)

Chapter 11
Open-Ended Elaborations in Creative Metaphor

John Barnden

Abstract This chapter looks at how a hearer or reader can understand metaphorical discourse, particularly when it is creative. The abilities of the ATT-Meta approach to creative metaphor understanding, implemented in an AI reasoning system, are discussed. A natural extension to metaphor generation is also briefly mentioned. As regards types of creativity, the chapter focusses on how ATT-Meta deals with open-ended elaborations (extensions) of familiar metaphorical views, although novel pairings of concepts in metaphor are also briefly discussed. A central feature of the approach is that it takes a (non-conventional) metaphorical utterance to imply a fictional or pretence scenario within which the elaboration takes place. The approach takes metaphorical mappings to act between fictions and reality (or other spaces outside the fiction), rather than between differentiated source and target domains as is usual in mapping-based metaphor theory. Another central feature is the adherence to an anti-analogy-extension thesis, which seeks to inhibit the extension of existing analogies to deal with elaborations. The approach lends itself naturally to the idea popular in Cognitive Linguistics that metaphor arises in thought as opposed to communication particularly. But the approach goes further in supporting a dramatic and disruptive version of this suggestion: namely that our thoughts can be intrinsically and creatively metaphorical in a way that cannot wholly be translated into non-metaphorical thoughts. This suggestion arises out of the anti-analogy-extension thesis: while analogy is involved in metaphor, and novel analogies are important in creativity, there is also a strong non-analogical side to creative metaphor.

J. Barnden (✉)
School of Computer Science, University of Birmingham, Birmingham, UK
e-mail: j.a.barnden@cs.bham.ac.uk

T.R. Besold et al. (eds.), *Computational Creativity Research: Towards Creative Machines*,
Atlantis Thinking Machines 7, DOI 10.2991/978-94-6239-085-0_11

11.1 Introduction

In metaphor in linguistic discourse, one subject matter is talked about in a way that can be analysed as using the resources of another subject matter. Metaphor often involves talking about aspects of the former subject matter as if they were aspects of the latter. For instance, the statement "The idea is buried deep in his mind" can be analysed as talking about the person's mind as if it were some sort of physical terrain in which things can be buried, and talking about the idea as if it were a physical object that can be buried. The person's mind (and its states, ideas, etc.) is the *target* subject matter and the physical objects and region constitute the *source* subject matter. We can also say the discourse is using a metaphorical view of the person's mind as a physical terrain and a metaphorical view of the person's ideas as physical objects. In such a view, target and source subject matters are paired with each other.

In the type of metaphor just discussed, an implicit similarity is assumed or created. But I also include simile, as in "The thought [acted] like a dark angry cloud",[1] as a case of metaphor. Here a likeness is explicitly stated.

What I call a metaphorical view is broadly similar to the much-researched notion of a conceptual metaphor [30], but I use a different, more intuitive term to avoid any particular alignment with particular theories of what conceptual metaphors are or how they work. Metaphorical views can in principle have any degree of familiarity for a particular understander. Competent speakers of English will have great familiarity with views of mental states as physical states (involving terrain, containers, etc), because of the prevalent use of such views in English discourse—see, e.g., Jäkel [28], and the metaphor databank developed in the ATT-Meta project.[2] On the other hand, the sentences "The writing desk is a penguin" or (relatedly) "The writing desk flapped its wings and attacked the bookcase" introduce or rest upon a metaphorical view that is likely to be novel to most hearers.[3]

To most people outside the academic discipline of metaphor, creative metaphor metaphor is probably a matter of an interesting, novel pairing of two subject matters. A good example, in Semino [40], is the metaphorical view of a migraine attack as being a matter of an animal moving about in the person's head (taken from Ian McEwan's novel *Atonement*). Another example, in Shakespeare's *Hamlet* (Act I Scene v), is a metaphor of Hamlet's father's ghost as a "mole" moving about beneath the earth. Those two examples are from literary sources, and that is no accident, as it has been claimed that in non-literary language such novel pairings are rare (see, e.g., Cameron [13], Nacey [38]). But metaphor is important and widespread

---

[1] Jolly, S. *Marigold Becomes a Brownie,* p. 44. London, U.K.: Blackie and Son—The Anytime Series (no date).

[2] http://www.cs.bham.ac.uk/~jab/ATT-Meta/Databank/.

[3] These sentences are invented variants of the famous literary example "Why is a raven like a writing desk?" from *Alice's Adventures in Wonderland*. They are novel. A search of the *Corpus of Global Web-Based English* (GloWbE) at http://corpus.byu.edu/glowbe/ (on 21 July 2014) found no relevant occurrences of "desk" or "desks" with "penguin" or "penguins" within seven words on either side of them.

not just in literary language but also in mundane fiction, news chapters, classroom discourse, advertisements, doctor/patient dialogue, blogs, device manuals, internet chat, and so on; and creative metaphor can be found in such everyday discourse as well. But, in such discourse, what appears to be much more common than novel pairings are phenomena such as the extension of familiar metaphorical views of things and the novel compounding of familiar metaphorical views. Indeed, Lakoff and Turner [31] claim that even in poetry, creative metaphor makes heavy use of extension and compounding. Extension will be clarified shortly. Compounding is illustrated by the example given by Deignan et al. [16, Chap. 9] of a chronic-pain sufferer describing himself both as being in a concrete straitjacket and as being in a black isolated space.

This chapter outlines and extends an approach to metaphor understanding called ATT-Meta [1, 3, 5, 6, 8] that has been strongly directed at accounting for how creative metaphor extensions and compounds can be handled. Because of space constraints, I do not address compounding here except for a brief mention late in the chapter, but ATT-Meta's approach to it is discussed in Barnden [8], extending Lee and Barnden [33].

There is an ambiguity in the word "extension" when applied to metaphor. One phenomenon it can describe is when a metaphorical view of something is used on multiple occasions over some possibly extensive swathe of discourse rather than just locally within a short sentence. But the sense of extension important in this chapter is a form of conceptual extension. It is a matter of exploiting some unusual aspect of the source subject matter of some familiar metaphorical view—unusual in the sense that that aspect is not normally exploited in uses of the view. For example, Semino [40] discusses a case of a pain being described as "viciously twisting" in the person's head. It is common to describe pain as "sharp" or "stabbing", using a familiar view of pain as the presence of an intrusive physical object that can cause pain, even when the pain is not in fact caused by any such object. But Semino shows that it is unusual to talk about vicious twisting when metaphorically describing pain. Such extensions are open-ended: there is no boundary to what a speaker might in principle say about the sharp object that is being twisted or otherwise manipulated, in the service of conveying useful information about the intensity and quality of the pain, and Semino and Deignan et al. (*loc. cit.*) give further creative examples.

To avoid the ambiguity of "extension" I will use the phrase "scenario elaboration" instead (or just "elaboration" for short). This is because some details of a scenario based on the source subject matter are being laid out, going beyond the basic source concepts used in the metaphorical views at hand. If a pain is metaphorically described as vicious twisting of a knife in the head, the source scenario is more elaborated than in just describing the pain as the presence of a knife, or yet more vaguely as stabbing. It is important to realize that what is being elaborated is a source scenario, not necessarily the source matter introduced by any one, specific metaphorical view. The elaboration might involve the source matter of *more* than one view, but might also introduce subject matter not belonging to *any* familiar metaphorical view. These points are missed by the term "extension", which is naturally taken to imply that a single, familiar metaphorical view is being elaborated. It is similarly missed by the notion of "entailments" (see, e.g., [29, pp. 384–389] of a metaphorical view.

I will be much concerned with the *open-endedness* of scenario elaboration. The following illustration is a rich but mundane example of creative metaphor using open-ended elaboration, from a story in a romance magazine found on a supermarket shelf:

(1) Sharon pulled herself out of her jeans, the words "How could he? How could he?" jumping about her wearied brain. Senseless, leaving her empty, cold, helpless. Another voice, angry and vindictive, shouted in her ear, 'Serves you right, you silly fool: play with fire and watch your life go up in flames. It was all so predictable'.[4]

This contains many metaphorical aspects, some of which take the form of conventional metaphorical wording (play with fire, watch, life go up in flames, empty, cold). "Conventional" metaphorical wording is frequently used, relatively fixed wording with an entrenched metaphorical meaning. Also, portraying thoughts as if they were spoken utterances is common, especially in written fiction, as evidenced for instance by the ATT-Meta databank's section on the metaphorical view called *Ideas As Internal Utterances* However, it is very unusual to talk about someone's thought as an utterance "jumping about" their brain. It is certainly very common for thoughts to be portrayed as animate beings.[5] But a search of GloWbE (see footnote 3) found no instances of anything jumping about a brain (or head, mind or skull) or jumping about in/inside a brain, etc. A search of the web delivered some instances of thoughts jumping about a brain.[6] But only two of those instances featured words jumping about the brain, one with a "book" doing the jumping, another with "stories" doing the jumping. No instances were found of a particular string of words jumping about a brain.

Another example is:

(2) The weather is settling into a drier frame of mind.[7]

It is conventional to talk of "settled" weather, and the "settling" in (2) is a minor elaboration conveying the process of the weather becoming settled. But talk of the weather being in a "drier frame of mind" is creative. It is unusual phraseology, and at the same time a highly effective and economical way to convey a subtle state of the weather. Note that (2) conveys not that the weather is necessarily going to be in a (settled) state that is consistently drier, but rather a state in which it has some *tendency* to be drier. This is just as when we say someone is in a "generous frame of mind" to mean that they are currently inclined to be generous, not that they are actually being consistently generous.

In (2) a (frame of) mind is being attributed to the weather, which is therefore being personified. Weather has historically often been personified (or deified), so that

---

[4] From magazine *My Story*, May 1995, p. 17. Gibraltar: Editions Press Ltd.

[5] Again see ATT-Meta databank, section on *Ideas As Persons or Other Animate Beings*.

[6] Eight instances on screen pages shown by Google searches conducted on 24th March 2014, using the search pattern "jump about * brain" and variants with "jumps", "jumping" and "jumped".

[7] From a weather report on BBC Radio 4, U.K., 7 a.m., 30 July 2003.

there is a source scenario in which a person (standing for the weather) is engaged in activity of some sort. But personification where a "frame of mind" is used to convey a particular state of the weather appears to be an unusual elaboration. A GloWbE search found no instances of "weather" within six words to the left of "frame of mind". A web search for "weather" being in a "frame of mind" reported four instances,[8] casting the weather as being in a "fickle", "friendly", "bi-polar", or "benevolent" frame of mind. These four examples are creative in the way they qualify the nature of the weather, by enriching the source scenario in an individual way. They also provide an important hint about the open-endedness of elaboration. In the case of "bi-polar", alternative wording that conveyed a mental state akin to bipolarity would lead to similar conclusions about the weather. For example, the weather being in an "up and down frame of mind" would work similarly to mean that the weather is tending to regularly move between two states. Equally, the "benevolent" case would be replaced by the weather being in a "frame of mind where it was inclined to dispense largesse over wide sections of the populace". There is no boundary round the way in which the frame of mind could be qualified. This gives great scope for metaphorical creativity.

Some other examples of open-ended and creative metaphor in mundane language will arise below.

The chapter looks mainly at how an understander (hearer or reader) understands metaphorical discourse according to the ATT-Meta approach, particularly when it is creative, rather than at creative metaphor generation, although a feature of ATT-Meta that helps with generation will be mentioned. Also, the approach leads to a new suggestion about the form metaphor may take within the mind (as opposed to within language, pictures, etc.) The suggestion is that, while analogy is involved in metaphor, and analogical thinking is extremely important in creativity, metaphorical thinking also has non-analogical aspects that are important in creativity.

A few words about ATT-Meta in general. The name ATT-Meta applies both to a theoretical account and the implemented AI reasoning system based on it. ATT-Meta is first and foremost a theoretical processing account of aspects of metaphor understanding. It can potentially inform psychological modelling of human metaphor understanding or intelligent computer programs for metaphor understanding, and has certain philosophical implications, some spelled out in Barnden and Wallington [10], for the nature of metaphor and metaphor understanding. Thus, in essence, the approach was developed in order to investigate the fundamental nature of metaphor. But the approach has been partially implemented in a computer program, a rule-based uncertain-reasoning system written in Prolog. The working system does not currently actually accept natural language sentences. Rather, it is a system for handling just the reasoning and source/target mapping actions that the ATT-Meta approach holds to be needed for handling a certain broad type of metaphor.[9] The system is not intended to be definitive as to how the approach should be implemented in computer

[8] In Google search result pages.

[9] However, in an ongoing EU project ATT-Meta is being combined with a natural language front-end for interpretation and generation of linguistic strings.

software—many other implementations of the broad principles of the ATT-Meta theoretical account could be envisaged.

The plan of the chapter is as follows. Section 11.2 explains the ATT-Meta approach and how it treats open-ended scenario elaborations. Section 11.3 explicates a crucial element of that treatment, namely an "anti-analogy-extension" thesis. Section 11.4 goes into further ways in which ATT-Meta handles creativity during metaphor understanding, and briefly discusses novel source/target pairings. It also touches on metaphor generation. Section 11.5 discusses the suggestion advertised above about the roles of analogy and metaphor in creative thinking. Section 11.6 concludes.

11.2 Open-Ended Elaboration and the ATT-Meta Approach

The ATT-Meta approach is mainly geared towards cases involving familiar metaphorical views, such as the view of the mind as a physical region. However, the approach is not much concerned with conventional metaphorical phraseology based on such views, as in "The idea was at the back of his mind". Rather, ATT-Meta it is geared towards open-ended elaborations that rest on familiar metaphorical views. This is best brought out by examples such as (1,2) and the following:

(3) The managers were getting cricks in their necks from talking up [to some people in power over them] and down [to the managers' subordinates].[10]

It is common for abstract control relationships, especially in organizational settings, to be metaphorically viewed in terms of relative vertical position of the people concerned. However, someone having a crick in their neck is not a matter addressed by this view, and no conventional metaphorical sense for "crick" appears in, for instance, the Chambers dictionary. Only one example was found in GloWbE, only one in the British National Corpus (BNC),[11] and only eight on the web, of metaphorical cricks in necks being used metaphorically to describe mental/emotional states in situations with no actual or potential turning of real heads. The BNC example was

> The draught from Microsoft's increasingly popular Windows is giving rival software firms a crick in the neck,

creatively exploiting the fact that a draught of air can cause a neck-crick, and that movement of an object can cause a draught. The GloWbE example was

> More likely, Romney as president would be a man with a strange crick in the neck, constantly looking over his right shoulder to see which pickup truck full of movement conservatives was about to run him over.

Annoying things and circumstances are often conventionally described as being a "pain in the neck" or just "a pain", and it might be possible to analyse (3) as resting

[10] Goatly ([24]: p. 162). The example is from the *Daily Telegraph* newspaper.

[11] Accessed via http://corpus.byu.edu/bnc/.

on a metaphorical view underlying these idioms, for example a view of *Annoying Item as a Pain*. However, (3) shows some elaboration in that the specific notion of a crick is introduced, and linked to the specific context-specific circumstance of the constant turning of the managers' heads in two opposite, physical directions. These cricks cause pain, emotional stress, difficulty in continuing such head-turning, and dislike of continuing it. Such feelings and so forth are in the source scenario, but I will show below how some are transferred into the target scenario (the situation actually described) as well. This generates conclusions such as that (a) the managers, in the target scenario, experience annoyance and other emotional stress, and (b) it is difficult for the managers to continue the conversations.

Point (b) is an illustration of how the (moderate) creativity of an example such as (3) rests not just in a degree of novelty but also in the way it economically conveys a range of useful information. (3) does not just convey (a), so the sentence is richer than if it had merely said that the managers were "getting annoyed" at their conversations. Annoyance does not in general imply difficulty of continuing (though it may imply reluctance to continue). However, having a crick in your neck not only causes pain but also *makes it difficult to continue turning your head,* and therefore difficult to continue the conversations.

The neck cricks begin to suggest the beneficial open-endedness of scenario elaborations in metaphor. The example could be further elaborated, creatively, by talking of the managers "massaging their necks", for instance. This can easily be understood as indicating that the managers are experiencing pains in their necks. Another elaboration would be to add a mention of pains in other parts of the managers' bodies, e.g. saying "getting cricks in their necks and cramps throughout their bodies", emphasizing the degree of overall discomfort. There is no limit to what could be said about what the managers are experiencing in the source scenario, with the intent of emphasizing their discomfort, difficulties and resulting mental states and feelings.

Now consider:

(4) I don't think strings are attached. If there are any they're made of nylon. I don't see them.[12]

The speaker was an African politician being interviewed about a new investment by China in mineral mining. He used the common metaphorical view of requirements or constraints as attached strings. However, the being-made-of-nylon is an elaboration in the sense that normal use of the metaphorical view does not concern itself with what the strings are made of. The elaboration appears to be creative in being both unusual and effective. No instance of "nylon" being used metaphorically was found in the BNC, whether or not in conjunction with "string[s]", and no metaphorical instance was found in GloWbE within seven words of "string[s]".[13] In the context of

[12] From *Newsnight* programme on TV channel BBC2, U.K., 3 July 2007. Plausible punctuation added.

[13] A web search using Google uncovered a case of the "Ramon y Gabriela" guitar duo having a "nylon strings attached" relationship. However, the nylon here is just a pun exploiting the nylon composition of the guitar strings the duo used, and does not make a comment about the abstract nature of the relationship.

the sentences, the nylon constitution of the putative strings helps to convey that the requirements, if present at all, are not readily noticeable, because of the translucency of nylon. Notice a subtle point here. The mention of the nylon constitution is not redundant, even though (4) also explicitly says that the speaker cannot see the strings. Without the mention of the nylon constitution of the strings, it could have been that the speaker could not see them because of, say, inattention on his part. But the nylon constitution suggests the nature of the strings themselves should be blamed instead.

These examples and others have been analysed under the ATT-Meta approach (see for example Barnden [2, 4, 6, 8]). Amongst them are two closely-related examples, (5) and (6), that I will analyse in detail in the following, to convey the nature of the approach. (6) preserves the essential quality, for the purposes of this chapter, of real-discourse example (7).

(5) One part of Mary was insisting that Mick was adorable.

(6) One voice inside Mary was insisting that Mick was adorable.

(7) Suddenly I was having second thoughts. About us, I mean. Did I really want to get married and spend the rest of my life with Mick? *Of course you do* one small voice insisted. *Are you quite sure about that?* another nudged.[14]

I take (5) to rest on two very general metaphorical views that are often used about the mind. First, there is the view of a person or a person's mind as containing persons with their own mental states. I call these the "subpersons" of the person. The subpersons are often portrayed as parts of the person as in (5). I call this view *Mind as Having Parts that are Persons* (elsewhere I have less accurately called it *Mind Parts as Persons*). Secondly, the subpersons may be portrayed as communicating in natural language. In such a case, the utterance also rests on a metaphorical view of *Ideas as Internal Utterances.*

(6) does not mention any person-like part of Mary. But the existence of the voice can be used to infer such a subperson by default, within the source scenario. It will therefore turn out, in the account below, that (6) conveys the same thing about Mary's mental state as (5) does. But an additional feature of (7) is the "small" qualification on the voice. I will discuss below what effect this has and how.

Now, I regard the uses of the notions of insisting and a voice as elaborations. The point of the *Mind as Having Parts that are Persons* is to relate the mental states of the real, whole person to the mental states of the subpersons, in particular to allow different subpersons to have different or even conflicting mental states. That relationship does not in itself make any use of the notion of a subperson "insisting" something, or indeed engaging in communication in any way, as I will make clearer below.

Furthermore, there is no bound on what sorts of elaborations make sense here. We can have any elaboration that, in the terms of the source subject matter, can have implications for the target. So, for example, suppose (5) had said "shouting" instead of "insisting" (cf. the shouting voice in (1)). The shouting has implications about the mental state of the subperson in question, for example that the subperson may be

[14] From magazine *My Story*, May 1995, p. 6/7. Gibraltar: Editions Press Ltd. Italics in original.

angry, as well as suggesting (as in the insisting case) that another subperson has said something contrary. The anger could then be transferred, as we will see, to become an implication that the real person has a motive to be angry. Again, if a sentence creatively mentioned a subperson as knocking another subperson unconscious, this would make sense because the latter subperson would no longer have conscious mental states and feelings, and therefore certain implications about the real, whole person would disappear.

A distinctive feature of ATT-Meta, compared to other approaches in cognitive linguistics, psychology and AI, is to handle metaphor through a *pretence* or *fiction* mechanism. I use a very broad notion of pretence here. It is akin and even perhaps identical to that involved in thinking counterfactually. In thinking through what would have happened had Obama lost the 2012 US presidential election, one mentally pretends that Obama did lose the election and then explores that pretend scenario. This broad notion of pretence in no way involves deceiving oneself or others of anything, or of physically acting a role. Under this weak notion of pretence, the metaphor understander pretends that what the metaphorical sentence literally says is true, and draws consequences from it using knowledge of the source subject matter. Those consequences are still a part of the pretence.

What I referred to above as a source scenario is more precisely the *pretended scenario*. It may then be that a consequence derived within the pretence—for example, that a string cannot easily be seen, in the case of (4)—can somehow be mapped in some modified form to become a proposition about the target scenario—e.g., that one cannot easily be aware of the constraint. Such a target-scenario proposition forms part of the meaning of the discourse for the understander.

Another way of putting it is that understanders use the literal meaning of the utterance to begin to construct a fictional scenario which they then fill out by means of inference and out of which they selectively extract information about the target. A pretended scenario is similar to a world as depicted by a fictional narrative. The ATT-Meta approach is therefore akin to fictionalist approaches to metaphor in Philosophy (e.g., Walton [44]), and to the use of imaginary worlds for poetry understanding (Levin [34]). See also Carston and Wearing [14] for a recent preliminary extension of Relevance Theory in a similar vein.

11.2.1 Metaphorical Views and Mappings in ATT-Meta

One major question is how information is transferred out of the pretended scenario into the target scenario (the reality scenario), possibly in modified form. Here ATT-Meta borrows in part from conceptual metaphor theory, while going beyond it. A conceptual metaphor consists of a set of mappings—or as I will say, *correspondences*—between aspects of the source subject matter and aspects of the target matter. These mappings constitute an analogy. The ATT-Meta approach broadly adopts this idea, though the correspondences are considerably different in form and function from those in conceptual metaphor theory and in analogy theory. One difference is that in the ATT-Meta approach there are two broad sorts of correspondence: (i) *view-specific* correspondences that are associated with particular metaphorical views, and (ii) *view-neutral mapping adjuncts* that apply by default in any case of metaphorical understanding and that build upon the effects of, and indefinitely extend the reach of, the correspondences in (i).

A metaphorical view in ATT-Meta involves a small number of very general, high level view-specific correspondences. In case of *Mind as Having Parts that are Persons*, only two correspondences appear to be needed for a large array of examples. They can intuitively be expressed as follows.

(C1) A person *having some motive* to believe/desire/intend/fear/like/... something metaphorically corresponds to *at least one subperson* of that person having a motive to believe/desire/intend/fear/like/... it.

(C2) A person believing/desiring/... something metaphorically corresponds to *every* metaphorical subperson of that person believing/desiring/... it.

If we know from an utterance such as (5) that some subperson believes something, then *a fortiori* that subperson presumably has a motive to believe it. Hence, via (C1), the overall, real person has a motive to believe it as well.

One main point of the *Mind as Having Parts that are Persons* view is that it allows different subpersons to have different beliefs (etc.) that conflict with each other. In such a case the real person has *motives* to believe various conflicting things, without actually *believing* any one of them. This sort of conflict is explicitly illustrated in (7), but as we will see, (5) and (6) implicitly convey conflict.

View-specific correspondences such as (C1) and (C2) are couched in ATT-Meta as *correspondence rules* that relate aspects of some source subject matter being used in a pretence to aspects of some target subject matter in the reasoning space surrounding the pretence. This surrounding space is usually the space concerning reality (the reality space), but pretences can be nested within other pretences and other sorts of space. Hence, correspondences in ATT-Meta serve to bridge, in general, between a pretence space and a surrounding space.

(C1) and (C2) are more precisely glossed as follows. X is a variable standing for some proposition. The "surround" is the space surrounding the pretence space. For simplicity here I restrict the rules to dealing with just the mental state of belief.

(8) IF (in the surround) P is a person AND (within the pretence) P has one or more subpersons
THEN
(in the surround) P's having some motive to believe X
CORRESPONDS TO
(in the pretence) at least one subperson having a motive to believe X.

(9) IF (in the surround) P is a person AND (within the pretence) P has one or more subpersons
THEN
(in the surround) P's believing that X
CORRESPONDS TO
(in the pretence) all the subpersons believing that X.

The IF part of such rules acts as an appropriateness condition or *guard*. During processing of a metaphorical utterance, it can become apparent that a person P is being viewed a having subpersons. This can happen, for instance, if a "part" of P is described as having thoughts. Then, the rules above can fire for P and any proposition X that may be salient. What the rule does is create the *individual correspondence* specified in the THEN part, for the *particular* person P and proposition X at hand.

11.2.2 The Process of Understanding (6)

The present subsection outlines how the ATT-Meta system processes (6). The system is a complex one based on a rule-based reasoning engine, implemented in Prolog, that performs a type of qualitatively-uncertain reasoning whose main feature is reasoning with defaults. It is also has a treatment of degrees with which situations can hold.

Taking sentence (6) literally, the mentioned voice insists that Mick is adorable. This fact about Mary is a premise used within the pretence, and is used to infer that (by default) there is a subperson inside Mary. Given the general default that when people claim things they believe them, the system can then infer that

(A) that subperson believes that Mick is adorable.
It follows by default that

(B) that subperson has a motive to believe that Mick is adorable.

Since Mary does have a subperson in the pretence, the above correspondence rule (8) applies, creating an individual correspondence between *Mary*'s having a motive to believe that Mick is adorable and *the subperson* having such a motive. Notice that this in itself is just a correspondence between two possible states of affairs. It doesn't say that these states of affairs obtain (in the reality space and pretence space respectively). But (B) does say that the state of affairs on the pretence side holds, so ATT-Meta creates the following proposition about reality:
(C) Mary has some motive for believing that Mick is adorable.

Thus, overall, a few simple inference steps lead from a within-pretence premise derived directly from (6) taken literally to a within-pretence proposition (B) that is mapped to become a within-reality proposition (C). See Fig. 11.1 (where the example is put into the present tense for simplicity).

But also the "insisting" in (6) can be used to infer that actually there is a subperson of Mary that believes that Mick is *not* adorable. This arises because of the real-world nature of "insisting". Typically, someone insists something because there is a conversation with a person who denies it. Thus, the presence of a subperson who claims that Mick is not adorable can be inferred by default. This is also shown in Fig. 11.1. Then, given again the general default that when someone claims something they believe it, this subperson presumably *believes* that Mick is not adorable. Hence, via a different application of correspondence rule (8), we get the result that Mary has a motive to believe that Mick is not adorable. Thus we get the effect that Mary has motives both to believe that Mick is adorable and to believe that he isn't. In addition, by means of correspondence rule (9) it can be inferred that Mary lacks the belief that Mick is adorable, since there is a subperson who lacks this belief.

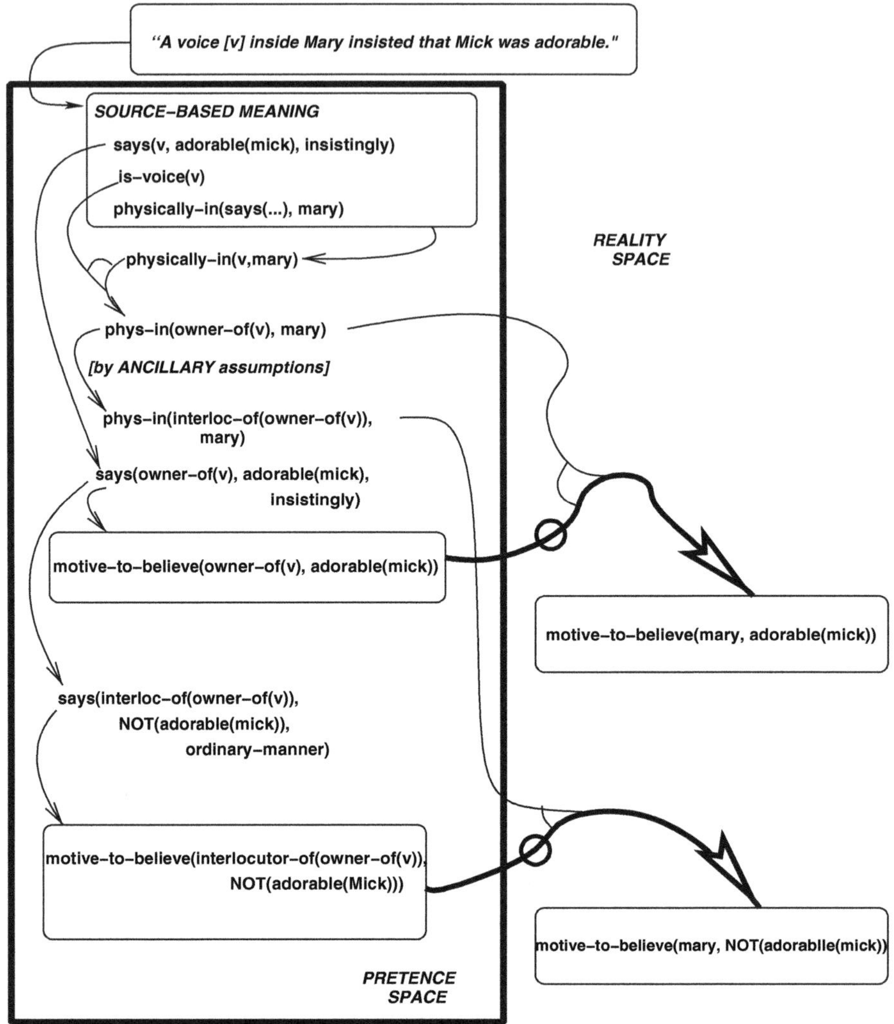

Fig. 11.1 Showing the processing for (a present tense version of) example (6). The heavily outlined box shows the pretence space. The *thick, circled arrows* crossing the box boundary from inside shows a mapping action, specifically one that arises from a view-specific correspondence rule (8) in the text, associated with the *Mind as Having Parts that are Persons* view. The *thin lines* joining the *thick arrow* show the dependence of the correspondence on the guard conditions in (8). Other *arrows* show ordinary inference steps

In summary so far, ATT-Meta is able to deal with open-ended elaborations through open-ended inference within the pretence, deriving conclusions (indirectly from the literal meanings of sentences) that can be transferred with the help of correspondence rules such as (8,9) to become statements about the target scenario. To emphasize this point further, I mentioned above a potential variant of (6) that uses "shouting" instead of "insisting". A possible line of reasoning would deliver all the consequences above plus an extra one that says that the shouting subperson is angry or alarmed. By (C1), we then get the implication that Mary has a motive to be angry or alarmed.

11.2.3 View-Neutral Mapping Adjuncts

Consider again the neck-crick example, (3). How can conclusions such as that the managers, in reality, are experiencing negative emotions, caused by the conversations, and find it difficult to continue their conversations, be created? Such conclusions arise within the pretence, but we need to transfer them to the surrounding space (reality space).

This is where *view-neutral mapping adjuncts* (VNMAs) come in. There are general qualities about source scenarios that are very often transferred in metaphor to the target scenarios no matter what the specific metaphorical view is. Amongst such qualities are the following:

- Emotional/attitudinal states, value-judgments, etc.
- Mental states, such as believing, intending, wanting.
- Time-Course, incl. starting, continuing, ending, immediacy, smoothness/intermittency, rates at which episodes occur, temporal relationships between episodes, etc.
- Causation, prevention, enablement, ability, attempting and tendency relationships, and related qualities such as effectiveness.[15]
- Ease/difficulty properties.
- Normal functioning (of a machine, organism, protective measure, structure, etc.)
- Modal qualities: possibility, necessity, obligation, ...
- Qualitative number/amount (little, much, few, many, more, ...)
- Uncertainty with which situations hold.
- Degrees to which situations hold.
- Complementation (the logical operation of switching between a situation holding and not holding).

As an illustration, to reflect the first two items in the above list, the ATT-Meta system contains the following rule:

(10) IF something P in a pretence CORRESPONDS TO something S in the surround
AND cognitive agent A in the pretence CORRESPONDS TO agent B in the surround
THEN
(in the pretence) agent A's bearing a particular mental/affective attitude towards P
CORRESPONDS TO
(in the surround) agent B's bearing the same attitude towards S.

In the neck-crick example, we can take P to be the conversations in the pretence and S to be the conversations in reality. They are actually the same conversations.

[15] Effectiveness has only recently been added, and is reported for the first time here.

Intuitively, something can exist in both spaces, but formally they are treated as separate but "corresponding" entities. We can take A to be the (set of) managers in the pretence and B the same managers in reality. Again, A "corresponds" to B. Thus, the rule delivers a correspondence between emotional distress of the managers about the conversations, in the pretence, and emotional distress of the managers about the conversations, in reality. The managers are indeed inferred to experience distress in the pretence, so the correspondence is used to infer that they experience it in reality as well.

The VNMA that handles causal relationships etc. takes the form of a rule like (10) that says that P causing Q in the pretence corresponds to R causing S in reality, if P corresponds to R and Q corresponds to S. As a result, the conversations causing the emotional distress in the pretence is inferred to correspond to their doing also in reality. Since the former casual link can be inferred to exist in the pretence, it is also assumed to exist in reality.

Note that there is nothing assumed in the causal chain in reality space to correspond to the neck-crick itself or the physical pain it causes. This is simply because the neck-crick itself does not correspond to anything in reality, nor does the physical pain. A causal link in the pretence between P and Q is only inferred to correspond to one in reality *if* P corresponds to something R in reality *and* Q corresponds to something S in reality. Similarly, (10) is parasitic on the existence of a correspondence between P and S and a correspondence between A and B.

Equally, the within-pretence difficulty for the managers of continuing with the conversations transfers to reality, because of VNMA rules handling time-course (a case of which is the continuation of a situation) and difficulty. For example, here is the rule for some aspects of the Time-Course VNMA.

(11) IF some situation P in a pretence CORRESPONDS TO some situation S in the surround
THEN
(in the pretence) P's having a specific qualitative temporal attribute
CORRESPONDS TO
(in the surround) S's having the same attribute.

The continuation of a situation is one case of a qualitative temporal attribute. So, because the conversations in the pretence correspond to the conversations in reality, the *continuation* of the conversations in the pretence corresponds to their continuation in reality, because of the Time-Course VNMA.

Then, given that these continuations correspond, and the managers correspond between pretence and reality, the difficulty the managers have in engaging in the continuations also corresponds, by the Ease/Difficulty VNMA.

Notice here how VNMAs can recursively build on each other: for example, continuations are made correspond by one VNMA, allowing another VNMA to make the difficulty of the continuations to correspond. If, further, an agent in the pretence were to have a mental/emotional state about this difficulty, then a corresponding agent in reality would be inferred to have it as well. Thus, the emotional/mental-state VNMA would have built on the Ease/Difficulty VNMA, which would have built on the Time-Course VNMA.

Note also that in all this, the only view-*specific* correspondences used are the very basic one of relative vertical position relationships in pretence corresponding to control relationships in reality, and the identity-like correspondences for the conversations and managers. *Thus, the meaning arising from (3) is generated mainly by VNMAs, not by view-specific correspondences.* Work on the ATT-Meta approach indicates that metaphorical utterances often get much, and in some cases all or almost all, of their effect via VNMAs rather than directly from view-specific mappings. The latter often merely provide a scaffold to allow VNMAs to handle the most important information. The treatment of a wide variety of examples in [2, 4, 6, 8] provides evidence for this claim.

However, VNMAs are merely default rules, so their implications can be defeated in specific circumstances by other evidence. This defeasibility applies also to the effect of view-specific rules.

As another example of the use of VNMAs, there was no treatment above of the effect of the "small" qualifier of the voice in example (7). Suppose this qualification is added to (6), to get

(12) One small voice inside Mary was insisting that Mick was adorable.

The effect of the smallness is quite subtle, but at least some of it can be captured by the notion that a small voice in a conversation is a relatively ineffective one (by default). So, the voice, and therefore also the mental states of the subperson whose voice it is, are relatively ineffective in the overall pretended scenario. Now, the subperson's motivational state corresponds under the *Mind as Having Parts that are Persons* view to Mary's having a motive to believe that Mick is adorable. So, the *effectiveness* of the subperson's motivational state in the pretended scenario corresponds by the Causation-etc. VNMA to the effectiveness of that motivational state of Mary's in the reality scenario. Moreover, as qualitative degrees to which situations hold carry across by the Degree VNMA, a relatively small degree of effectiveness in the pretended scenario maps to a relatively small degree of effectiveness in the reality scenario.

Notice that *the pretend subpersons themselves do not correspond to anything in reality*. It is merely that certain things about their mental states correspond to certain things about the real person's mental states. This point leads into the next section.

11.3 The Anti-Analogy-Extension Thesis

In the ATT-Meta approach, as in conceptual metaphor theory, metaphor is based on familiar analogies. An ATT-Meta metaphorical view is largely realized as a set of entrenched correspondence rules, and VNMAs are additional correspondence rules.[16] Nevertheless, a key point about the ATT-Meta approach can be called the

[16] The realization of a view can also involve *ancillary assumptions* [8]. These are default assumptions about the nature of the pretended scenarios involved in a metaphorical view. Figure 11.1 shows one

Anti-Analogy-Extension Thesis. This says that open-ended elaboration in metaphor should *not*, normally, be handled by extending already known analogies. ATT-Meta seeks to get away with the least amount of analogy possible, *contra* other theories such as Structure-Mapping Theory [11, 18, 22] which assume that the task is to maximize the extent of (structurally well-founded) analogy.

I call both the view-specific correspondence rules and the view-neutral mapping adjuncts *standard* analogical mapping rules. They are part of an understander's long-term knowledge. However, suppose an understander is faced with a metaphorical sentence where there are *as yet unparalleled* within-pretence items, such as the nylon in (4), and the sub persons, voices, smallness, insisting and nudging in (5, 6, 7, 12). Such items are ones that are not mappable into the surrounding space by any standard mappings. Then, the Anti-Analogy-Extension Thesis says that, unless there are special pressures to act otherwise, understanding does *not* seek to provide parallels in the surrounding space for the as-yet-unparalleled items. Rather, the understander tries to connect these items to within-pretence content that *can* be converted via standard mappings. This is on the theoretical principle that, typically, the unparalleled items are proposed by a speaker not as individually standing for aspects of the target scenario being addressed, but rather to build a pretended scenario that holistically illuminates the target side using knowledge of metaphor that the understander is expected already to have.

In particular, there is no need at all to propose that for examples like (5) and (6) the mentioned parts and voices correspond to identifiable aspects of the real person, or to propose that there is some internal, real mental action that can be clearly held to correspond to the action of insisting in the sentence. Rather, the mentions of parts, a voice and insisting in (5, 6) are *merely* tools towards constructing a rich pretended scenario, which in turn conveys in an economical, accessible and vivid manner the possession of a particular sort of mental state by Mary.

Similarly, in the neck-crick example (3), the cricks and resultant physical pain have no parallel in reality. The cricks are only there to convey emotional distress, difficulty in continuing the conversations, etc. In the nylon constraints example (4), the speaker is not trying to refer to possible real features of the putative abstract constraints that correspond to nylon, to being made of nylon, or the general being-made-of relationship. The being-made-of-nylon is only there to convey information such as that the speaker cannot see the constraints and to suggest it is not his fault.

Although the nylon itself has no parallel, the nature (whatever it is) of an item in a pretence is taken to correspond to the (possibly different) nature of a corresponding item in the surrounding space. Since the nature of the strings in the pretence causes the politician not to see them, their nature in reality causes him not to be aware of them in reality (by the Causation-etc. VNMA). But there is no need at all to *analyse* the nature of the constraints in order to find items analogous to being-made-of and

(Footnote 16 continued)
application of such assumptions. In brief, it is assumed that if an agent has a subperson that is involved in a conversation, then there is an interlocutor who is another subperson involved in the conversation.

to nylon. Those within-pretence aspects are not paralleled in reality, even though the overall nature of the strings is, as an unanalysed unit.

Consider also the following example:

> And do you remember that one of the Vault's SPECIAL FEATURES was A DARK, DUSTY, COBWEBBY CORNER WHERE UNWANTED THOUGHTS AND FEELINGS COULD BE DUMPED? Well, that's where you can push the no-longer wanted, unrealistic and overly negative thoughts too![17]

There is no need for an understander to find a parallel for the mentioned corner or the qualities of being cobwebby, dusty and dark. Instead, these aspects of the pretended scenario lead to within-pretence inferences about such matters as the relative inaccessibility and hiddenness of the negative thoughts (metaphorically viewed as physical objects) once the thoughts have been "pushed" into the "corner".

The example is moderately novel not only in the notion of a "Vault" in the mind but also in the use of "cobwebby" (whereas mention of pushing ideas, or of parts of the mind such as corners and recesses, is common). This word may be unusual in any mode of use, judging by the mere 29 occurrences in GloWbE. None of the uses are similar to that in the example above, but a few uses of cobwebby corners, cellars, etc. of the mind can be found in other sources.[18]

The Anti-Analogy-Extension Thesis goes hand in hand with a form of holism about the pretended scenarios and the metaphorical sentences leading to them. The pretended scenario is to be regarded not as having a detailed analogy to a reality scenario (or a scenario in some other sort of surrounding space) but rather something that *holistically* conveys information about a reality scenario. This conveying is, to be sure, done by the action of correspondences that pick on specific aspects of the pretended scenario. But the ultimate intent here is to transfer information, not specify an analogy. And any specific aspect of the pretended scenario that is grabbed by a correspondence may be the result of inference over large amounts of information within the scenario. In particular what this means is that there may be no specific part of the metaphorical sentences that can be said to correspond to a given aspect of the reality scenario (although this can happen in simple cases of metaphor). For example, going back to (6), the fact that Mary lacks the belief that Mike is adorable does not correspond to any one aspect of (6) but rather to the whole of (6).

Another work that emphasizes both frequent holism of metaphor and the lack of need for (or indeed the frequent undesirability of) analogy-extension is Langlotz's [32] treatment of idioms, including metaphor-based ones.

---

[17] From excerpt found in http://books.google.co.uk of: Kate Collins-Donnelly (2014), *Banish Your Self-Esteem Thief: A Cognitive Behavioural Therapy Workbook on Building Positive Self-Esteem for Young People*, p. 127. Jessica Kingsley Publishers.

[18] For instance, three such uses in http://cardkiwi.com/words/cobwebby/cobwebby-sentence-examples.

11.4 How ATT-Meta Helps with Creativity

There is no limit to the amount or type of inference that can take place in a pretence, so that indefinitely rich elaborations can be handled with the aid of suitable reasoning mechanisms and enough general knowledge of the source subject matter(s). The Anti-Analogy-Extension Thesis gives both speaker and hearer the freedom to process elaborations without a general imperative to establish new analogical linkages: the thesis seeks instead to rely on standard mappings, whether view-specific or view-neutral. But there are some further features of the ATT-Meta approach that help with open-ended elaborations.

First, the ATT-Meta approach gives a major role to goal-directed reasoning. Although the descriptions of reasoning above are couched as moving forward from premises towards conclusions, the process is actually typically assumed to proceed in a goal-directed way.[19] That is, there is some goal or issue that the system is trying to address, and reasoning steps are attempted towards that end. For example, in the case of (7), the context raises the question of whether to get married to Mick. Given the presence of view-specific correspondence rules (8,9), this can be converted into the goal of investigating, in the pretence, which subpersons want, or have a motive to want, the marriage. By a process of backwards chaining through inference rules, it is discovered that there is a subperson who does have this motive and one that has a motive not to want it. This then rolls forward via the relevant view-specific correspondence to become the conclusion that Mary herself both has a motive to want the marriage and a motive not to. Goal-directed reasoning is an extremely powerful tool for combatting the notorious indeterminacy and context-sensitivity of metaphorical meaning (see, e.g., [41]). Suitably deployed it can guide metaphor understanding towards uncovering meaning that is relevant to the context, thereby helping creative metaphor to have a useful meaning—cf. the Sternberg and Lubart's [42] definition of creativity as "the ability to produce work that is both novel (i.e., original, unexpected) and appropriate (i.e., useful, adaptive concerning task constraints"). See Barnden [6] for more on this.

The ATT-Meta system currently has no facility for analogy discovery. That is, new correspondence rules cannot at present be created in the system. However, there is already a way in which ATT-Meta can cope to some extent with novel pairings of target and source in metaphor. This is because, just as with open-ended elaborations, novel pairings can work partly or even wholly by means of standard correspondence rules (view-specific and view-neutral), rather than new ones. For instance, suppose someone says in a fit of metaphorical creativity, "My TV set is a pile of rotting mongoose tails that've been lying round in the sun for a few days". It may be clear from context that the speaker is merely trying to convey her feeling of disgust towards the TV set in question. This feeling gets transferred from pretence to reality because of a VNMA. There is no need for the hearer to find a more detailed analogy between the TV set and the mentioned tail-pile, and indeed it would be waste of effort to try

[19] The overall ATT-Meta approach allows non-goal-directed as well as goal-directed reasoning to be used. However, the implemented ATT-Meta system can at present only do goal-directed reasoning.

to find one. Similarly, recall the real example in the Introduction of a migraine being described as a moving animal in the head. Much of the point is about the time-course of the migraine developing and about the fact that the progress cannot be resisted. Such content can be carried over from pretence to reality by VNMAs. For example, difficulty of resistance amounts to difficulty of preventing continuation, and here the difficulty, prevention and continuation are handled by three VNMAs.

But the ATT-Meta approach could be expanded to include the creation of new correspondence rules using methods discussed in the analogy literature (e.g., Falkenhainer et al. [18], Indurkhya [27]). If this were done then the approach would actually allow two different ways in which an explicit metaphorical pairing could be processed, such as in "Yesterday was Christmas Day for newspaper editors"[20] or similarly "Yesterday was a feast day"—when yesterday was not a special day. One of these ways, the one already available, and suitable in some circumstances, would be for such a sentence to be understood as a pretended equation (the day is Christmas Day) or pretended categorization (the day is a feast day). Then, within the pretence, an inference might be made that the day was very pleasurable, and this could be carried over into reality, with the help of VNMAs, to create the conclusion that the real day was very pleasurable. The other understanding approach, suitable in other circumstances, would not be to put an equation or categorization into the pretence, but rather to try to discover an analogy between the target item (the real day) and the source item (Christmas Day, or a feast day, as an aspect of the pretence). The analogy would be couched initially in newly-constructed correspondence rules about the specific day, but might later be generalized to cover any day.

This potential double approach to explicit novel pairings resonates with and is closely analogous to a prevalent concern in the psychological literature on metaphor. Here there is dispute between the categorization theory (or class-inclusion or property-attribution theory) and the so-called comparison theory (see, e.g., [11, 23]). Considerations in that dispute about when categorization or comparison are appropriate could illuminate the question of when ATT-Meta should seek a new analogy across the pretence/surround boundary and when it should use a within-pretence categorization or equation.

ATT-Meta's handling of creativity benefits from the fact that metaphorical views are not reified in ATT-Meta: they are not themselves explicitly represented. This allows great freedom in the nature of the correspondence rules. Metaphorical views are merely a theoretical abstraction from the guards in the individual correspondence rules possessed by an understander using the approach. Consider again correspondence rule (8). Because the guard requires there to be a person (in the surrounding space) that (within the pretence) has subpersons, the rules can be thought of as belonging to the *Mind as Having Parts that are Persons* view. So can rule (9). But there is no explicit connection between these rules, other than the fact that they have the same guard, or any labelling of the rules as belonging to any view. Thus, the use of correspondence rules does not require a prior, explicit decision that some particular metaphorical *view* is in play. Rather, what needs to be inferred is certain

[20] On BBC Radio 4, *Today* programme, 4 November 2005.

aspects of the pretended scenario and surrounding scenario (e.g. that the person has subpersons). Thus, correspondence rules come into play opportunistically during the understanding process, as a seamless part of the overall reasoning activities involved in understanding.

Furthermore, it may be that guards of different correspondence rules might merely overlap rather than being identical, or might merely be similar in some other way. There may only a rough way to theoretically categorize rules as regards what metaphorical views they are in, and it may difficult to say that a given metaphorical view is reflected in any clearly delineated set of correspondence rules.

The flexibility of the contents of pretences in ATT-Meta is boosted by an avoidance of any restriction on correspondence rules as regards how different the subject matters they link are. In conceptual metaphor theory, it is often claimed or tacitly assumed that metaphorical mappings cross between delineated, qualitatively very different "domains". This certainly respects the intuition that in much metaphor the source and target subject matters are often completely different types of thing, as when an idea or mind is viewed as being a physical creature (see ATT-Meta databank cited above). But in Barnden [7] I argue for scepticism, shared with other authors such as Haser [25, p. 32ff], about the scientific utility of the notion of "domain" in describing what metaphor fundamentally is or how it works. There is at most a general tendency for the target and source to be completely different sorts of thing, and much metaphor does not obey this tendency.

Indeed, *Mind as Having Parts that are Persons* radically fails to obey it. There, both the target and the source subject matter involve people, thoughts, utterances, etc. Also, there are simpler metaphors like "Japan—The Britain of Asia?"[21] and "Tuesday is honorary Monday this week" (heard in conversation) where target and source cannot be said to be completely different sorts of thing. Of course the target and the source are always *different* (a person does not contain a conversation amongst subpersons, Japan is not Britain) but things can be different while being the same sort of thing, intuitively. Ironically, even some famous, much-discussed examples of conceptual metaphors as *Life as Journey* and *Argument as War* [30] fail to exhibit the domain difference that is usually claimed. Literal journeys are an important part of life itself; and wars importantly contain, and may even sometimes be based on, literal arguments. In sum, it is quite common for there to be a strong overlap between target and source subject matters.

Accordingly, in ATT-Meta, correspondences (mappings) are *not* by definition between subject matters or domains that have a radical qualitative difference. Rather, by definition they are between the contents of a pretence and contents outside the pretence. ATT-Meta makes no assumptions whatever as to how close or distinct, or how disjoint or overlapping, the two subject matters are, and there is no use of subject matter distinctions anywhere in the approach. Thus, there are no artificial limits on what can go in the pretence, or on what source subject matter can be used to illuminate a given target.

---

[21] http://www.jref.com/forum/all-things-japanese-26/japan-britain-asia-17745/.

In ATT-Meta, a transfer of information from pretence into surrounding space is just a reasoning step like any other, and an individual correspondence created by a correspondence rule is just a proposition like any other. (Recall that a correspondence rule such as (8) creates an individual correspondence when it fires, such as: Mary's having a motive to believe that Mike is adorable corresponds to at least one subperson having a motive to believe this.) Further, correspondence rules just sit with other rules in the system, firing when able to. The overall reasoning engine that handles the firing of rules makes no distinction between correspondence rules and other rules. In addition, while reasoning actions occur in different spaces, there is no regimentation of when reasoning steps happen in different spaces, except as is dictated by the natural linkages between the outputs of some rules and inputs of others. The actions in different spaces are interleaved in whatever way is convenient for the goals for reasoning. Thus, as already hinted, the peculiarly metaphorical aspects of ATT-Meta's reasoning (reasoning within pretences; and the action of correspondences to transfer information) are seamlessly woven into the overall reasoning endeavour. This gives tremendous flexibility especially when dealing with more elaborate cases of metaphor, and especially when a metaphorical scenario is developed over a multi-sentence or larger region of discourse. It is typical in such cases for reasoning to be needed in the ordinary terms of the target subject matter, not just metaphor-based reasoning, partly because metaphorical stretches can be separated by literal language.

In the ATT-Meta approach (and implemented system) metaphorical correspondences cause information to be transferred in *both* directions between a pretence and the surrounding space. I call the surround-into-pretence direction the "reverse" direction. The motives for this are detailed in Barnden et al. [9]. In brief, one reason is to try to maximize the degree of consistency between what is being inferred within the pretended scenario and what is being inferred within the surrounding space's scenario, and another is to allow useful enrichment of the pretended scenario based on information independently known about the surround. For example, in a *Mind as Having Parts that are Persons* example, if we know that the real person believes X then it may be useful to infer that each subperson believes X too, by virtue of correspondence rule (9). This could allow important inference about different subpersons' thoughts to take place. This case of reverse transfer is a simple instance of the "metaphorization" of information in the surrounding space. A more elaborate case, concerning a creative metaphor of doubts and grievances moving in and out of a conversation as though they were independent-minded cats, is discussed in Barnden et al. [6, 9].

At the time of writing, ATT-Meta is being developed towards metaphor generation as well as understanding. The reverse transfer capability puts ATT-Meta in an advantageous position in this respect.

11.5 Metaphor in Thought

Within the Cognitive Linguistics field, it is typical to think of metaphor as something that is somehow fundamental in the mind, not just in communication and external expression, and in particular to think of many concepts, particularly abstract ones, as in some way structured by metaphor (i.e., by being the target of metaphorical mappings with source concepts). See Vervaeke and Kennedy [43] and Murphy [36] for a discussion of some of the main points here. One reason for the hypothesis is that metaphor occurs in media other than language, such as in graphical media (notably advertisements, comic strips, etc.), gesture, and music [15, 20, 21]. One might try to account for this in a number of ways, but one parsimonious option is that metaphor is inherently a mental as opposed to purely communicative or externally-expressive phenomenon. I will take the point to basically be that, when thinking but not externally communicating about some subject matters, we are at least sometimes mentally using metaphorical mappings between those subject matters and suitably-related source subject matters. This mental activity may be unconscious.

Various features of ATT-Meta, notably the Anti-Analogy-Extension Thesis, suggest interesting possibilities about metaphor within the mind. That thesis leads to an especially strong claim: namely, that portions of a metaphorical thinking episode may not individually have any translation into non-metaphorical thoughts within the person's mind. This is because extensive areas within a metaphorical pretence may not have any analogical correspondence to the target subject matter, but rather just serve indirectly to support those limited aspects of the pretence that are in analogical correspondence to the target. Open-ended elaboration of metaphor could exist in mind just as much (or more than) in language and other external expression, and it could creatively lead to new thoughts about the target that are relatively unconstrained by analogy between source and target.

For example, someone thinking (but not communicating) about the managers in (3) may mentally develop the pretended scenario in creative ways as above, such as imaging pains in many parts of the managers' bodies, not just their necks, imagining the managers massaging themselves, contorting themselves, etc. These could have consequences about the intensity of the emotional states, their longevity and difficulty of eradication, and the desires of the managers. These conclusions can be mapped to reality. But most of the pretended scenario is *not* mapped. Similarly, when Bill is thinking about an idea as physical object in, say, John's mind, metaphorically conceived as a physical terrain, Bill could think of an idea being buried under the floor of an ocean and eaten away by creatures there, as a way of considering the lengthy inaccessibility, hiddenness and deterioration of the idea, without thinking that the ocean and the creatures correspond to anything about John's mind. There is no more reason to think that anything like a comprehensive analogy for a pretended scenario is found in the case of metaphor that is purely in the mind than that such an analogy can be found in the case of metaphor in discourse.

The approach thus suggests a creative use of metaphor within the mind that goes beyond the typical idea that the way that metaphor helps creativity is through novel analogies being created between subject matters. That type of metaphorical creativity is important, but so is the open-ended elaboration of an existing analogy without extending it.

11.6 Conclusions and Final Remarks

This chapter has indicated how the ATT-Meta approach can handle, from an understander's viewpoint, certain broad and important types of creativity in metaphor. The chapter has concentrated on the case of open-ended pretended-scenario elaborations. Even very creative elaborations and novel pairings work in important part by engaging standard analogical parallels (view-specific or view-neutral), rather than by extension of an existing analogy through the addition of new correspondence rules. Elsewhere (Barnden [8], extending Lee and Barnden [33]) I have shown that the ATT-Meta approach can handle a variety of ways in which metaphorical views can be compounded (mixed). One, parallel, type of compounding involves mixing of different subject matters within one pretence, or use of multiple pretences side by side. Another, serial, type (i.e., chained metaphor) is handled by having pretences nested within other pretences. This is all implemented in the ATT-Meta system.

The chapter has also mentioned a natural extension to metaphor generation, by virtue of the reverse-transfer capability of ATT-Meta, which is a significant measure for suitably enriching pretended scenarios even during understanding. Given the open-endedness of the way in which pretended scenarios can be developed, the stage is set for creative generation as well as understanding.

ATT-Meta's emphasis on the Anti-Analogy-Extension thesis is distinctive compared to most other work on metaphor. Also, ATT-Meta's emphasis on the following topics from the previous section is distinctive compared to conceptual-metaphor theory: use of view-neutral correspondences as well as view-specific ones; opportunistic use of correspondences through guard satisfaction; intertwining of mapping with other reasoning; non-reification of metaphorical views; rejection of a domain-difference assumption; and reverse transfer. However, the emphasis on view-neutral mapping resonates with the transfer of higher-order structure in Structure-Mapping Theory (Gentner [22]) and owes a considerable debt to the AI work on metaphor by Carbonell [12].

At a suitably high level the ATT-Meta approach bears some strong similarities to those of Hobbs and Narayanan [26, 39], and somewhat less so to the MIDAS system of Martin [35]. The differences are explained in Barnden [5]. But the closest point of contact with other work, and one very relevant to the topic of creativity, is with Conceptual Integration theory, or Blending [19]. Reasoning within the pretence is broadly like the elaboration of a blend space. The relationship to blending is especially close because of something not yet mentioned: namely that a pretence can opportunistically use information from the surrounding space, much as a fictional story about, say, Sherlock Holmes can use real information about London. A pretence can also blend different source subject matters together, as needed for example for some types of metaphor compounding. However, as well as some technical differences, there is perhaps a significant difference in underlying philosophy between the ATT-Meta approach and blending. The core notion in ATT-Meta is that of a pretence or fiction. Blending of knowledge from different aspects of life is an important opportunity that fiction provides, but it is merely a subordinate one. It is a special case of the fact that fiction can depart arbitrarily from reality. Indeed, much open-ended metaphor does not rely in any important way on blending as such, as opposed to the development of a fictional scenario in the quite straightforward terms of some familiar source subject matter.

Various types of creativity in metaphor have not been mentioned at all so far. One is the way that source subject matters can be unrealistically distorted. A simple, practical example of this is provided by Musolff's example [37] of the EU as a "marriage" between countries. It is a strange marriage, in that there are far more partners than usual even in cultures that allow polygamy or polyandry, and there is *no* involvement of sex, whether intragender or intergender! Yet, we somehow understand the metaphor without any difficulty or (usually) any sense of shock, irony or amusement. ATT-Meta can cope with such distortion, which can be seen as another type of pretended scenario elaboration, partly through downplaying analogy.

A fiction-based approach to metaphor facilitates creativity in many ways, including through facilitating free elaborations and distortions, liberated moreover by policies such as the downplaying of analogy. Also, there has been extensive work on various forms of fictionalism in philosophy [17]. This suggests that fictionalist approaches to many phenomena, not just metaphor, could usefully be studied in A(G)I.

The free intertwining of metaphorical processing and general reasoning is essential in the ATT-Meta approach, and signals a way in which metaphor and creativity need to be worked more closely into theories in an AGI vein as well as into psychological and philosophical theories. Also, the chapter has advocated serious consideration of the idea that metaphor is fundamental in some thinking, in a way that goes far beyond the popular idea that some concepts are structured though mappings to others, and the popular idea in creativity theory that much creativity rests on analogy. The fundamentality of metaphor in (some) thinking is amplified by a type of holism: a frequent *lack* of analogy between specific metaphorical thoughts and any specific aspects of reality.

Acknowledgments This work was supported in part by an EU grant, FP7-PEOPLE-2012-IIF project 330569 ("GenMeta"). I am grateful to Andrew Gargett for help with data collection, and to an anonymous reviewer for helpful comments.

References

1. Barnden, J.A.: Uncertainty and conflict handling in the ATT-Meta context-based system for metaphorical reasoning. In: Akman, V., Bouquet, P., Thomason, R., Young, R.A. (eds.) Modeling and Using Context: Third International and Interdisciplinary Conference (CONTEXT 2001). Lecture Notes in Artificial Intelligence, pp. 15–29. Springer, Berlin (2001a)
2. Barnden, J.A.: Application of the ATT-Meta metaphor-understanding approach to various examples in the ATT-Meta project databank. Technical Report CSRP-01-02, School of Computer Science, The University of Birmingham (2001b)
3. Barnden, J.A.: Artificial intelligence, figurative language and cognitive linguistics. In: Kristiansen, G., Achard, M., Dirven, R., Ruiz de Mendoza Ibáñez, F.J. (eds.) Cognitive Linguistics: Current Applications and Future Perspectives, pp. 431–459. Mouton de Gruyter, Berlin (2006a)
4. Barnden, J.A.: Consequences for language learning of an AI approach to metaphor. In: Salazar, J., Amengual, M., Juan, M. (eds.) Usos Sociales del Lenguaje y Aspectos Psicolingüísticos: Perspectivas Aplicadas, pp. 15–57. Universitat de les Illes Baleares, Palma, Mallorca (2006b)
5. Barnden, J.A.: Metaphor and artificial intelligence: why they matter to each other. In: Gibbs Jr, R.W. (ed.) The Cambridge Handbook of Metaphor and Thought, pp. 311–338. Cambridge University Press, Cambridge (2008)
6. Barnden, J.A.: Metaphor and context: a perspective from artificial intelligence. In: Musolff, A., Zinken, J. (eds.) Metaphor and Discourse, pp. 79–94. Palgrave Macmillan, Basingstoke (2009)
7. Barnden, J.A.: Metaphor and metonymy: making their connections more slippery. Cognit. Linguist. **21**(1), 1–34 (2010)
8. Barnden, J.A.: Mixed metaphor: its depth, its breadth, and a pretence-based approach. In: Gibbs, Jr, R.W. (ed.) Mixed Metaphor. John Benjamins, Amsterdam (in press)
9. Barnden, J.A., Glasbey, S.R., Lee, M.G., Wallington, A.M.: Varieties and directions of inter-domain influence in metaphor. Metaphor Symb. **19**(1), 1–30 (2004)
10. Barnden, J.A., Wallington, A.M.: Metaphor and its unparalleled meaning and truth. In: Burkhardt, A., Nerlich, B. (eds.) Tropical Truth(s): The Epistemology of Metaphor and Other Tropes, pp. 85–121. De Gruyter, Berlin (2010)
11. Bowdle, B.F., Gentner, D.: The career of metaphor. Psychol. Rev. **112**(1), 193–216 (2005)
12. Carbonell, J.G.: Metaphor: an inescapable phenomenon in natural-language comprehension. In: Lehnert, W., Ringle, M. (eds.) Strategies for Natural Language Processing, pp. 415–434. Lawrence Erlbaum, Hillsdale (1982)
13. Cameron, L.: Metaphor in proasic and poetic creativity. In: Swann, J., Pope, R., Carter, R. (eds.) Creativity in Language and Literature: The State of the Art, pp. 69–82 (2011)
14. Carston, R., Wearing, C.: Metaphor, hyperbole and simile: a pragmatic approach. Lang. Cognit. **3**(2), 283–312 (2011)
15. Cienki, A., Mueller, C. (eds.): Metaphor and Gesture. John Benjamins, Amsterdam (2008)
16. Deignan, A., Littlemore, J., Semino, E.: Figurative Language, Genre and Register. Cambridge University Press, Cambridge (2013)
17. Eklund, M.: Fictionalism. In: Zalta, E.N. (ed.), The Stanford Encyclopedia of Philosophy (Fall 2011 Edition) (2011) http://plato.stanford.edu/archives/fall2011/entries/fictionalism/
18. Falkenhainer, B., Forbus, K.D., Gentner, D.: The structure-mapping engine: algorithm and examples. Artif. Intell. **41**(1), 1–63 (1990)
19. Fauconnier, G., Turner, M.: Rethinking metaphor. In: Gibbs Jr, R.W. (ed.) The Cambridge Handbook of Metaphor and Thought, pp. 53–66. Cambridge University Press, Cambridge (2008)
20. Forceville, C.: Non-verbal and multimodal metaphor in a cognitivist framework: agendas for research. In: Kristiansen, G., Achard, M., Dirven, R., Ruiz de Mendoza Ibáñez, F.J. (eds.) Cognitive Linguistics: Current Applications and Future Perspectives, pp. 379–402. Mouton de Gruyter, Berlin (2006)
21. Forceville, C.J., Urios-Aparisi, E. (eds.): Multimodal Metaphor. Mouton de Gruyter, Berlin (2009)

22. Gentner, D.: Structure-mapping: a theoretical framework for analogy. Cognit. Sci. **7**(2), 95–119 (1983)
23. Glucksberg, S.: Understanding Figurative Language: From Metaphors to Idioms. Oxford University Press, New York (2001)
24. Goatly, A.: The Language of Metaphors. Routledge, London (1997)
25. Haser, V.: Metaphor, metonymy and experientialist philosophy: challenging cognitive semantics. Mouton de Gruyter, Berlin (2005)
26. Hobbs, J.R.: Metaphor and abduction. In: Ortony, A., Slack, J., Stock, O. (eds.) Communication from an Artificial Intelligence Perspective: Theoretical and Applied Issues, pp. 35–58. Springer, Berlin (1992)
27. Indurkhya, B.: Metaphor and Cognition: An Interactionist Approach. Kluwer, Dordrecht (1992)
28. Jäkel, O.: The metaphorical concept of mind: mental activity is manipulation. In: Paper No. 333, General and Theoretical Papers, Series A, Linguistic Agency, University of Duisburg, D-4100 Duisburg, Germany (1993)
29. Lakoff, G.: Women, Fire and Dangerous Things. University of Chicago Press, Chicago (1987)
30. Lakoff, G., Johnson, M.: Metaphors We Live By. University of Chicago Press, Chicago (1980)
31. Lakoff, G., Turner, M.: More than Cool Reason: A Field Guide to Poetic Metaphor. University of Chicago Press, Chicago (1989)
32. Langlotz, A.: Idiom Creativity: a Cognitive-linguistic Model of Idiom-representation and Idiom-variation in English. John Benjamins, Amsterdam (2006)
33. Lee, M.G., Barnden, J.A.: Reasoning about mixed metaphors with an implemented AI system. Metaphor Symb. **16**(1&2), 29–42 (2001)
34. Levin, S.R.: Metaphoric Worlds. Yale University Press, New Haven, CT and London (1988)
35. Martin, J.H.: A Computational Model of Metaphor Interpretation. Academic Press, San Diego (1990)
36. Murphy, G.L.: On metaphoric representation. Cognition **60**, 173–204 (1996)
37. Musolff, A.: Metaphor and Political Discourse: Analogical Reasoning in Debates About Europe. Palgrave Macmillan, Basingstoke (2004)
38. Nacey, S.: Metaphors in Learner English. John Benjamins, Amsterdam (2013)
39. Narayanan, S.: Moving right along: a computational model of metaphoric reasoning about events. In: Proceedings of the National Conference on Artificial Intelligence (AAAI'99), pp. 121–128. AAAI Press (1999)
40. Semino, E.: Metaphor, creativity, and the experience of pain across genres. In: Swann, J. Pope, R., Carter, R. (eds.) Creativity in Language and Literature: The State of the Art, pp. 83–102 (2011)
41. Stern, J.: Metaphor in context. Bradford Books, MIT Press, Cambridge, MA (2000)
42. Sternberg, R.J., Lubart, T.I.: The concept of creativity: prospects and paradigms. In: Sternberg, R.J. (ed.) Handbook of Creativity, pp. 3–15. Cambridge University Press, Cambridge (1999)
43. Vervaeke, J., Kennedy, J.M.: Conceptual metaphor and abstract thought. Metaphor Symb. **19**(3), 213–231 (2004)
44. Walton, K.: Metaphor and prop oriented make-believe. In: John, E., Lopes, D.M. (eds.) Philosophy of Literature-Contemporary and Classic Readings: An Anthology, pp. 239–247. Blackwell, Oxford (2004)

Chapter 12
Poetry Generation with PoeTryMe

Hugo Gonçalo Oliveira and Amílcar Cardoso

Abstract PoeTryMe is a platform for the automatic generation of poetry, with a versatile architecture that provides a high level of customisation. The user can define features that go from the poem configuration and the line templates, to the initial seed words that will define a generation domain, and also the generation strategy. In this chapter, we introduce PoeTryMe's architecture and describe how we used it to generate poetry in Portuguese, using natural language processing resources for this language as well as patterns that denote semantic relations in human-created poetry. After presenting the resources used with PoeTryMe, the problem of poetry generation is tackled incrementally, as our decisions are explained and illustrated, step-by-step. In the end, the objective features of the poems generated by the implemented strategies are compared, while the best-scoring poems are shown.

12.1 Introduction

Natural language generation [30] is a sub-field of artificial intelligence and computational linguistics, which aims at developing computer programs capable of producing text that is understood by humans. Biographies [20] and weather forecasts [3] are some of the text genres that have been generated automatically. Although with singular features, the generation of text that can be seen as creative, such as story narratives [5], jokes [31] or poetry [11, 24], also fits in this sub-field.

In this chapter, we address the automatic generation of poetry, by presenting PoeTryMe, a generic platform designed for this task. Given a generation grammar and a set of relation instances, PoeTryMe generates grammatically correct and meaningful sentences. Those may then be combined according to a pre-defined strategy, towards the production of a poem, with certain features. PoeTryMe has a versatile architecture that provides a high level of customisation and can be the starting point

H. Gonçalo Oliveira (✉) · A. Cardoso
Department of Informatics Engineering, CISUC, University of Coimbra, Coimbra, Portugal
e-mail: hroliv@dei.uc.pt

A. Cardoso
e-mail: amilcar@dei.uc.pt

T.R. Besold et al. (eds.), *Computational Creativity Research: Towards Creative Machines*,
Atlantis Thinking Machines 7, DOI 10.2991/978-94-6239-085-0_12

for the development of different poetry generation systems. Several of its components can be changed: the base semantics, represented as relation instances; the sentence templates, included in the generation grammars; the generation strategies, that select and organise the lines to include in the poem; and, of course, the poem configuration, which sets the target number of stanzas, lines and their size. The combination of all these components enables the generation of a diverse set of poems, thus contributing to a positive perception of creativity.

We start this chapter by referring some work on the automatic generation of poetry, including two groupings for this task. Then, we present an overview on the architecture of PoeTryMe, followed by a detailed description of our approach to the automatic generation of poetry, in Portuguese. Each decision of our approach is described step-by-step, from an enumeration of the resources used and the process for acquiring line templates, to the implemented generation strategies and the contextualization of the poems. Before concluding, we compare the implemented strategies regarding several features, including the diversity of the poems they produce, the time they take to produce them, and the score of the produced poems, which considers how they match a target metre as well as the rhymes they use. Together with the comparison numbers, the results of each strategy are illustrated with a generated poem, which gives an idea of what we can do with PoeTryMe.

12.2 Automatic Generation of Poetry

The automatic generation of poetry is a complex task, as it involves several levels of language (e.g. phonetics, lexical choice, syntax and semantics) and usually demands a considerable amount of input knowledge. What makes this task more interesting is that some of those levels do not have to be strictly addressed.

On the one hand, writing poetic text does not have to be an extremely precise task [10], as several rules, typically present in the production of natural language, need to (or should) be broken [24]. On the other hand, poetry involves a high occurrence of interdependent linguistic phenomena where rhythm, metre, rhyme and other features like alliteration and figurative language play an important role. For instance, it is ok to have a not very well-defined message, in a trade-off for a pleasant sound given by a highly regular metre.

In this section, we present two groupings of poetry generation systems, proposed in the literature. One of them considers the applied techniques and another the properties of the generated text.

12.2.1 Poetry Generation Techniques

Regarding the followed approaches and applied techniques, Gervás [12] groups poetry generation systems into five categories:

- **Template-based**: systems that just fill templates of poetry forms with words that suit syntactic and/or rhythmic constraints.
- **Generate-and-test**: random word sequences are produced according to formal requirements, that may involve metre or other constraints.
- **Evolutionary**: systems that rely on evolutionary computation techniques.
- **Case-based reasoning (CBR)**: existing poems are retrieved, considering a target message, and then adapted to fit in the required content.
- **Stochastic language modelling**: text is generated according to a learned probabilistic model.

In fact, several poetry generation systems use templates, but most of them go further and combine them with other techniques (e.g. [7, 33]). Agirrezabal et al.'s [1] approach is mostly based on part-of-speech (POS) templates – POS-tag sequences are first learned from human-written poems, and then used as line templates. When those templates are merely filled with random words, though consistent with the target POS, the resulting text is easily recognised as machine-generated.

Manurung's chart system [23], WASP [10] and the generate-and-test strategy of Tra-la-Lyrics [16, 17] fall into the generate-and-test category. In the chart system, sentences are logically represented by first-order predicates describing the input semantics, and charts are used to generate natural language strings that match a given stress pattern. Before adding the result of a new rule to the chart, its stress pattern is checked for compatibility with the target pattern. Only results with compatible patterns are used, ensuring that the generated text satisfies the pattern. WASP is a forward reasoning rule-based system that aims to study and test the importance of the initial vocabulary, word choice, verse pattern selection and construction heuristics, regarding the acceptance of the generated verses and complete poems. Tra-la-Lyrics generates text based on the rhythm of a song melody, given as input. Using the sequence of strong and weak beats as a rhythmic pattern, the task of generating song lyrics is very similar to the generation of poetry. In the generate-and-test strategy, grammatical sentences are produced and then scored according to their suitability to a given metre and rhythmic pattern.

POEVOLVE [22] and McGonnagall [24, 25] rely on evolutionary approaches. POEVOLVE generates limericks according to a model that takes the real process of human poetry writing as a reference. In McGonnagall, the poem generation process is formulated as a state space search problem using stochastic hill-climbing. There, a state in the search space is a possible text with all its underlying representations, and a move can occur at any level of representation, from semantics to phonetics. The search model is an evolutionary algorithm encompassing evaluation and evolution.

ASPERA [11] and COLIBRI [8] are CBR poetry generation systems. They are forward reasoning rule-based systems that, given a prose description of the intended message and a rough specification of the type of poem, select the appropriate metre and stanza, generate a draft poem, request modification or validation by the user, and update their database with the information of the validated verse.

Ray Kurzweil's Cybernetic Poet [21] and Babieri et al.'s [2] lyrics generator are examples of stochastic approaches. They are both trained on a selection of poems/lyrics by a seed author. Probabilistic language models are learned and then used for generating text in the style of the seed author. In Barbieri et al.'s system, the models include constrains for rhyme, metre, syntax and semantics.

In addition to the previous, alternative techniques have been proposed for this task including constraint programming [33], or a blackboard multi-agent system [26]. The truth is that, as more systems are developed for the automatic generation of poetry, the line between these categories is becoming thinner and there are, of course, overlapping approaches. The combination of different approaches, some focused on the form and others on the content, is related with the degree of articulation [12] of a poetry generation system.

12.2.2 Properties of Generated Poetry

In his dissertation on the automatic generation of poetry, Manurung [24] affirms that poetic text must hold all the three properties of meaningfulness (M), grammaticality (G) and poeticness (P). Of course, this does not cover all kinds of poetry, as it leaves out some alternative types, including concrete poetry, but these properties were a starting point for much work on this area. Manurung adds that poetry must: (i) convey a conceptual message, meaningful under some interpretation; (ii) obey linguistic conventions prescribed by a given grammar and lexicon; and (iii) exhibit poetic features. Having this in mind, he divides poetry generation systems according to the properties held by the text they generate. To accommodate additional poetry generation systems, we refined and expanded Manurung's categories with an additional group, thus resulting in the following categories:

- **Word salad**: random words are concatenated together, without following grammatical rules, therefore not holding any of the properties.
- **Features-aware** (P): generated text exhibits poetic features (e.g. metre, rhyme), but there are no syntactical nor semantic constraints.
- **Features+Syntax-aware** (P+G): besides exhibiting poetic features, the choice of words follows a (pre-defined) grammatically correct form.
- **Poetry generation** (P+G+M): generated text exhibits poetic features, obeys syntactic rules and the choice of words follows some semantic constraints.

The random words strategy of Tra-la-Lyrics [16, 17] falls into the features-aware category. In this strategy, the selection of words is only constrained by the target metre and, when in the end of a line, by their termination, which should rhyme with previously used terminations. The choice of words does not follow any grammatical rules nor semantic constraints, so the generated text does not convey a precise meaning. The WASP system [10], POEVOLVE [22], and the generative grammar strategy of Tra-la-Lyrics [16, 17] are features+syntax-aware systems, because they generate text that is grammatically correct, but do not have any kind of semantic constraints nor concerns.

ASPERA [11], COLIBRI [8] and McGonnagall [24, 25] fall in the last category, as they satisfy the three properties. In the first two systems, words are combined according to the syntax of the language and should make sense according to a prose message provided by the user. When occurring at the end of lines, words may be additionally constrained by the strophic form (e.g. rhymes). McGonnagall's goal state is precisely a text that satisfies all the three properties. However, after several experiments, Manurung et al. [25] admit that it is difficult to produce both semantically coherent text in a strict agreement to a predefined metre.

In the last years, a series of poetry generators have been developed. Although not producing poems towards a precise conceptual message, most of them constrain the selection of words according to their meaning. These, which would also fit in the last category, include systems that produce syntactically correct text in different forms of poetry, such as haikus [27, 34] or song lyrics [2, 28, 29], in different languages, including Tamil [28, 29], Chinese [35], Finnish [33], or Basque [1]. To define a starting point for semantics, most systems start generation from a given theme or set of seed words [27, 34, 35]. This constrains the space of possible generations in a way that the poem should use these exact words, or others semantically related. The choice of relevant words may be achieved either by exploring models of semantic similarity, extracted from corpora [29, 33–35], with the help of lexical-semantic knowledge bases [1, 27, 28], or both [7]. As for systems that learn a language model [2], it is expected that model already holds the three properties.

In addition to poetic features such as rhyme and metre, the Full-FACE system by Colton et al. [7] generates poems according to the mood for a certain day, given by newspaper articles. Poems are produced with line templates, collected from the articles and from short phrases in the Web. Among other constraints, the selection of words considers the sentiment they transmit, as given by a polarity lexicon. Furthermore, this system provides comments, supporting the choices made (e.g. mood, used sentences, aesthetic measures), which contextualize the poem and are useful for more objective evaluations of the results.

12.3 PoeTryMe

PoeTryMe is not a poetry generation system, but more like a poetry generation platform, on the top of which different approaches for poetry generation can be implemented. It relies on a modular architecture (see Fig. 12.1), which enables the independent development of each module. This architecture provides a high level of customisation, depending on the needs of the system and ideas of the user. It is possible to define the semantic relation instances to be used, the sentence templates of the generation grammar, the generation strategy and the configuration of the poem. In this section, the modules, their inputs and interactions are presented.

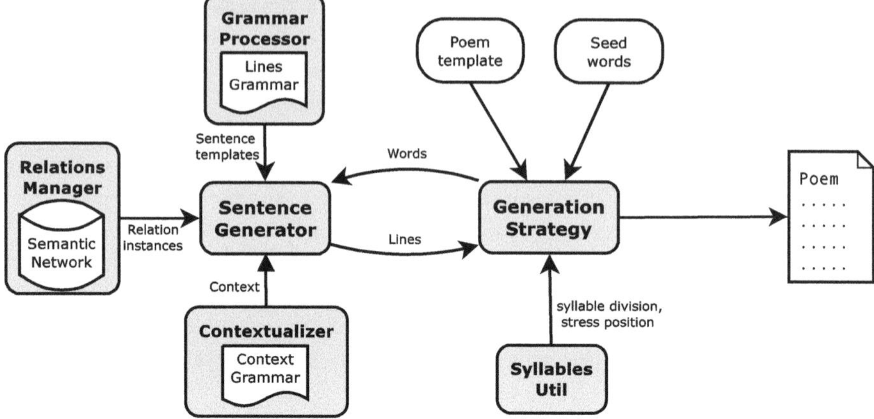

Fig. 12.1 PoeTryMe's architecture

```
#haiku
estrofe{verso(5);verso(7);verso(5)}

#sonnet
estrofe{verso(10:A);verso(10:B);verso(10:B);verso(10:A)}
estrofe{verso(10:A);verso(10:B);verso(10:B);verso(10:A)}
estrofe{verso(10:C);verso(10:D);verso(10:C)}
estrofe{verso(10:D);verso(10:C);verso(10:D)}
```

Fig. 12.2 Templates with the structure of a haiku and a sonnet with a rhyme pattern

12.3.1 Generation Strategies

A Generation Strategy organises sentences according to some heuristics, such that they suit, as much as possible, a target template of a poetic form, and that they exhibit certain features. A poem template contains the poem's structure, including the number of stanzas, the number of lines per stanza and the number of syllables of each line. Some templates might as well use a symbol for denoting the target rhyme for each line. Figure 12.2 shows the representation of poem structure templates, for generating a haiku (5-7-5) and a sonnet (14*10-syllable verses). There, the Portuguese word *estrofe* indicates a stanza and *verso* indicates a line. There is no rhyme pattern specified for the haiku, but each line of the sonnet has a symbol that results in the following rhyme pattern: *ABBA ABBA CDC DCD*.

Each strategy uses the Sentence Generator module to retrieve natural language sentences, which might be selected as poem lines. For each generation, a set of seed words may be provided and used to narrow the set of possible generations, this way defining the generation domain. To evaluate the suitability of a sentence towards a predefined metre, Syllable Utils provides several syllable-related operations, including syllable division, stress and rhyme identification (for rhymes).

We recall that an instantiation of the Generation Strategy does not generate sentences. It just follows a plan to select the most suitable sentences for each line. Selection heuristics might consider features like metre, rhyme, coherence between lines or other, depending on the desired purposes.

12.3.2 Sentence Generator

The Sentence Generator is the core module of PoeTryMe's architecture. It is used to generate meaningful natural language sentences, with the help of:

- A **semantic graph**, managed by the Relations Manager. Its nodes are words and its edges are labelled according to a relation predicate. A tuple $t = \{node_1, predicate, node_2\}$ establishes a relational instance;
- Generation **grammars**, processed by the Grammar Processor, which contain textual renderings for the (chart) generation of grammatical sentences that express semantic relations.

The generation of a sentence starts by selecting a random relation instance, e.g. $t = \{word_1, predicate, word_2\}$, from the semantic graph. Then, a random rendering for the predicate of t is retrieved from the grammar. After inserting the arguments of t in the rule body, the resulting sentence is returned. A third module, the Contextualizer, keeps track of the instances that were used to generate each line and may be used to explain the choices made.

12.3.2.1 Relations Manager

The Relations Manager is an interface to the semantic graph. It may be used, for instance, to retrieve all words related to another, or to check if two words are related.

In order to narrow the space of possible generations, a set of seed words is provided to the Relations Manager. This set will define the generation domain, used for selecting a subgraph from the main semantic graph. There, the relation instances should either contain one of the seed words or somehow related words. More precisely, the subgraph will only contain relation instances with words that are at most δ nodes far from a seed word, where δ is a neighbourhood depth threshold. It is also possible to define a surprise factor, ν, interpreted as the probability of selecting instances one level further than δ.

The number of seed words is open, and it can be enlarged with the top n relevant words for those seeds. For this purpose, the PageRank [4] algorithm is run in the full semantic graph. Initial node weights are uniformly distributed across the seed words, while the rest of the nodes have an initial weight of 0. After 30 iterations, nodes will

HYPERNYM-OF → a <arg1> like a <arg2>
HYPERNYM-OF → <arg2> is a delicious <arg1>
HYPERNYM-OF → <arg2> before <arg1>

Fig. 12.3 Chart parser example rules

be ranked according to their structural relevance to the seeds. The *n* highest-ranked nodes are selected.

12.3.2.2 Grammar Processor

The Grammar Processor is an interface for the generation grammar. Similarly to Manurung [23], to perform chart generation, the Grammar Processor uses a chart-parser in the opposite direction. The body of the rules should consist of natural language renderings of semantic relations and there must be a direct mapping between the relation names, in the graph, and the name of the head rules, in the grammar. Besides the simple terminal tokens, that will be present in the poem without any change, the Grammar Processor supports terminal tokens that indicate the position of the relation arguments (*<arg1>* and *<arg2>*), to be filled by the Sentence Generator. This way, given a relation predicate, the Grammar Processor can retrieve one (or several) renderings for a given relation instance.

PoeTryMe uses an implementation of the Earley [9] chart-parsing algorithm[1] that analyses sentences according to input grammars, which are editable text files. A very simple example of a valid rule set, with three hypernymy patterns, is shown in Fig. 12.3. These rules could be used to generate sentences such as: *a tool like a hammer*, *mango is a delicious fruit*, *man before animal.*

12.3.2.3 Contextualizer

The ability to explain how its artefacts are created is an important feature of a creative system. PoeTryMe provides this feature by keeping track of all the relation instances that originated each line. These can later be used to contextualize the poem by indicating the relation instances used to form the lines and how they are connected to a word in the generation domain. The context can be a mere list of relation instances or, if a grammar is provided specifically for this purpose, it can be a natural language piece of text.

[1] Parser available from http://code.google.com/p/pen/.

12.4 An Incremental Approach to Poetry Generation in Portuguese

Using PoeTryMe's platform, we devised a plan for the automatic generation of poetry, in Portuguese. This section starts by presenting the lexical resources we used together with PoeTryMe for this task. Moreover, given all the features that add value to this kind of creative text, we present our approach to poetry generation step-by-step. Instead of tackling all features at once, we describe how we coped with each feature, towards our main goal. Different approaches, targeting different features, originated different generation strategies, also described in this section.

12.4.1 Lexical Resources

CARTÃO [14] is a public lexical knowledge base for Portuguese, extracted automatically from three Portuguese dictionaries.[2] It contains about 330,000 semantic relation instances, held between words, which establish a semantic graph. A relation instance, represented as *<arg1 PREDICATE arg2>*, indicates that one sense of the word in the first argument (*arg1*) is related to one sense of the word in the second (*arg2*), by means of a relation identified by PREDICATE. For example:

`animal HIPERONIMO_DE cão`

(*animal* HYPERNYM_OF *dog*)

CARTÃO includes relations as synonymy, hypernymy, part-of, causation, purpose and property-of, amongst others. The name of the semantic relation also defines the part-of-speech (POS) of its arguments.

SilabasPT[3] is a tool for syllabic division and stress identification in Portuguese. It was developed for generating text based on rhythm, in the scope of Tra-la-Lyrics [16, 17], but it is independent, and can be integrated with other applications. Our implementation of the Syllables Util module, used for evaluating poem lines according to their size, relies on SilabasPT.

LABEL-LEX[4] is a lexicon of Portuguese, with 1,5 million inflected word forms, automatically generated from about 120,000 lemmas. For each form, it provides information such as the lemma, the POS and other morphological information. It was used for handling the inflection of nouns and adjectives (number and gender).

[2] CARTÃO is available from http://ontopt.dei.uc.pt/index.php?sec=downloads.

[3] SilabasPT is available from http://code.google.com/p/silabaspt/.

[4] LABEL-LEX is available from http://label.ist.utl.pt/pt/labellex_pt.php.

12.4.2 Learning Renderings for Semantic Relations

Whilst, with PoeTryMe, we can write our own grammar rules manually, this may take too much time and have an unbalanced coverage. We thus automated this task by exploiting human-created Portuguese poetry.

It is a well-known fact that semantic relations can be expressed in running text by discriminating patterns, typically used to discover new relations (see e.g. Hearst [19]). Therefore, we discovered line patterns automatically, by exploiting available poetry. The discovering process consisted of the following steps:

1. Collect and tokenize all the lines of the exploited poems;
2. For each relation instance $t = (w1, r, w2)$ in CARTÃO, select all lines (or pairs of lines) containing both words $w1$ and $w2$;
3. For each original line selected, replace $w1$ and $w2$ respectively with the terminal tokens indicating the first and second relation argument (<*arg*1> and <*arg*2>).
4. Add each sentence as a grammar rule, whose name is the same as the relation predicate r. This rule can be seen as a generic rendering for relations of type r.

A total of 4,107 renderings were discovered, after exploiting the following textual collections:

- Poems in *Versos de Segunda*, a web portal dedicated to Portuguese poetry.[5] These included mostly classical forms of poetry, especially sonnets and other poems that followed a strict metre, rhythm and rhyme pattern.
- Portuguese song lyrics, transcribed in the scope of project *Natura*.[6] As lyrics tend to follow the rhythm of the song, these poems tend to have a higher degree of freedom, concerning their form, as compared to strict forms of poetry.

Table 12.1 shows examples of the relations used, example arguments, and automatically discovered patterns, used as renderings for the relations. We included rough translations of the patterns.

We should add that, in order to deal with inflected nouns and to keep number and gender agreement in the generated sentences, before discovering the patterns, we added the number and the gender of the noun and adjective arguments to the relation predicate name. For example, the instance {*destino* synonym-of *futuro*} was changed to {*destino* ms-synonym-of-ms *futuro*} (both masculine, singular), while the instance {*versos* part-of *quadras*} was changed to {*versos* mp-part-of-fp *quadras*} (masculine, plural and feminine, plural). For the sake of clarity, this information was not included in Table 12.1. We recall that the number and gender information was obtained from LABEL-LEX. An alternative for this procedure, would be to associate inflection information to the argument tokens, similarly to what Agirrezabal et al. [1] do.

---

[5] *Versos de Segunda* is hosted at http://users.isr.ist.utl.pt/~cfb/VdS/zlista.html (as of December 2013).

[6] Project *Natura* is hosted at http://natura.di.uminho.pt/~jj/musica/lista_transcricoes.html (as of December 2013).

Table 12.1 Automatically discovered renderings, then included in the grammars

Type	POS	Example args.	Example rendering

Synonym-of	noun,noun	*destino,futuro*
		
		(destiny,future)
			`não sei que <arg1> ou <arg2>`
			`compete à minha angústia sem leme`
			(i don't know what < 1 > or < 2 > concerns my distress)

	adj,adj	*quebrada,rota*
		
		(broken,ragged)
			`<arg1> a espada`
			`já <arg2> a armadura`
			(< 1 > the sword already < 2 > armor)

Antonym-of	adj,adj	*possível,impossível* *(possible,impossible)*	`tudo é <arg1>, só eu <arg2>` (everything is < 1 >, only I'm < 2 >)

Hypernym-of	noun,noun	*mágoa, dor*
		
		(sorrow,heartache)
			`e a própria <arg2>`
			`melhor fora <arg1>`
			(and even the < 2 >, better had been < 1 >)

Part-of	noun,noun	*versos,quadras* *(lines,blocks)*	`as minhas <arg2> têm três <arg1>` (my < 2 > have three < 1 >)

Causation-of	noun,noun	fogo,queimadura
		
		(fire,burn)
			`e com o <arg1>,`
			`fiz a <arg2> que me leva`
			(and with the < 1 >, I made the < 2 > that takes me)

	verb,noun	*dor,doer* *(pain,to_hurt)*	`é <arg2> que desatina sem <arg2>` (it's < 2 > that flips out without < 2 >)
Purpose-of	noun,noun	*arma,munição* *(weapon,ammunition)*	`com <arg2> sem <arg1>` (with < 2 > without < 1 >)

	verb,noun	*taça,beber*
		
		(cup,to_drink)
			`<arg1> para <arg2>`
			`junto à perturbada intimidade`
			(< 1 > for < 2 > near the disturbed intimacy)

Has-quality	noun,adj	*certeza,certo*
		
		(certainty,sure)
			`eu que não tenho nenhuma <arg1>`
			`sou mais <arg2> ou menos <arg2>`
			(I, who has not any < 1 >, am more < 2 > or less < 2 >)

Property-of	adj,noun	*letal,morte* *(letal,death)*	`a <arg2> é branda e <arg1>` (the < 1 > is mild and < 2 >)

12.4.3 Filling Poem Templates

Using the learned renderings and the semantic graph, we may generate a broad range of random sentences, which can work as poem lines. We recall that sentences are generated from two related words. A pattern is retrieved for their relation type and the words are put in this pattern, which results in an alternative way to denote their relation. This guarantees that most sentences are semantically coherent.

In order to organise sentences into a target poetic form, we just have to generate a number of sentences equal to the number of lines in a given template. For instance, using the first four renderings of Table 12.1 and their original example arguments, we could generate the following block of four:

não sei que destino ou futuro compete à minha angústia sem leme
quebrada a espada, já rota a armadura
tudo é possível, só eu impossível
e a própria dor melhor fora mágoa

Using the same renderings, but different relation arguments, a variation of the previous poem can be generated by replacing those arguments (underlined). For example, the following block is obtained with the relation instances then listed:

não sei que cuidado ou zelo compete à minha angústia sem leme
próspera a espada, já brilhante a armadura
tudo é benigno, só eu perigoso
e a própria raiva, melhor fora doença

- *cuidado* synonym-n-of *zelo* (care, zeal)
- *próspera* synonym-adj-of *brilhante* (prosperous, shining)
- *benigno* antonym-adj-of *perigoso* (benign, dangerous)
- *doença* hypernym-of *raiva* (disease, rabies/rage)

Although, when alone, the lines are syntactically and semantically coherent, there is not a semantic connection between different lines. This kind of text is like a line-salad, in a sense that coherent lines are randomly put together in a poetic form.

12.4.4 Setting a Generation Domain

In order to generate more closely related lines, we set some boundaries on the semantic graph, and limit the generations to a subgraph, which will roughly define a domain. The subgraph is defined in terms of provided seed words and their neighbourhoods. To illustrate this, Fig. 12.4 shows part of the subgraph with the words related to the Portuguese word *poesia* (poetry).

With just this subgraph and our grammars, a great number of sentences can be generated, and different poems created, including the following block:

não soubemos de poesia nem harmonia
mais lírica que a poesia pura
só poeticamente em poesia de criar seca
que descubro as artes por dizer poesia os meus lábios

While it is still not safe to say that the poem conveys a precise meaning, all the sentences are related to the seed word. This way, one or several global meanings tend to emerge. In this case, a possible interpretation would be a poet, looking for the best way to recite poetry.

But the previous example is not very interesting, because all the lines use the word *poesia*. To have more diversity, a set of seed words can be given, instead of

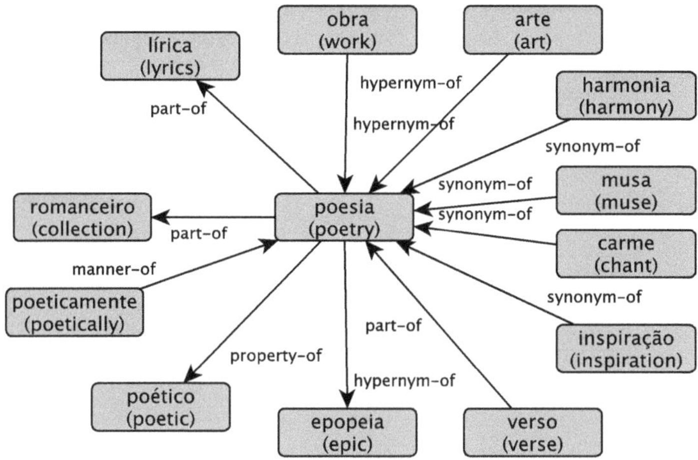

Fig. 12.4 Part of the semantic graph established by the words related to the word *poesia*

just one. Alternatively, structurally relevant words could be selected by applying the PageRank algorithm to the semantic graph. With our graph, the top-10 relevant words for *poesia* would be:

- *musa, pessoa, arte, poeta, poético, poeticamente, obra, piério, lírica, alba*[7]
  (muse, person, art, poet, poetic, poetically, work, Pierian, lyric, alba)

With these, more diverse blocks can be generated, including:

> *já sobre os dantescos poetas pende*
> *que maneiras e artes são de árduas empresas*
> *entre pessoa e o passeante*
> *com tantas liras e poesias inventadas*

To extend the previous list, we could also set $\delta >= 2$. For $\delta = 2$, the *poesia* subgraph would become something like Fig. 12.5 where we have omitted the higher level relation names. It is clear that this graph is structured on words but not senses. Therefore, there will be relations mixing different senses of polysemous words. For instance, the word *obra* might either denote a work of art, thus the relation with the bible and repertoire, or a construction, thus the relation with building. Moreover, the graph includes figurative meanings, which explain the relations between poetry, inspiration, muse and harmony, as well as the relation between inspiration and flame. While this could be problematic in other tasks, we believe that it makes poetry generation even more interesting.

With higher depths, the generated poem tends to be less focused on the domain, as in the following, where *poesia* is the only seed, with $\delta = 3$.

---

[7] Noun and adjective inflections, also included in our semantic graph, were excluded from this list.

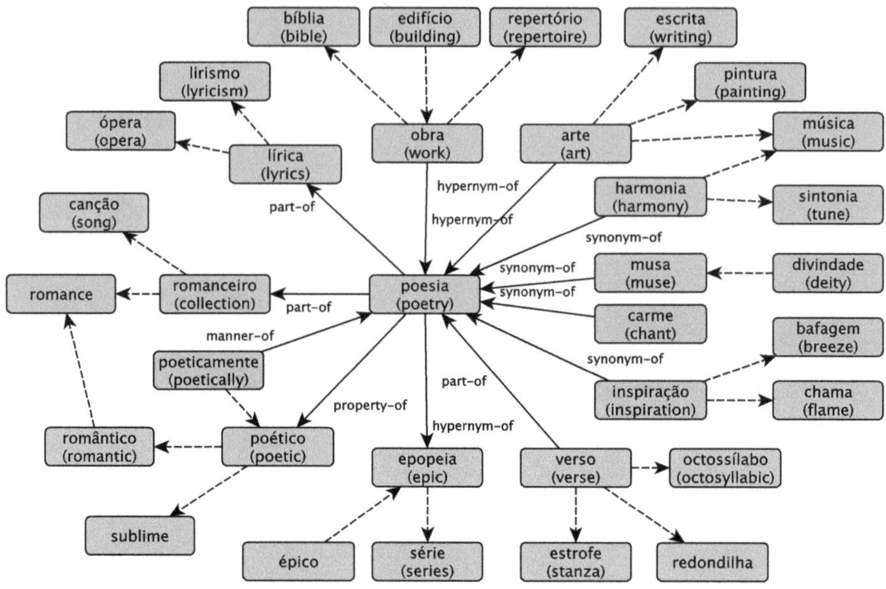

Fig. 12.5 Part of the semantic graph surrounding the word *poesia*, with 2 levels deep

e à sensação de que tudo é sonho como pintura pictórico por dentro
matinas ofícios anos breves
muitos a vida e em terra sagrada e sublime
crematística meu brasil crematístico

Therefore, we prefer using $\delta = 1$ with a surprise factor, for instance $\nu = 0.02$. This way, only a minority of out of domain relation instances is selected. Moreover, instead of only one seed word, we can use several, such as: *poesia* (poetry), *arte* (art), *escrever* (writing), *máquina* (machine), *computador* (computer) and *automático* (automatic). This set defines the domain used in all the illustrative poems in the rest of this section.

Of course, this set could still be extended by applying PageRank and selecting the top-10 relevant words: *pessoa* (person), *aparelho* (apparatus), *dispositivo* (device), *instrumento* (instrument), *inconsciente* (unconscious), *indivíduo* (individual), *automatizar* (automatise), *mecânico* (mechanical), *maquinal* (synonym of mechanical) and *automaticamente* (automatically). Yet, to have higher control on the domain, we did not apply PageRank in the examples shown here.

12.4.5 Towards the Exhibition of Poetic Features

We have shown how, given a domain, a broad range of grammatically correct sentences can be generated, and how they are used to fill poem templates. At this stage, the property of grammaticality is held by the generated text, while the relation instances used and the domain boundaries provide an approximation to meaning. But poetic text should exhibit certain features, such as regular metre and rhymes.

The next step deals with the evolution of the generated text, towards a poem with a stronger presence of those features. For this purpose, we developed three generation strategies, each based on an evaluation function that scores lines regarding how well they exhibit the target feature. The strategies are briefly described as follows:

- **Evolutionary (Ev):** an initial population of n poems is generated with the basic strategy. Each poem is scored with the evaluation function. New generations of poems consist of those with the best evaluation among the population, new poems created with the basic strategy, and some resulting from crossing two random poems in the population—a descendent poem has the same configuration and lines selected from both of its ancestors. In the end, the best scoring poem of the last generation is returned. Evolution stops either after a predefined number of generations or after m consecutive generations with the same best score.
- **Plain generate-and-test (G&T):** for each line in a poem template, n random sentences are generated, one after each other. The best scoring line is selected. All unused sentences are indexed and can be picked later, if a new line needs exactly the same number of syllables.
- **Generate-and-test + Evolutionary (G&T+Ev):** similar to the evolutionary approach except that poems in the initial population are created using the plain G&T, typically with a lower n.

A comparison of poems generated with these strategies is provided in Sect. 12.5. In the remaining of this section, we describe how the poems can evolve towards the desired metre and the presence of rhymes. All the examples were obtained with the G&T approach, because this strategy generates poems with the correct metre more consistently.

12.4.5.1 Matching the Metre

For matching the metre, it is enough to have an evaluation function that, for each generated sentence, computes the absolute difference between their number of syllables and the number of syllables in the target line of the template — the lower the evaluation, the better the sentence is. Syllables Util is used to count the number of syllables of each sentence and to identify its last stress. For a poem with n lines, this evaluation function is formally:

$$Score_{metre} = \sum_{i=1}^{n} |\#syllables(sentence_i) - \#syllables(target_i)|$$

To illustrate how metre is matched, we present a block of four, with 8 syllables per line. It was obtained with the G&T strategy, with $n = 100$, $\delta = 1$, $\nu = 0.01$.

desenhos retratos carretas
é da poesia das obras

pasmei na arte das poéticas
sempre a expor e a pôr

12.4.5.2 Forcing Rhymes to Occur

While the metre is matched with the previous function, there are still no rhymes, another important feature in poetry. For this purpose, the evaluation function is changed to include a bonus for rhymes (ρ) and a penalty for lines ending with the same word (σ). The former is an additional score for pairs of lines in the same stanza with the same termination, while the latter minimises the bias towards the selection of the same word when matching a rhyme. The final score of a poem thus becomes the following:

$$Score_{poem} = Score_{metre} + (\#rhyme\_pairs * \rho) - (\#repetitions * \sigma)$$

An important difference between the G&T and the Ev strategies is that, in G&T, bonuses and penalties are given for each sentence, at generation time. As for Ev, a global score is attached to the full poem.

To increase the probability of rhymes, a higher number of generations should be considered. Moreover, in the G&T we added a progressive multiplier π that increases the maximum number of generations (n) for lines of higher order in a stanza, where rhyme is expected. This is such that, for the ith line, $n_i = n + (n * \pi * i)$.

Rhymes are not always guaranteed for every line, as they depend not only on the number of generations, but also on the domain words and on the available patterns in the grammar. Yet, using our resources with G&T, a substantial amount of lines ends up in rhyme. The following block of four, with two pairs of rhyming lines, was obtained with the G&T and $n = 750$, $\pi = 0.35$, $\delta = 1$, $\nu = 0.01$, $\rho = 2$, $\sigma = 1$.

e a arte navegação
arte de grande criação
sistema por apriorismos
estilo por minimalismos

12.4.5.3 Forcing a Rhyme Pattern

If it is nice to have rhymes in random lines of the poem, it is probably more interesting to control the lines where they occur. Therefore, we developed a variation of the G&T strategy where: (i) the rhyme bonus is only added if the rhyme is in a predefined target place; (ii) there is an additional multiplying factor ϕ for generating lines in rhyming position.

Still, even with a high ϕ factor, the rhyming pattern is less often matched than simple rhymes, especially with the domain we defined earlier. The following is a block of four with 8 syllables per verse and rhyming pattern *ABAB*, obtained with the G&T strategy, with the following properties: $n = 750$, $\pi = 0.35$, $\phi = 15$, $\delta = 1$, $\nu = 0.01$, $\rho = 2$, $\sigma = 1$.

que vai depor vai assentar
com arte santo e manobra
na estaca me amparar
eu fiquei obrante e obra

12.4.6 Contextualizing the Poem

We took advantage of PoeTryMe's contextualizer to generate comments that describe the relation instances used for creating each verse. When the instances do not contain a domain word, their connection to the domain is also provided. To have a more friendly format, more easily understood by human readers, we manually developed a grammar to express semantic relations as text. It is similar to those for the poem lines, but simpler and with more objective sentences.

Figure 12.6 shows an 'English sonnet' (4-4-4-2), obtained with the G&T strategy, without any given rhyme pattern and with $n = 750$, $\pi = 0.35$, $\delta = 1$, $\nu = 0.01$, $\rho = 2$, $\sigma = 1$. Also, for the generation of this poem, we added two words to our previous domain, namely: *criar* (create) and *gerar* (generate). All the lines rhyme, but it does not have the maximum possible score because the 4th and 6th lines have one more syllable than the target 10. The same figure includes the contextualization of the sonnet and its rough translation. Since two of the sonnet lines are repeated, there are just 12 sentences in the contextualization.

12.5 Comparing Different Generation Strategies

It is not possible to show all the capabilities of PoeTryMe in just a few pages. While comparing the more objective features of our generation strategies, this section gives an idea of what we can do with it.

For each generation strategy, we produced 100 sonnets, with standard parameters. In each generation, a domain was set, with 2 to 10 random words of the graph. For comparison convenience, we used the same neighbourhood parameters, $\delta = 1$ and $\nu = 0.01$, and we set the same bonuses and penalties $\rho = 3$ and $\sigma = 1$. As a sonnet has 14 lines, the best possible evaluation would be -21.

Figures 12.7, 12.8, 12.9, and 12.10 have some numbers regarding the diversity and the evaluation of the poems obtained respectively with the Ev, G&T, G&T+Ev strategies and also for the rhyme-pattern variation of G&T (G&T-rp). About the diversity, each figure reports on the distinct patterns used in the 100 poems, from the total 4,107. This includes the number of distinct patterns, the average (Avg) and standard deviation (StDev) of occurrences, and the number of times the most frequently selected pattern was used. As for evaluation, they report the average and the standard deviation of the sonnet's score. They also present the average time for generating one sonnet,[8] and they are illustrated with the best-scoring sonnet, its score and domain.

We did the same exercise with a baseline strategy, where there is no evolution. The best sonnet is not shown because, as expected, it is not very interesting. But we leave the baseline numbers for comparison purposes (see Table 12.2).

Comparatively to the other strategies, the baseline generates poems much faster and uses more diverse patterns, in a trade-off for substantially lower scores. On diversity, these numbers show that, with the current settings, there is not a pattern which is constantly used. Each selected pattern is, on average used between 1.7 and 2.2 times, depending on the strategy and the most frequent pattern was never used more than 20 times in the 100 poems. Ev (Fig. 12.7) stands out as the strategy using more distinct patterns, but its average scores and generation time suggest this strategy is not the best choice for this task. Although G&T+Ev (Fig. 12.9) generated the best poem of this exercise, averages show that the plain G&T (Fig. 12.8) is the strategy that, more consistently, produces poems with higher scores. Moreover, G&T runs faster.

Poem:

vou fundar uma vida fundadora
outras de máquina perfuradora
duma certa sensação de magia
não te peço a arte cavalaria

da grande expressão da cega arte
e com seus desenhos e arte se parte
os gerativos que querem gerar
os turvadores que querem causar

os gerativos que querem gerar
os turvadores que querem causar
o que foi utensílio de trelhos
verdes preparos verdes aparelhos

eu tenho uma musa poesia
passa a proporção que harmonia

Contextualization:

- **fundadora diz-se do que fundar, criar é o mesmo que fundar** (founder is one who founds, to create is the same as to found)
- **perfuradora é um tipo de máquina** (drill is a kind of machine)
- **magia é um tipo de sensação, magia é um tipo de arte** (magic is a kind of sensation, and a kind of art)
- **cavalaria é um tipo de arte** (cavalry is a kind of art)
- **arte é um género de expressão** (art is a sort of expression)
- **desenhos são arte** (drawings are art)
- **gerativos diz-se do que gerar** (generatives are those who generate)
- **turvadores diz-se do que causar, causar é o mesmo que gerar** (muddying are those who cause something, to cause is the same as to generate)
- **trelhos são utensílio, máquina é uma espécie de utensílio** (churns are implements, machine is a kind of implement)
- **aparelhos são preparos, aparelhos são máquina** (devices are preparations, and are machines)
- **poesia é o mesmo que musa** (poetry is the same as muse)
- **proporção é equivalente a harmonia, harmonia é igual a poesia** (proportion is equivalent to harmony, harmony is the same as poetry)

Fig. 12.6 Sonnet and its contextualization

[8] Measured in a MacBook Pro with a Intel i7 CPU and 8GB of RAM, running MacOS X 10.8.

Table 12.2 Figures of 100 sonnets generated with the baseline strategy

Diversity	
Distinct patterns used	1,138
Avg(uses/pattern)	1.05
StDev(uses/pattern)	2.23
Most used pattern	41 times
Evaluation	
Avg(score)	115.68
StDev(score)	27.80
Avg(generation time)	0.03 s

When trying to generate a poem with a predefined rhyme pattern, G&T (Fig. 12.10) produced a sonnet scored with −15, but the average score of the 100 generated poems is just −1.6, which shows that it generally results in poems with few rhymes.

Properties	
Strategy	Ev
Maximum generations	50
Reproduction probability	0.4
Crossing probability	0.5
Remain probability	0.3
Stop after stagnation	8 generations
Diversity	
Distinct patterns	824
Avg(uses/pattern)	1.50
StDev(uses/pattern)	1.73
Most used pattern	18 times
Evaluation	
Avg(gen. time)	94.01 seconds
Avg(score)	26.54
StDev(score)	8.83
Best poem	
Score	11
Domain	*podriqueira* (laziness) *punhal* (dagger) *progredimento* (progress)

tão franzinos os fez e tão débeis
esbate as linhas fracas e lânguidas
sempre a ir e a progredir
podriqueiras fontes o débeis manavam

de progredimentos e progressos
tempos punhais e gomedares
tu vais por dolo e sofisma
bago de punhal que tocou meu gládio

sugestões que o dolo braço já fizera
podriqueiras débeis mentirosas
preguiças glórias calaças palmas

depois que em feitos verdes e débeis
esbate as linhas froixas e fracas
podriqueiras glórias débeis palmas

Fig. 12.7 Figures of 100 sonnets with the evolutionary strategy, and the best-scoring sonnet

Properties	
Strategy	G&T
Generations/line	800
Progressive multiplier (π)	0.35
Diversity	
Distinct patterns	593
Avg(uses/pattern)	1.95
StDev(uses/pattern)	2.00
Most used pattern	14 times
Evaluation	
Avg(gen. time)	3.89 seconds
Avg(score)	-6.25
StDev(score)	3.32
Best poem	
Score	-17
Domain	*doçal* (variety of grape/pear) *naceja* (South American snipe) *revalorizar* (to revalue) *dissolúvel* (dissoluble)

que a uma naceja ou narceja
eu tenho uma narceja naceja
de segunda pessoa da videira
dos doçais ou das águas sem pereira

a pereira em todas as doçais
as videiras pasmadas de folhais
dos meus divinos doçais de videira
das doçais que trouxeram macieira

eu tenho uma videira parreira
nos fortes doçais na grande macieira
as videiras pasmadas de doçais

dos doçais ou das águas sem videira
eu tenho uma videira parreira
a pereira em todas as doçais

Fig. 12.8 Figures of 100 sonnets with the G&T strategy, and the best-scoring sonnet

Properties	
Strategy	G&T+Ev
Generations/line	3
Reproduction probability	0.4
Crossing probability	0.5
Remain probability	0.3
Stop after stagnation	8 generations
Diversity	
Distinct patterns	517
Avg(uses/pattern)	2.25
StDev(uses/pattern)	2.34
Most used pattern	14 times
Evaluation	
Avg(gen. time)	36.25 seconds
Avg(score)	-5.23
StDev(score)	3.31
Best poem	
Score	-18
Domain	*abezerrado* (like a calf) *versar* (to rhyme)

sou mau versejador mas sei versar
circunspeccionar e ponderar
versar em frente ir a estudar
tratar em frente ir a combinar

circunspeccionar e ponderar
vai abordar jeito p'ra versar
quero versar e volver rosa branca
ponderar a dor, pensar a labuta

sou mau versejador mas sei versar
minha mãe manda-me volver a volta
constar a dor, consistir a labuta

vai praticar jeito p'ra versar
sempre a consistir e a constar
usar em frente ir a praticar

Fig. 12.9 Figures of 100 sonnets with the G&T+Ev strategy and the best-scoring sonnet

We should also stress that the domains of these specific poems were defined by random words. Therefore, although the generated sentences should be semantically coherent and related to a domain word, when compared to a manually defined domain, it will be harder to have semantic coherence across lines.

Properties	
Strategy	G&T-rp
Generations/line	800
Progressive multiplier (π)	0.35
Rhyme pos. multiplier (ϕ)	10
Diversity	
Distinct patterns	590
Avg(uses/pattern)	1.99
StDev(uses/pattern)	2.0
Most used pattern	20 times
Evaluation	
Avg(gen. time)	45.13 seconds
Avg(score)	-1.61
StDev(score)	2.83
Best poem	
Score	-15
Domain	*africano* (african) *odiento* (hateful) *cangalhas* (packsaddles) *parlamentário* (negotiator) *festivo* (merry) *batido* (whipped) *debilitante* (debilitating) *débito* (debit) *gatunar* (to steal)

mais testante que a pessoa pura
tão triviais os fez e tão batidos
dos gastos feitos grandes e comidos
jubiloso e ovante sou eu

de ti há um festo pela largura
do louro deus calcados e batidos
um dos batidos ministros vencidos
mais testante que a pessoa pura

de todo o navio de cargueira
tudo é natural só eu forçada
uma janota e taful terceira

das cangalhas que trouxeram puxada
duma certa pessoa de veleira
abarcou-me a festa churrascada

Fig. 12.10 Figures of 100 sonnets with rhyme pattern *ABBA ABBA CDC DCD*, obtained with the G&T strategy, together with the best-scoring sonnet

12.6 Concluding Remarks

This chapter presented PoeTryMe, which can be seen as an exploratory creativity system acting in the complex conceptual space of poetry. One of its distinctive characteristics is definitely the wide range of possible parameterisations, which make it highly customisable and thus a powerful creative support tool. PoeTryMe, originally described in 2012 [13], may be used with different lexical resources and generate poems with different configurations. Generation grammars may be edited manually, in order to cover new linguistic constructions. Generated sentences can be selected and organised according to different criteria, in what we called the generation strategies. Since PoeTryMe's earlier instantiations, significant improvements were made, including an improved G&T strategy, where rhymes are more common, and also the new contextualizer module.

In this context, we described an incremental approach to poetry generation, in Portuguese, using PoeTryMe as a starting point. We detail each step we took towards poetry generation and the main generation strategies implemented—evolutionary (Ev), generate-and-test (G&T), and their combination (G&T+Ev). In the end, the most objective features of their results were compared, together with illustrative examples of the best scoring poems.

PoeTryMe was conceived having in mind the idea that, for a computational system to be accepted as creative, it must convey a perception of creativity into its public, i.e., the idea that creativity lies "in the eyes of the beholder" [6]. PoeTryMe follows this path in two ways: first, its modular architecture makes it adaptable to a wide range of disparate situations, including the exploration of different generation strategies (possibly considering different features), adaptation to an evolving world by learning, exploitation of different lexical and semantic resources, and possible adaptation to other languages; second, it explains the process behind each new poem, which is particularly relevant for ruling out the idea of "randomness".

According to the techniques used, our baseline implementation falls into simple template-based generation, while the others, although using templates, are closer to evolutionary and generate-and-test approaches. As for the generated poems, they are grammatically correct and, when the generation strategy considers the metre and the rhyme, they exhibit poetic features. Also, despite using templates, the richness of the lexical-semantic resources and of the poetry collection exploited provide an interesting degree of variation in the generated lines. Semantics is coherent in each line and the poem uses words of a pre-defined semantic domain. Although minor issues might arise regarding the connection of two contiguous lines, a meaning tends to emerge from the full poem, possibly assisted by the provided contextualization. Given their semantic concerns, we can say that all our strategies, but the baseline, fall into Manurung's poetry generation category.

Although our approach targeted Portuguese, it should be emphasised that PoeTryMe's architecture is flexible enough for generating poetry in other languages, as long as we have the following language-specific components: (i) a semantic graph, which could be obtained, for instance, from a wordnet-like resource; (ii) line templates, which could be learned from any corpus of poetry in the target language; (iii) a syllable division tool; and (iv) a lexicon with morphological information. In fact, PoeTryMe has recently been adapted to Spanish [18] and, in the future, other languages might be targeted. We have also been working on the generation of poetry with a predefined sentiment orientation (e.g. positive or negative), using available polarity lexicons [15]. Other future directions might involve adding other kinds of relations to the semantic network (e.g. word associations, as in [33]) and embracing current trends in poetry generation, such as generating poetry based on a given piece of text (e.g. news, as in [7, 32]).

References

1. Agirrezabal, M., Arrieta, B., Astigarraga, A., Hulden, M.: Pos-tag based poetry generation with wordnet. In: Proceedings of the 14th European Workshop on Natural Language Generation, pp. 162–166. ACL Press, Sofia, Bulgaria, August 2013
2. Barbieri, G., Pachet, F., Roy, P., Esposti, M.D.: Markov constraints for generating lyrics with style. In: Proceedings of 20th European Conference on Artificial Intelligence (ECAI), Frontiers in Artificial Intelligence and Applications, vol. 242, pp. 115–120. IOS Press (2012)
3. Belz, A.: Automatic generation of weather forecast texts using comprehensive probabilistic generation-space models. Nat. Lang. Eng. **14**(4), 431–455 (2008)

4. Brin, S., Page, L.: The anatomy of a large-scale hypertextual web search engine. Comput. Netw. **30**(1–7), 107–117 (1998)
5. Bringsjord, S., Ferrucci, D.A.: Artificial Intelligence and Literary Creativity: Inside the Mind of BRUTUS, A Storytelling Machine. Lawrence Erlbaum Associates, Hillsdale (1999)
6. Cardoso, A., Veale, T., Wiggins, G.: Converging on the divergent: the history (and future) of the international joint workshops in computational creativity. AI Mag. **30**(3), 15–22 (2009)
7. Colton, S., Goodwin, J., Veale, T.: Full FACE poetry generation. In: Proceedings of 3rd International Conference on Computational Creativity. ICCC 2012, pp. 95–102. Dublin, Ireland (2012)
8. Díaz-Agudo, B., Gervás, P., González-Calero, P.A.: Poetry generation in colibri. In: Proceedings of 6th European Conference on Advances in Case-Based Reasoning (ECCBR 2002), pp. 73–102. Springer, London (2002)
9. Earley, J.: An efficient context-free parsing algorithm. Commun. ACM **6**(8), 451–455 (1970). Reprinted in Grosz et al. (1986)
10. Gervás, P.: WASP: evaluation of different strategies for the automatic generation of spanish verse. In: Proceedings of AISB'00 Symposium on Creative and Cultural Aspects and Applications of AI and Cognitive Science, pp. 93–100. Birmingham (2000)
11. Gervás, P.: An expert system for the composition of formal spanish poetry. J. Knowl.-Based Syst. **14**, 200–201 (2001)
12. Gervás, P.: Computational modelling of poetry generation. In: Artificial Intelligence and Poetry Symposium, AISB Convention 2013, University of Exeter, United Kingdom, April. The Society for the Study of Artificial Intelligence and the Simulation of Behaviour (2013)
13. Gonçalo Oliveira, H.: PoeTryMe: a versatile platform for poetry generation. In: Proceedings of the ECAI 2012 Workshop on Computational Creativity, Concept Invention, and General Intelligence, C3GI 2012, Montpellier, France, August 2012
14. Gonçalo Oliveira, H., Pérez, L.A., Costa, H., Gomes, P.: Uma rede léxico-semântica de grandes dimensões para o português, extraída a partir de dicionários electrónicos. Linguamática **3**(2), 23–38 (2011)
15. Gonçalo Oliveira, H., Cardoso, A.: Using a generic poetry generation system to produce song lyrics with sentiment. In: Proceedings of 3rd International Workshop on Computational Creativity, Concept Invention, and General Intelligence (C3GI), PICS, Prague, Czech Republic, August 2014. collocated with ECAI (2014)
16. Gonçalo Oliveira, H., Cardoso, F.A., Pereira, F.C.: Exploring different strategies for the automatic generation of song lyrics with tra-la-lyrics. In: Proceedings of 13th Portuguese Conference on Artificial Intelligence, EPIA 2007, pp. 57–68. APPIA, Guimarães, Portugal (2007)
17. Gonçalo Oliveira, H., Cardoso, F.A., Pereira, F.C.: Tra-la-lyrics: an approach to generate text based on rhythm. In: Proceedings of 4th International Joint Workshop on Computational Creativity, pp. 47–55. IJWCC 2007, London (2007)
18. Gonçalo Oliveira, H., Hervás, R., Díaz, A., Gervás, P.: Adapting a generic platform for poetry generation to produce Spanish poems. In: Proceedings of 5th International Conference on Computational Creativity, ICCC 2014. Ljubljana, Slovenia, June 2014
19. Hearst, M.A.: Automatic acquisition of hyponyms from large text corpora. In: Proceedings of 14th Conference on Computational Linguistics, COLING'92, pp. 539–545. ACL Press (1992)
20. Kim, S., Alani, H., Hall, W., Lewis, P.H., Millard, D.E., Shadbolt, N.R., Weal, M.J.: Artequakt: generating tailored biographies with automatically annotated fragments from the web. In: Proceedings of ECAI 2002 Workshop Semantic Authoring, Annotation and Knowledge Markup, SAAKM 2002, pp. 1–6 (2002)
21. Kurzweil, R.: Cybernetic poet. http://www.kurzweilcyberart.com/poetry/rkcp_overview.php
22. Levy, R.P.: A computational model of poetic creativity with neural network as measure of adaptive fitness. In: Proceedings of the ICCBR-01 Workshop on Creative Systems (2001)
23. Manurung, H.: A chart generator for rhythm patterned text. In: Proceedings of 1st International Workshop on Literature in Cognition and Computer (1999)
24. Manurung, H.: An evolutionary algorithm approach to poetry generation. Ph.D. thesis, University of Edinburgh (2003)

25. Manurung, R., Ritchie, G., Thompson, H.: Using genetic algorithms to create meaningful poetic text. J. Exp. Theor. Artif. Intell. **24**(1), 43–64 (2012)
26. Misztal, J., Indurkhya, B.: Poetry generation system with an emotional personality. In: Proceedings of 5th International Conference on Computational Creativity, ICCC 2014, Ljubljana, Slovenia, June 2014
27. Netzer, Y., Gabay, D., Goldberg, Y., Elhadad, M.: Gaiku: generating haiku with word associations norms. In: Proceedings of the Workshop on Computational Approaches to Linguistic Creativity, CALC'09, pp. 32–39. ACL Press (2009)
28. Ramakrishnan, A., Devi, S.L.: An alternate approach towards meaningful lyric generation in Tamil. In: Proceedings of NAACL HLT 2010 2nd Workshop on Computational Approaches to Linguistic Creativity, CALC'10, pp. 31–39. ACL Press (2010)
29. Ramakrishnan, A., Kuppan, S., Devi, S.L.: Automatic generation of Tamil lyrics for melodies. In: Proceedings of the Workshop on Computational Approaches to Linguistic Creativity, CALC'09, pp. 40–46. ACL Press (2009)
30. Reiter, E., Dale, R.: Building Natural Language Generation Systems. Cambridge University Press, New York (2000)
31. Ritchie, G., Manurung, R., Pain, H., Waller, A., Black, R., O'Mara, D.: A practical application of computational humour. In: Proceedings of 4th International Joint Workshop on Computational Creativity, pp. 91–98. London (2007)
32. Toivanen, J.M., Gross, O., Toivonen, H.: The officer is taller than you, who race yourself! using document specific word associations in poetry generation. In: Proceedings of 5th International Conference on Computational Creativity, ICCC 2014, Ljubljana, Slovenia, June 2014
33. Toivanen, J.M., Järvisalo, M., Toivonen, H.: Harnessing constraint programming for poetry composition. In: Proceedings of the 4th International Conference on Computational Creativity, ICCC 2013, pp. 160–167. The University of Sydney (2013)
34. Wong, M.T., Chun, A.H.W.: Automatic haiku generation using VSM. In: Proceeding of 7th WSEAS International Conference on Applied Computer and Applied Computational Science (ACACOS'08), Hangzhou, China (2008)
35. Yan, R., Jiang, H., Lapata, M., Lin, S.-D., Lv, X., Li, X.: I, poet: automatic chinese poetry composition through a generative summarization framework under constrained optimization. In: Proceedings of the Twenty-Third International Joint Conference on Artificial Intelligence, IJCAI'13, pp. 2197–2203. AAAI Press (2013)

Chapter 13
From MEXICA to MEXICA-Impro: The Evolution of a Computer Model for Plot Generation

Rafael Pérez y Pérez

Abstract In this chapter I describe the engagement-reflection computer model of creative writing and provide details on how this model can be employed to implement agents capable of generating individual or collaborative plots. I emphasize the role of evaluation as part of the creative process and suggest some criteria that an agent must satisfy to generate c-creative products.

13.1 Introduction

> Jaguar Knight was an inhabitant of the great Tenochtitlan. Princess was an inhabitant of the great Tenochtitlan. A bad spirit took Jaguar Knight's soul provoking Jaguar Knight to become intensely jealous of Princess. Jaguar Knight tried to scare Princess by pretending that Jaguar Knight wanted to kill Princess with a lance. But instead, Jaguar Knight stumbled and wounded itself. Princess, knowing that Jaguar Knight's life was at risk, did not try to cure Jaguar Knight. In this way, Princess expected Jaguar Knight's death. Princess's state of mind was very volatile and without thinking about it Princess charged against Jaguar Knight. Suddenly, Jaguar Knight and Princess were involved in a violent fight. Princess threw some dust in Jaguar Knight's face. Then, using a dagger, Princess perforated Jaguar Knight's chest. Imitating the sacred ceremony of the sacrifice, Princess took Jaguar Knight's heart with one hand and raised it towards the sun as a sign of respect to the gods. Princess got so depressed that committed suicide.
>
> MEXICA

Narratives are an essential part of human beings. They are present in all cultures and have been recognized as essential social and cognitive tools (e.g. see Eubanks [1]). In this chapter I describe a computer model of narrative generation. It is based on Sharples account of creative writing (1999) [2]. According to his account, creative writing consists of cycles of engagement and reflection guided by constraint. During engagement "the writer devotes full attention to creating a chain of associated ideas and turning them into text". ([3], p. 143). During reflection the writer reviews

R. Pérez y Pérez (✉)
Divisi6n de Ciencias de la Comunicación y Diseño, Universidad Aut6noma Metropolitana, Cuajimalpa, Mexico
e-mail: rperez@correo.cua.uam.mx
http://www.rafaelperezyperez.com

T.R. Besold et al. (eds.), *Computational Creativity Research: Towards Creative Machines*, Atlantis Thinking Machines 7, DOI 10.2991/978-94-6239-085-0_13

the material produced, explores and transforms possible options, produces plans and constraints which guide furthers periods of engagement. These ideas originate the Engagement-Reflection Computer-Model of narrative generation. The model is implemented in a program called MEXICA; it produces plots of stories about the Mexicas, the inhabitants, in centuries past, of what is now México City. (These people are sometimes inaccurately called the Aztecs.) Its main purpose is the production of novel, coherent and interesting plots that generate useful knowledge (see the definition of all these terms later in this chapter). That is, a creative agent generates at least two types of outputs: a final product, e.g. a plot, and novel and useful knowledge that increases its expertise, in this case the capacity of the program for the generation of narratives.

Thus, computerized creativity (c-creativity) occurs when, besides producing interesting, coherent and novel outputs, and as a result of performing (a routine that represents) a creative process, a computer agent generates knowledge that:

- *does not explicitly exist in its original knowledge-base*
- *is significant for the produced output*
- *is available for the generation of future outputs* ([4, 5])

The main purpose of this chapter is to offer a general description of our project in narrative generation and to illustrate how the infrastructure created around the storyteller can be used to advance the study of c-creativity. This document provides a general description of: MEXICA, our plot generator; the Evaluator, an agent that grades MEXICA's generated plots; and MEXICA-impro, a system where two computer agents collaborate to produce as a team a narrative.

The rest of the document is organized as follows. Section two describes how the Engagement-Reflection Computer–Model for plot generation was implemented in MEXICA. Section three describes the model for plot evaluation. Section four describes the model for collaborative plot generation. The chapter ends with some conclusions about the project and some comments about the field of computational creativity in general.

13.2 MEXICA: The Engagement-Refection Computer-Model for Plot Generation

The Engagement-Reflection cycle forms the backbone of the computer model, which has been implemented in a program called MEXICA (for details see Pérez y Pérez [6], Pérez y Pérez and Sharples [5] and Pérez y Pérez [7]). This program is divided in two main parts: the first transforms a database into appropriate knowledge structures in memory. The second makes use of such structures to develop novel stories.

13.2.1 Preparing the Information

MEXICA employs two files to create its knowledge base: a dictionary of story-actions and a file of Previous Stories. Both are text files provided by the user of the system.

The dictionary of story-actions specifies all possible actions that can be performed by a character in a tale. The definition includes the name of the action, the number of characters involved in it and the set of preconditions and post conditions of performing such an action. As a distinctive characteristic, all preconditions and post conditions in MEXICA are defined in terms of emotional links and tensions between characters. Emotional links have an intensity and a valence; some examples are: character A is very grateful towards Character B; Character A hates Character B; Character A is in love with character B; Character A envies Character B. Examples of tensions are: the life of Character B is at risk; the health of Character B is at risk; character A is a prisoner; Character A is dead. Thus, the precondition to perform the action A proposes B is the emotional link A is in love with B; the consequences of the action A murders B is that character B is dead (a tension) and that all friends of B that are aware of the murder hate A (an emotional link).

Notice how some of the preconditions and post conditions depend on cultural aspects. For instance, if Character A rises his voice towards Character B, in some cultures B might not care about it while in other cultures B might feel really upset. In MEXICA, this situation can be represented by changing the intensity of a negative emotional link. Thus, preconditions and post conditions are classified in two types: logical and social. Logical conditions do not depend on cultural settings; for example, the death of a character. Social conditions depend on cultural characteristics, for example, an emotional response to a specific situation. In this way, the user of the system has the flexibility to include some cultural aspects in his definition of story actions.

The different events occurring in a story when it unfolds modify the circumstances or context surrounding the story world. In this research, such circumstances are called Story-Context. Thus, Story-Contexts are structures which represent the current story-world state of affairs in terms of emotional links and tensions between characters. For instance, imagine a plot where Princess suffers an accident and then Jaguar knight heals Princess. When the first action occurs the Story-Context registers that the life of Princess is at risk (a tension); then, when the second action is performed, the Story-Context is updated: the tension life at risk is deactivated and Princess is very grateful towards Jaguar knight (an emotional link). Thus, the Story-Context is a very dynamic structure that is refreshed each time an action is performed; it can become a very complex construction comprising several emotional relations between characters.

The Story-Context encodes the core events performed in the tale in terms of emotional links and tensions between characters. In that way, it represents as a single structure the consequences of a sequence of actions. As it will be explained later, this characteristic is essential for the model.

A set of tales–denominated Previous Stories–forms the material used to create the structures representing MEXICA's knowledge and experience. The format employed to define the Previous Stories is Character-1 Action Character-2, e.g. Jaguar Knight kills Tlatoani. That is, MEXICA does not employ natural language; so, a story is defined as a sequence of actions performed by characters. The purpose of the Previous Stories is to provide information about how the world works; they are considered as well constructed narratives and the user decides their number and topics. Like in the case of the Dictionary of Story-Actions, the Previous Stories allow expressing some cultural characteristics. These characteristic will be illustrated in the next section.

13.2.2 Construction of the Knowledge-Base

Employing the dictionary of story-actions and the Previous Stories, MEXICA creates three types of knowledge structures: Abstract Representation, Tensional Representation and Concrete Representation.

The Abstract Representation encodes part of the knowledge necessary to retrieve an appropriate following action during story development. To build these structures MEXICA performs a detailed analysis of the Previous Stories: for each action in each tale MEXICA gets the actual Story-Context, re-represents such a context in more abstract terms (all characters are substituted by variables) and stores it in memory as a new structure known as Contextual-Structure; finally, it is linked to the following action performed in the tale. In this way, MEXICA establishes a relation between Contextual-Structures (abstract representations of story-world contexts) and logical next actions to be performed. If we employ again the imaginary plot where Princess suffers an accident and then Jaguar knight heals Princess, the process to build a Contextual-Structure works as follows. After action 1 is performed the system updates the Story-Context registering the tension "The life of Princess is at risk". Then, the system re-represents the context by substituting the character Princess for the variable A; so, the Contextual-Structure records "The life of character A is at risk". Finally, the next action in the story, in this case Jaguar knight heals Princess, is linked to such a structure and characters are also substituted by variables. In this way, the system registers that when the life of character A is at risk, something logical to do is that Character B heals A (where A and B are variables that can be instantiated by any character). This process is repeated until the story ends. The same sequence of steps is repeated for each of the previous stories. Contextual-Structures are employed during plot generation to progress a story.

The use of emotional links and tensions as a way to organize knowledge provides a great flexibility. For instance, picture a Previous Story that includes a scene where Jaguar knight fancies Princess and then he serenates her; now, imagine another tale where Jaguar knight fancies Princess and then he feels intimidated by her and runs away. MEXICA records in its knowledge base that when a character A fancies character B, some logical actions to perform are that A serenates B or that A feels intimidated and runs away from B. That is, the same context can offer different—and sometimes contrary—ways to progress a story; but, nevertheless, all of them make sense. Notice that if different Previous Stories generate the same context, the system only generates one Contextual-Structure and associates all possible next actions to it. So, one Contextual-Structure might include several possible next actions to be performed.

In the same way, if the Previous Stories are alike and therefore generate similar Story-Contexts, the knowledge base will include similar Contextual-Structures, each one with its own set of possible next actions to be performed. That is, the system registers multiple options to progress the story under related circumstances. This characteristic is exploited during plot generation.

The content of the Previous Stories determines what a "logical next action to perform" is. For instance, if in the Previous Stories characters respond to other characters in an aggressive way when they are in a bad mood, MEXICA records that being aggressive is a logical action to perform when someone is irritable. In this way, Previous Stories encode cultural characteristics that influence the development of new plots. Thus, Contextual-Structures represent cultural common-sense knowledge.

In this way, the Abstract Representation establishes the universe of all possible events that MEXICA can retrieve from memory during engagement.

The Tensional Representation is employed to document the dramatic tension in the tale: when conflicts arise the tension grows; when conflicts are sorted out the tensions decrease. In MEXICA all tensions have associated a numeric value; each time an action is performed the system registers the value of all active tensions. In this way, it is possible to obtain a graphic (see Fig. 13.1). MEXICA classifies a narrative as interesting when it includes increments and decrements of tensions. Thus, a story is recounted in a correct manner when it follows the classical Aristotelian structure of a story: development, climax and resolution (see Clayton [8] , p.p. 13–15) (see Fig. 13.1). Tensions play several important roles in MEXICA. The Tensional Representations of the Previous Stories are employed as guidelines during plot generation. In this way, the system knows when to incorporate new conflicts and when to work them out. Finally, as it will be explained later, tensions are also employed during the evaluation of a finished plot.

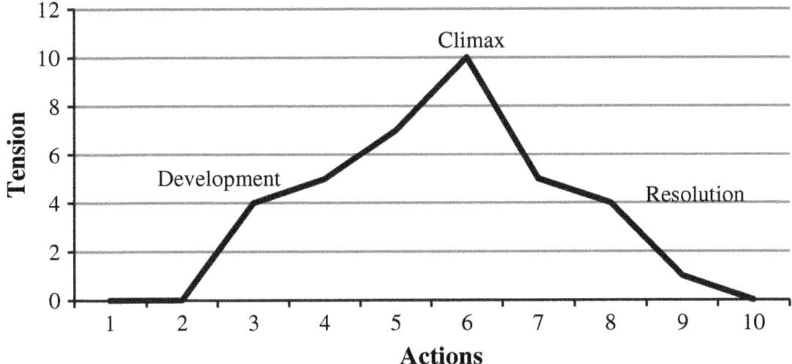

Fig. 13.1 A graphic of the dramatic tension in a story (Tensional Representation)

The Concrete Representation comprises the dictionary of Story-Actions and the set of Previous Stories in memory. It is used during reflection to break an impasse (see explanation in the next section).

The Abstract, Tensional and Concrete Representations represent the experience and beliefs of the agent. They are structures that characterize different aspects of the Story-Actions and the Previous Stories at different levels of abstraction.

13.2.3 The Engagement-Reflection Cycle

MEXICA generates stories through cycles of engagement and reflection. During engagement an initial action is performed producing a Story-Context. The system looks in memory for a Contextual-Structure that matches such a context and retrieves the set of possible next actions associated to it. One of those actions is selected as the next event in the story; the action is executed by the system, the Story-Context is updated, and the cycle starts again. During reflection MEXICA evaluates if the material produced during engagement satisfies the requirements of coherence, novelty and interest. As a result of such an evaluation MEXICA produces a set of constraints called guidelines; their purpose is to influence the production of material during engagement.

An impasse is declared when the system cannot match any Contextual-Structure in memory and therefore cannot continue developing the tale. Then, it switches to reflection to try to break the impasse. The following lines provide some details of these processes.

13.2.3.1 The Engaged State

Constraints, rather than goals or predefined story structures, guide the production of material during the Engaged State. "A writing episode starts not with a single goal, but with a set of external and internal constraints... As the writing progresses, constraints provide the tacit knowledge to guide the writing process" ([3], p. 142–143).

During engagement MEXICA performs the following tasks:

1. The consequences of an initial action chosen by the user produce an initial Story-Context.
2. This context is used as cue to probe memory; the goal is to match a Contextual-Structure *equal or similar* to the current Story-Context and then retrieve the set of possible next actions associated to it.
3. The system eliminates from such a set all those actions that do not satisfy the guidelines.
4. The system selects between the remaining actions one as the next event in the tale. The action is performed and the Story-Context updated.
5. Go to step 2.

Engagement stops after three actions have been retrieved (this is a parameter modifiable by the user) or when an impasse is declared.

The model includes a crucial feature that allows generating novel and interesting narratives: the possibility of matching a Contextual-Structure that is *equal or similar* to the current Story-Context (see step 2 of the engagement cycle). If MEXICA only matches Contextual-Structures that are equal to the current Story-Context then the output will be a plot that already exists in the Previous Story. That is, the system only reproduces ("remembers") what it already knows. If MEXICA only matches Contextual-Structures that are vaguely similar to the Story-Context the resulting story probably will be original but incoherent. Thus, the key is to find a balance between these two poles. MEXICA includes routines that manage this situation; they allow matching structures that are similar enough to the Story-Context in order to keep the coherence of the tale and at the same time retain the flexibility of the process. This characteristic allows MEXICA to produce sequences of actions that are original but also relatively consistent, like in a daydream style. During reflection the system fixes all incongruous situations produced during engagement. Thus, it is important that the agent's knowledge base includes several similar Contextual-Structures because in that way the system has more options to produce novel and interesting plots. So, if the agent's experience is rich the system has more possibilities of generating appealing outputs.

13.2.3.2 The Reflective State

During Reflection MEXICA performs the following tasks:

1. Evaluation of novelty and interestingness of the material produced during engagement.
2. Verification of the coherence.
3. If necessary, breaking an impasse triggered during engagement.

(1) During reflection MEXICA evaluates two aspects of the material produced during engagement: novelty and interest. Each time MEXICA performs an evaluation the guidelines are updated. Guidelines can be pictured as messages sent from the Reflective State to the Engaged State that attempt to contribute to the generation of novel and interesting narratives.

A story is considered novel when it is not very similar to any of the tales in the file of Previous Stories. If the system detects that the story in progress is too similar to any of the Previous Stories the novelty-guideline can be set as follows: "during engagement only employ actions that have not been used frequently in the previous stories" (c.f. Turner [9]).

A story is considered interesting when it includes, at least, an introduction, climax and resolution (see Fig. 13.1); so, in order to assess the interestingness of the story in progress MEXICA analyses its Tensional Representation. For instance, if the story in progress is too "flat" and it is necessary to increase the Tension to the reader the following interestingness-guideline can be set: "during engagement only employ actions that intensify the conflict". So, after evaluation the system might decide to increase, decrease or hold the current tension in the story in progress.

(2) The purpose of preconditions is to avoid that a story in progress contains a sequence of actions that does not satisfy common-sense knowledge (e.g. it is not possible to cure someone unless that person is wounded or ill). During reflection MEXICA checks that all actions' requirements in the story in progress are satisfied. When the system detects a deed with unfulfilled preconditions, it searches the dictionary of story-actions and chooses one whose post conditions solve the problem. Then, the agent inserts it just before the unsatisfied event. The added action might also have unfulfilled preconditions that need to be sorted out. Thus, whole episodes can be included to satisfy the preconditions of a single action.

(3) It is possible that during engagement MEXICA declares an impasse; i.e. the story-world context of the story in progress cannot match any structure in memory. In this case, MEXICA attempts to break the impasse employing the "plagiarist" tactic: it looks in the Previous Stories for a similar situation and copies the way it continues hoping that in this way the impasse will break.

13.2.4 Some Considerations About MEXICA

MEXICA does not work with natural language; it generates sequences of actions, i.e. a plot. The following lines show an example of MEXICA's output (for the sake of clarity some texts have been modified):

Jaguar knight is introduced in the story.
Princess is also introduced in the story.
Jaguar knight got intensely jealous of Princess
Jaguar knight simulated that he would hurt Princess but instead he hurt himself
Princess decided not to cure Jaguar knight
Princess attacked Jaguar knight
Jaguar knight and Princess fought
Princess killed Jaguar knight
Princess committed suicide

The system has the possibility of substituting the produced sequence of actions with texts predefined by the user; the story at the beginning of this chapter is an example.

The interplay between engagement and reflection and the way they influence each other constitute the essence of the model. MEXICA obtains from its knowledge base information that shapes the ER cycle (parameters that control internal processes depend on such information); likewise, the outputs produced by the system can be incorporated into the agent's knowledge base. Thus, all the elements that comprise the model are connected. These relationships, together with the use of emotional links and tension between characters to represent knowledge structures, give the MEXICA project its singularity.

One of the most intriguing experiences associated to the development of computer models like the one described in this document is what I refer to as the *unforeseen dynamics of the system.* Characteristics that become visible only during the implementation of the model and that, at the least, force you to reflect on aspects that were not considered during the design of the model but play essential roles in it.

An example is the instantiation of characters. The selection of the number of characters participating in a tale and the decision about how they conduct in the story-world affect the coherence, interestingness and novelty of a plot. For instance, some experiments have shown that if during plot generation the system starts to incorporate more and more characters the narrative becomes incoherent. Here is another interesting example. One of my favorite plots generated by MEXICA describes how Princess is kidnapped by Enemy and then she is rescued by Eagle Knight. As a consequence she falls in love with the knight and then, surprisingly, she decides to kill him. This plot is interesting because none of the Previous Stories describes a situation where a character kills her lover; this action provides its singularity to the tale. It emerges because of the way MEXICA instantiates characters. The following lines explain it. Sometimes, during plot generation, MEXICA retrieves actions that are not fully instantiated (an explanation of how this situation arises can be found in Pérez y Pérez [10]). The agent has some routines to sort out this problem. One of them, inspired by the work on improvisation by Johnstone [11], instructs the system to reincorporate characters. So, if a not instantiated action is retrieved from memory, the agent attempts to instantiate it using one of the characters that already exists in the narrative. In this example, during engagement, after the princess falls in love with the knight, the agent gets as the next action to perform "Princess kills Someone" (the variable Someone represents an undefined character); the system selects who will participate in the action by reincorporating the only character available in the tale: the knight. This is how Princess ends killing Jaguar Knight. Notice that if the system had introduced a new character to be murdered, the story would have taken a completely different twist and probably would not have been as interesting as it is.

There are thousands of pages written about emotions; however, it is hard to find works that focus on studying the role of emotions as a way to link ideas during the creative process (c.f. Gelernter [12]). The MEXICA project offers a computational version of this activity that illustrates the potential and flexibility of such representation. For instance, a sequence of actions can be converted into one Contextual-Structure. In this way, the events that occurred previously in the tale influence what happen next, a fundamental requirement in a model of creativity. It is possible to associate to one Contextual-Structure several possible next actions to be performed that make sense but drive a story into completely different directions.

13.3 A Computer Model for Plot Evaluation

The engagement-reflection model considers evaluation as an integral part of the creative process. The current version of MEXICA is capable of performing two types of assessments: one during plot generation and the other after the story has been finished. The first of these processes evaluates the story in progress; its purpose is to contribute to advance the narrative. The second values the final product; its goal is to determine if the system's output might be classified as c-creative. I refer to a computer agent that is capable of assessing a final product as Evaluator. This section describes MEXICA's features as an Evaluator.

MEXICA has the capacity of assessing the narratives it generates or the narratives generated by other agents, as long as they are represented in terms of sequences of actions that follow the format employed in the Previous Stories. There are three main aspects that the model considers:

(1) As explained earlier, I believe that an essential aim of creativity is the generation of expertise that is useful for the creative process itself. I claim that the same principle is valid during the assessment of a narrative. Thus, a computer model of evaluation must consider if the Evaluator, as a result of the assessment it performs, incorporates new knowledge structures into its knowledge base. Otherwise, what is the point of considering new material?
(2) As explained earlier, in MEXICA a story is recounted in a correct manner when it follows the classical Aristotelian structure of a story: development, climax and resolution (or setup, conflict and resolution). The Tensional Representation plays a fundamental role in this part of the process because it allows analysing these characteristics within a plot (see Figs. 13.1 and 13.2).
(3) It is necessary to verify if the sequence of actions that comprise a story satisfies common sense knowledge, i.e. to check if all preconditions are satisfied.

In order to perform the evaluation MEXICA weighs up seven characteristics. In some cases the process consists in calculating a value for each feature and then

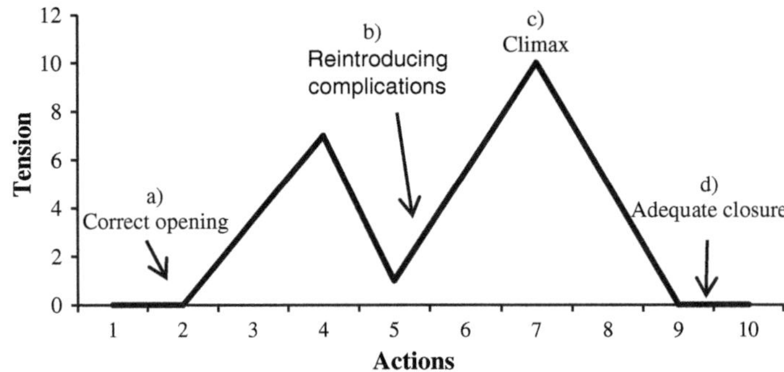

Fig. 13.2 Employing the Tensional Representation to evaluate a plot

comparing it against some references; in others, it involves establishing how much some attributes are present within the plot. The system rates each of these aspects; possible values range between zero and one. With all these information the program calculates a final result. Details of the whole evaluation model can be found in Pérez y Pérez [13]. The following lines describe the seven characteristics employed by the current version of the Evaluator.

1. Opening: A story has a correct opening when at the beginning there are no active dramatic tensions in the tale and then the tension starts to grow (see Fig. 13.2a).
2. Closure: A story "should display an overall integrity and closure, for example with a problem posed in an early part of the text being resolved by the conclusion" ([5]). Thus, the closure of a plot is adequate when all the dramatic tensions in the story are sorted out when the last action is performed (see Fig. 13.2d).
3. Climax: In the Tensional Representation the climax is characterised as the highest peak in the figure. In order to evaluate it, the system gets the average value of all Previous Stories' climax and employs it as a reference (see Fig. 13.2c).
4. Reintroducing Complications: In this work, we support narratives that appear to end and then new problems for the characters arise (i.e. where all tensions are solved and then they rise again). We refer to this situation as reintroducing-complications. MEXICA calculates the average number of complications that are reintroduced in the Previous Stories and employs it as a reference (see Fig. 13.2b).
5. Novel Contextual-Structures: A new story generates new knowledge when it generates structures that did not exist previously in the knowledge base of the system and that can be employed to build novel narratives.
6. Original Value: Besides calculating the number of novel Contextual-Structures, it is necessary to determine how original they are with respect to the information that already exists in the knowledge base.
7. Preconditions: All actions' preconditions, which represent common sense knowledge, must be satisfied. MEXICA verifies if this requirement is fulfilled.

After evaluating the plot MEXICA generates a report which includes the final result as well as some comments explaining the reasons that lead the agent to give that particular mark. The following lines show the evaluation of MEXICA's story introduced in Sect. 13.2.4.

EVALUATION OF THE STORY

This story is really good! I like it! Congratulations!

Here are some comments about your work that I hope will be a useful feedback.

COHERENCE

The story is very logical; all actions are nicely integrated and form a coherent unit.

It has a clear closure where all conflicts are worked out. Nice!

INTERESTINGNESS

The text has a good introduction.

The story reaches a nice climax with a good amount of tension. This is an important characteristic of a good narrative. Great!

The enjoyable closure of the story makes the reader to connect nicely with the plot.

There are surprising events that make the story appealing. I enjoyed that!

NOVELTY

The plot is inventive. I find the way the story unravels original!

My evaluation of your story is ->100/100

These are predefined texts. They are chosen based on the assessment of the seven features described earlier.

13.4 Generation of Collective Narratives

Creative processes take place in very complex environments. In order to have a better understanding of these phenomena it is necessary to build computer models that at least combine a social dimension and a cognitive dimension (c.f. [14]). The interaction between agents provides a useful social setting. Thus, the development of collaborative systems where agents—each one capable of producing individually a final product—work together in the construction of interesting outputs might contribute to the study of creativity.

Employing MEXICA as a base, we decided to create a model for the generation of collective narratives named MEXICA-impro ([15, 16]). The system is formed by two MEXICA agents working together to develop as a team a story plot. Agent 1 is called the Leader and agent 2 is called the Follower. The Leader starts the improvisation and decides when it finishes. The Leader generates material through one complete E-R cycle and then cues the Follower to continue the narrative. Then, the Follower takes the material generated so far and progresses the story through one complete E-R cycle and then cues the Leader to continue the narrative, and so on.

The collaboration between agents follows some rules inspired by human improvisation. Because "Strictly speaking, improvisation is making it up as you go alone" ([17], p. 14), collaborative MEXICA agents do not employ any explicit planning, predefined story-structure or characters' goal during collective plot generation. Furthermore, neither the Leader nor the Follower can modify the sequence of actions that the other agent generates, or change anything that they have previously produced. Thus, the initial contribution of the Leader shapes the unraveling of the plot.

The MEXICA-impro project has an important goal: the stories generated by the collaborative agents cannot be developed by any of them alone. In this way, the following sentence complements the previous definition of c-creativity:

> If the narrative generated by the collaborative agents cannot be developed by any one of them alone, and when it is incorporated as part of the Previous Stories that narrative generates original structures within their knowledge base that eventually can be employed by the agents to produce new stories, we refer to such a story as a collectively-creative narrative ([4]).

It is possible to achieve this goal by providing each agent with different knowledge-bases. We have considered three different ways of doing this.

The first has to do with preconditions and post conditions. As explained earlier, actions' (pre and post) conditions can be classified as logical and social. The former are the same for all agents; however, the latter might change between participants in a collective narrative. Thus, if agents' pre and post conditions are different then the requirements to perform an action might vary between them as well as the consequences of it. As a result, each participant will generate different Contextual-Structures from the same sequence of actions. This situation opens interesting questions about how this characteristic influences the interaction between the narrative's contributors and how it shapes the features of the final product.

In the second option, the number of actions that each agent includes in its knowledge base might be different. That implies that during plot generation one or both of them might have to deal with unknown situations. So, collaborators must be prepared to affront these conditions.

In the third option, the agents' set of Previous Stories might be dissimilar. They can diverge in two aspects: the narratives' themes or their number. Regarding themes, one can provide each agent with a set of completely different tales, e.g. a set of physical violent Previous Stories and a set of love and tacky Previous Stories. As a result, their Contextual-Structures contain different information that might or might not complement each other. With respect to their number, the amount of Contextual-Structures depends on the quantity of Previous Stories. Thus, given a sequence of actions, one agent might have more choices to progress a narrative than the other. Of course, all these alternatives can be combined. So, MEXICA-impro has a great potential to explore different aspects of the interaction between collaborative creative agents.

So far, we have only tested the model employing two different sets of Previous Stories: violent narratives versus love narratives (see Pérez y Pérez [4]). The work concentrates on analysing how the agents' knowledge-bases are affected when the new plots generated by MEXICA-impro are incorporated as part of the Previous Stories. To analyse the results we employ what we refer to as a Knowledge-Map (K-Map) (see Fig. 13.3). K-Maps allow comparing the number and features of existing KB-Structures in memory through time. Elements in the map are grouped by its characteristics. The vertical axis indicates the number of tensions and the horizontal

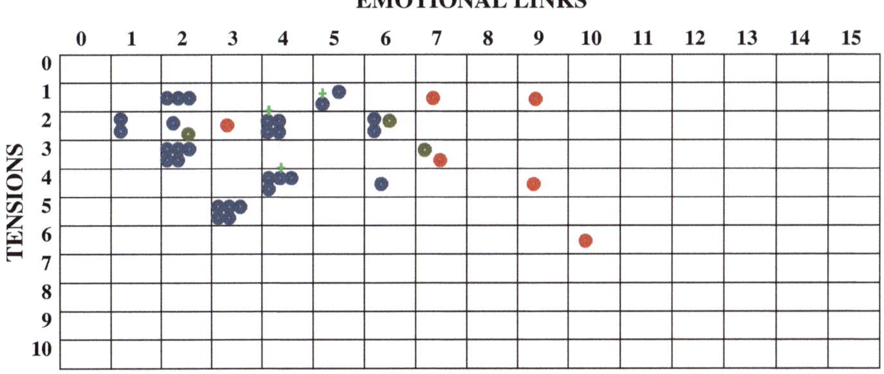

Fig. 13.3 A knowledge map

axis indicates the number of emotional links. I employ coordinates of type (Number of tensions, number of emotional links) to point out specific positions. For instance, (4, 3) refers to the position (4 tensions, 3 emotional links). Each entrance in the map represents the number of CK-Structures in that position.

The following lines provide a hypothetical example to illustrate how to utilise the map. In Fig. 13.3 the blue circles represent the Contextual-Structures generated after processing the Previous Stories. They are situated close to each other within the Map. As explained earlier, the reason of this proximity is the alikeness in the content of the Previous Stories; as a consequence, the Contextual-Structures have similar characteristics and therefore they are located in the same or adjacent areas. The similarity between narratives also provokes that some of such structures include several possible next actions to be performed. These two characteristics provide an important flexibility to the system during plot generation.

The green circles in Fig. 13.3 represent the new knowledge structures produced after incorporating into the system a narrative generated by MEXICA-impro; as a main characteristic, such a tale is similar to those in the Previous Stories. As expected, only few new Contextual-Structures are created and they tend to be located in populated blue areas within the K-Map. In the same way, some new possible next actions to be performed are added to existing structures; this is represented in Fig. 13.3 by a cross on top of some blue circles.

The red circles represent the new knowledge structures generated after incorporating into the system a second narrative generated by MEXICA-impro; as a main characteristic, such a tale is dissimilar to those in the Previous Stories. When a new tale is unrelated to the Previous Stories several new Contextual Structures are created and they tend to be located in isolated regions within the map. In most cases, they have associated only one next possible action to be performed.

Intuitively, one might wish that our computer agent generates narratives as different as possible to those in the Previous Stories. After all, one of the main characteristics of creativity is novelty. However, that is not necessarily the best scenario in a long term. It is true that the isolated Contextual-Structures that emerge from singular narratives allow MEXICA-impro to register in its knowledge base unknown situations. That is, the agent is increasing its knowledge-base. Certainly, that is a good thing to do. However, the lack of different possible next actions to be performed and the absence of similar Contextual-Structures around it make the use of isolated structures very rigid and predictable. Some experiments have shown that the use of one isolated structure pushes the system to employ other isolated structures producing as a final result the reproduction of tales that already exist in the Previous Stories. Thus, in order to improve its capacity to produce original outputs, MEXICA-impro requires mechanisms that allow locating new structures in isolated areas within its K-Maps and then populate them. I refer to this process *as Full Expansion of Knowledge*. The interaction of agents with different knowledge bases seems useful for this purpose. Such expansion brings to mind what some artists report when they are preparing themselves to produce a new piece of art. For example, Pablo Picasso generated several similar sketches in preparation to paint the Guernica; Gabriel García Márquez reports that he spent several years researching about dictators in Latin

America before writing *El Otoño del Patriarca*; and so on. Within the framework provided by MEXICA-impro the Full Expansion of Knowledge might be achieved by running several times the system and constantly interchanging the role of Leader and Follower between the two agents. This solution needs to be tested. But the important point here is to illustrate the type of research that a system like MEXICA-impro allows.

13.5 Conclusion

The Engagement-Reflection computer-model of writing provides detailed descriptions of every process involved in the model along with meticulous accounts of the relationships that such processes establish with each other. This in depth description of even the smallest mechanisms implicated in the system is a typical characteristic of this type of models. The requirement of expressing in such specific terms all the concepts behind an idea forces the researcher to ponder aspects that otherwise he never would have considered. The development of the ER-Computer-Model based on Sharples' general account of writing illustrates this situation.

MEXICA has proved to be a flexible tool. Because the Dictionary of Story-Actions and the Previous Stories are defined in text files by the user, it is simple to modify the agent's knowledge base. In the same way, it is possible to observe every activity within the system. The program also generates all kind of reports that permit analyzing the data produced. Thus, it is possible study the so called Unforeseen Dynamics of the System. All these characteristics provide lots of freedom during the testing of the system and allow a deep understanding of the model.

However, all this flexibility comes with a price to be paid. Computer models focus on specific aspects of the phenomenon that they represent. In that sense they are limited. For instance, MEXICA only generates plots rather than stories; it cannot explicitly manipulate objects within a story; the system cannot include suspense, flashbacks, parallel events; it does not take into consideration the effects of the environment during the creative process; and so on. The list of potential components that might influence the development and evaluation of a narrative and that MEXICA does not consider is big. Although this might be seen as a drawback I believe it constitutes one of the main advantages of computer models: it allows reducing to practicable dimensions what otherwise would be unmanageable.

It is necessary to recognize that we are dealing with a phenomenon that we hardly understand. So, it makes sense to decompose this massive complex problem into more accessible pieces that humans can deal with, like computer models. Even so, the task is still enormously difficult. That is what makes it interesting.

So, rather than focusing on what is missing we should concentrate on critically reflecting on the concepts introduced by the model, establish its scope—have a clear understanding of its strong points and limitations—and, based on that information, suggest novel directions to progress the research. That is the way a computer model of creativity must be understood and used.

In this chapter I have described three of the works that constitute the project on narrative generation: the individual plot generator, the Evaluator and the collective plot generator. Together, they combine more than 15 years of research. The way the project has progressed shows some of the diverse courses that research on narrative generation can follow. It also illustrates the value of building new models on top of previous works.

This chapter can be summarized as follows. The ER Computer-Model provides plausible explanations about how important aspects of the creative process of writing operate. It represents the creative process as a continuous cycle between automatic generation of material during engagement and the evaluation, and if it is necessary the modification, of such material during reflection. The evaluation process produces a set of guidelines that constrains the production of material during engagement. In this way, stories emerge as a result of the ER-cycle; that is, there are no predefined story structures or explicit character goals. The system first generates sequences of actions and then goes back to check that all actions' preconditions are satisfied. These characteristics distinguish MEXICA from other computer models of writing.

An extra singularity is the organization of memory in Contextual-Structures; such structures are comprised by emotional links and tensions between characters. As far as I know, there is no other model that employs emotional links and tensions between characters to represent the state of affairs of a narrative. It turns out that this representation provides great flexibility during plot generation. These results make me wonder about the essential role of emotions to make sense of the world and on their utility to deal with endless circumstances through life.

There are "external" components of computer models that are hardly mentioned but, nevertheless, play a significant role in it. They are components that "we take for granted", so to speak. MEXICA offers an opportunity to reflect on them. An example is the information that the agent exploits to construct its knowledge base and which is supplied by the user: the pre and post conditions, the story actions and the previous stories. Based on the ideas expressed in this text, it would be useful to work on developmental agents that allow observing how emotional links and tensions emerge during development, how they are associated to particular actions, and how the cognitive abilities necessary to represent them appear. Some efforts in that direction have been done (e.g. see Aguilar and Pérez y Pérez [18]).

Evaluation is a recurrent topic in meetings about computational creativity. Despite its relevance we hardly understand how it works. Human evaluation is too complex; it depends on individual knowledge and experience, cultural and social traditions, and even personal mood. That is why it is so hard to model it. Although some authors have contributed in this area (e.g. [19–22]) much more work is desirable. The main goal of the Evaluator is to help to understand better this key process. MEXICA and MEXICA-impro provide a useful framework for this task.

The model for the generation of collaborative narratives includes a cognitive dimension and a social dimension. Because each agent has its own knowledge base and the capacity of producing interesting plots, their interaction allows exploring the links between cognitive and social aspects during plot generation. We need to enhance the social dimension. We are starting by enriching the communication between agents.

What is next? Currently we are working on improving the cognitive abilities of MEXICA (e.g. see Morales and Pérez y Pérez [23]) to produce more elaborate plots where features like suspense can be included; we are interested in incorporating explicit cultural representations, like social norms, and study their influence in plot generation and evaluation (see Guerrero-Román[24]); we are experimenting with visual narratives (see Pérez y Pérez and Morales [25] and Pérez y Pérez [26]) and, of course, we want to enrich the Evaluator and the collaboration between agents.

In the literature one can find important works on plot generation (e.g. [9, 27–29]); nevertheless, much more research is needed. We need novel ideas in order to tackle the problems we are facing. Initiatives like SLANT ([30]), where three well recognised systems collaborate to produce stories that neither of them could produce alone, are important.

Narrative generation is one of the most challenging areas within computational creativity. There is much work left to do. If I were asked to explain what computer models of creativity are useful for, this would be my answer:

They are tools for reflection that challenge our deep-rooted ideas about how we create.

I hope that through this chapter you have confronted some of yours.

References

1. Eubanks, P.: Poetics and narrativity: how texts tell stories. In: Bazerman, C., Prior, P. (eds.) What Writing Does and How it Does It. LEA, Mahwah (2004)
2. Sharples, M.: How We Write: Writing as Creative Design. Routledge, London (1999)
3. Sharples, M.: An account of writing as creative design. In: Levy, C.M., Ransdell, S. (eds.) Science of Writing: Theories, Methods. Individual Differences and Applications, pp. 127–148. Erlbaum Press, Mahwah (1996)
4. Pérez y Pérez, R.: A Computer-based Model for Collaborative Narrative Generation (under review)
5. Pérez y Pérez, R., Sharples, M.: Three computer-based models of storytelling: BRUTUS. MINSTREL and MEXICA. Knowl. Based Syst. **17**(1), 15–29 (2004)
6. Pérez y Pérez, R., Sharples, M.: MEXICA: a computer model of a cognitive account of creative writing. J. Exp. Theor. Artif. Intell. **13**(2), 119–139 (2001)
7. Pérez y Pérez, R.: Employing emotions to drive plot generation in a computer-based storyteller. Cognit. Syst. Res. **8**(2), 89–109 (2007)
8. Clayton, J.J.: Introduction: on fiction. The Heath Introduction to Fiction. Heath and Company, Lexington (1996)
9. Turner, S.R.: MINSTREL: a computer model of creativity and storytelling. Ph.D. Dissertation, University of California, Los Angeles (1993)
10. Pérez y Pérez, R.: MEXICA: a computer model of creativity in writing. Doctoral Dissertation, University of Sussex (1999)
11. Johnstone, K.: IMPRO Improvisation and the Theatre. Methuen Drama, London (1989)
12. Gelernter, D.: The Muse in the Machine. Fourth Estate, London (1994)
13. Pérez y Pérez, R: The three layers evaluation model for computer-generated plots. In: Proceedings of the Fifth International Conference on Computational Creativity, Ljubljana, Slovenia (2014) (http://computationalcreativity.net/iccc2014/proceedings/)

14. Sun, R.: Prolegomena to integrating cognitive modeling and social simulation. In: Sun, R. (ed.) Cognition and Multi-Agent Interaction, pp. 3–26. Cambridge University Press, Cambridge (2006)
15. Pérez y Pérez, R., Castellanos, V., Ávila, R., Peñalosa, E., Negrete, S.: Mexica-impro: ideas para desarrollar un modelo computacional de improvisación. CIENCIA Ergo Sum, Vol. 18 Número 1, pp. 35–42 (2011)
16. Perez y Perez, R., Negrete, S., Peñaloza, E., Castellanos, V., Ávila, R. and Lemaitre, C.: MEXICA-Impro: a computational model for narrative improvisation, In: Proceedings of the International Conference on Computational Creativity, pp. 90–99. Lisbon, Portugal (2010)
17. Halpern, C., Close, D., Johnson, K.: Truth in Comedy: The Manual of Improvisation. Meriwether, Colorado Springs (2001)
18. Aguilar, W., Pérez y Pérez R.,: Computer model of a developmental agent to support creative-like behavior. In: Proceedings Creativity and (Early) Cognitive Development: A Perspective from Artificial Creativity, Developmental AI, and Robotics, AAAI Spring Symposium, Technical Report SS-13-02, pp. 8–13. AAAI Press (2013)
19. Jordanous, A.: A standardised procedure for evaluating creative systems: computational creativity evaluation based on what it is to be creative. Cognit. Comput. **4**(3), 246–279 (2012)
20. Pease, A.; Winterstein, D.; and Colton, S. (2001). Evaluating machine creativity. In: Weber, R. and von Wangen-heim, C. G. (eds.) Case-Based Reasoning: Papers from the Workshop Programme at ICCBR 01Vancouver, Canada, pp. 129–137
21. Peinado, F., Francisco, V., Hervás, R., Gervás, P.: Assessing the novelty of computer-generated narratives using empirical metrics. Mind. Mach. **20**(4), 565–588 (2010)
22. Ritchie, G.: Some empirical criteria for attributing creativity to a computer program. Mind. Mach. **17**, 76–99 (2007)
23. Morales-Palafox, E., Pérez y Pérez, R.: Razonamiento analógico: una herramienta en la creación de narrativas. Res. Comput. Sci. **55**, 3–13 (2012)
24. Guerrero-Román, I., Pérez y Pérez, R.: Social Mexica: a computer model for social norms in narratives. In: Proceedings of the Fifth International Conference on Computational Creativity, Ljubljana, Slovenia (2014) (http://computationalcreativity.net/iccc2014/proceedings/)
25. Pérez y Pérez, R., Morales, N., Rodríguez, L.: Illustrating a computer generated narrative. In: Proceedings of the Third International Conference on Computational Creativity, pp. 103–110. Dublin (2012)
26. Pérez y Pérez R., González de Cossío, M., Guerrero, I.: A computer model for the generation of visual compositions. In: Proceedings of the Fourth International Conference on Computational Creativity, pp. 105–112. Sydney, Australia (2013)
27. Gervás, P., Díaz-Agudo, B., Peinado, F., Hervás, R.: Story plot generation based on CBR. Knowl. Based Syst. **18**, 235–242 (2005)
28. Meehan, J.: TALE-SPIN. In: Shank, R.C., Riesbeck, C.K. (eds.) Inside Computer Understanding: Five Programs plus Miniatures, pp. 197–226. Erlbaum Press, Hillsdale (1981)
29. Riedl, M.: Narrative planning: balancing plot and character. Ph.D. dissertation, Department of Computer Science, North Carolina State University, Raleigh (2004)
30. Montfort, N., Pérez y Pérez R., Harrell, F., Campana, A.: Slant: a blackboard system to generate plot, figuration, and narrative discourse aspects of stories. In: Proceedings of the Fourth International Conference on Computational Creativity, pp. 168–175. Sydney, Australia (2013)

Chapter 14
Handle: Engineering Artificial Musical Creativity at the "Trickery" Level

Simon Ellis, Alex Haig, Naveen Sundar G, Selmer Bringsjord, Joe Valerio, Jonas Braasch and Pauline Oliveros

Abstract We embrace a "middle-standard" view of creativity in AI, according to which the driving goal is to engineer computational systems able to "trick" humans into regarding them to be human-level creative. We then report upon three versions of our system of this type in the realm of music: Handle. One of the important hallmarks of our engineering is a commitment to exploiting the power of formal computational logic within the highly intuitive domain of music; accordingly, coverage of the still-incomplete but fast-maturing *music calculus* $\mathcal{M}$ is included.

S. Ellis (✉) · A. Haig · J. Valerio
Department of Computer Science, Rensselaer Polytechnic Institute (RPI), Troy, NY 12180, USA
e-mail: elliss5@rpi.edu

A. Haig
e-mail: katadh@sdcglobal.net

J. Valerio
e-mail: valerj3@rpi.edu

N. Sundar G
Department of Computer Science & Department of Cognitive Science, Rensselaer Polytechnic Institute (RPI), Troy, NY 12180, USA
e-mail: naveensundarg@gmail.com

S. Bringsjord
Department of Cognitive Science, Department of Computer Science, and Lally School of Management & Technology, Rensselaer Polytechnic Institute (RPI), Troy, NY 12180, USA
e-mail: Selmer.Bringsjord@gmail.com

J. Braasch
School of Architecture, Rensselaer Polytechnic Institute (RPI), Troy, NY 12180, USA
e-mail: braasj@rpi.edu

P. Oliveros
Department of the Arts, Rensselaer Polytechnic Institute (RPI), Troy, NY 12180, USA
e-mail: olivep@rpi.edu

T.R. Besold et al. (eds.), *Computational Creativity Research: Towards Creative Machines*, Atlantis Thinking Machines 7, DOI 10.2991/978-94-6239-085-0_14

14.1 Introduction and Plan

We briefly articulate a "middle-standard" position on machine creativity as measured against human creativity (§14.2), and report on the status of our decade-long attempt to concretize that position in the realm of music, specifically in the form of a multi-talented intelligent agent: Handle. Our middle-standard position falls between what we see as two extreme positions on machine creativity: on the one hand, Cope's position, which (for reasons we briefly explain below) is acutely "low-standard," and on the other, a position articulated by Bringsjord et al.: a "high-standard" one according to which it's probably impossible that any standard computing machine could be creative.

After placing our stake in the "middle ground," we summarize three versions of Handle, one for each of three types of muscial creativity from among the full array we ultimately seek to reach. Our ultimate goal is simply stated: For every significant and determinate type of musical creativity seen in the associated human sphere of music, and every level reached by humans in each type, Handle will match the type and level—in the "middle-standard" sense of 'match.'

As to the three types discussed in the present, short chapter: First is Handle as conductor (Handle$_{cond}$), introduced at C3GI 2012 in Montpellier, France (§14.5.1). Handle$_{cond}$ is designed to be able to interpret and provide feedback on a human performance of a score-based, classical solo piece of music. Second is Handle as jazz musician (Handle$_{jazz}$), summarized in Sect. 14.5.2, engineered to able to join humans in the production of improvised, "free" jazz. Third is Handle as film composer (Handle$_{film}$); here our engineering, still very active and more suggestive than mature, centers around the composition of music to suitably accompany narrative expressed cinematically. Handle$_{film}$ is summarized in Sect. 14.5.3.

One of the important hallmarks of the Handle project is an unwavering commitment to exploiting the power of formal computational logic within the highly intuitive domain of music. Accordingly, *the music calculus*, $\mathscr{M}$, a first-order sorted modal logic, has been invented as a means of modeling certain aspects of musical creativity and cognition. A significant portion of the sequel (viz., §14.3) is devoted to the presentation of parts of this calculus.

We turn now, as planned, to staking out our aforementioned "middle standard."

14.2 Our "Middle-Standard" View of Creativity

In earlier work, Bringsjord, joined by Ferrucci and Bello, argues that while machines can't be *genuinely* creative, at least in the literary sphere they can nonetheless be engineered to *seem* to be creative [8]. This two-part position is partly philosophical in nature (based as it is upon a priori reasoning), and partly engineering-centred (based as it is upon producing a computational artifact capable of generating compelling short-short stories [8]). On the philosophical side, in order for a machine to be genuinely creative (*creative*$_G$), it would need to pass the so-called "Lovelace Test"

(LT), which entails that what the machine does cannot be anticipated by the designer of this machine [8]. On the engineering side, it's enough for the storytelling machine to trick human readers, in Turing-testing-style, into believing that the stories produced by this machine were produced by creative humans (*creativity*$_T$). The Handle project is, like Brutus, based on a direct analogue of this two-part position: viz.,

P1 Computing machines can't be genuinely creative in the musical sphere. This means that no AI system or agent can be a genuinely creative conductor or composer. Nonetheless . . .

P2 from an AI-engineering point of view, (a) it's enough to aim for a machine conductor or composer able to trick human listeners, in Turing-testing-style, into believing that the music produced/guided by this machine was produced by genuinely creative humans; and (b) such a creative$_T$ conductor and/or composer can in fact be engineered within the foreseeable future.

The work described herein is of course directly in line with P1 and P2, and is intended to empirically demonstrate, eventually, the truth of P2 (b).

Our "middle-ground" approach to mechnical creativity differs radically from the approach advanced by Cope, a longtime researcher of the first rank working in the intersection of AI and musical creativity, who abides by a "lower" definition of creativity. To confirm this, we need only turn to Cope's *Computer Models of Musical Creativity* [12], where he tells us that for him creativity is merely "[t]he initialization of connections between two or more multifaceted things, ideas, or phenomena hitherto not otherwise considered actively connected" (Cope 2005, 11). Immediately after giving this latitudinarian definition, Cope provides a series of examples of his brand of creativity in action. His last example is the solving of the following puzzle:

> I have three sons whose ages I want you to ascertain from the following clues. Stop me when you know their ages. One, the sum of their ages is thirteen. Two, the product of their ages is the same as your age. Three, my oldest-in-years son weighs sixty-one pounds.
>
> Stop, says the second man, I know their ages.
>
> What are their ages?

Under the assumptions that: (i) the second man is an adult, and hence—in our culture—at least 21 years of age; (ii) the second man couldn't deduce the answer after the second clue; and (iii) the second man knows his own age, it's possible to provide an outright proof that the correct answer is 2, 2, and 9. In an informal nutshell here, the reasoning runs as follows: Of the permutations of three numbers n, m, and k that sum to 13 and have a product that's at least 21, the only two that produce the same product (36) are: 1, 6, 6 and 2, 2, 9. Since in the former case there is no oldest, we are left with the latter as the only possibility. Since, using standard formalisms in logic-based AI [6], we have engineered a machine able to find and certify a formal proof of the argument just given, it's clear that a theorem-prover-based program able to solve this problem would not be creative$_G$. The reason is that the designer of such a computer program wouldn't be surprised in the least when a formal proof expressing the argument is found. In addition, such a program wouldn't be creative$_T$, for the simple reason that cracking such puzzles is precisely the kind of thing humans *expect* computers to be able to do, while humans, save for a select few trained in formal logic, have quite a bit of trouble with such puzzles.

Despite the fact that we anchor our r&d to "trickery," we do make one assumption about the internals of a creative$_T$ machine, and one general assumption about the epistemic context of any computational artifact that we produce.

Regarding internals, we assume that the computing machine that is a candidate for creativity$_T$ have at least the minimum representation-and-reasoning power of quantified epistemic logic. This minimality condition, and the methodology that accompanies it, have been defended elsewhere [5]. The condition reflects Bringsjord's affirmation of Piaget's seminal position that mature, general-purpose human cognition (including, most prominently, problem-solving) consists of processes operating largely on formulas having at least the expressivity of formulae in full first-order logic [17]. Given the affirmation of the condition in question, the present work reflects a desire to engineer machines that are *creative*$_{T+}$: they are both creative$_T$ *and* their internal processing conforms to Piagetian concepts of general intelligence and creative problem-solving. For more on machines that embody the first condition, see [10].

What is our second assumption? We assume that computational creativity cannot be formalized and engineered without yet another nod in the direction of logic, and specifically epistemic logic. To see our second assumption, consider a person/agent A who has produced an artifact a through some creative process p. Any formal model of A's cognition before, during, and after the creative process should by our lights have sufficient representational capacity to let us conclude, or at least consider, whether:

1. A *knows* that A itself has produced a in a creative manner.
2. A *believes* that a has not been produced by any other agents.
3. A *knew* what existed and what did not exist before it started the process that produced a.
4. A *desired* to create an artifact to fulfill some need.
5. A *intended* to engage in a series of actions making up the process p.

The above (informal) micro-theory is admittedly incomplete, but does seem unavoidable, and suggests that higher cognition surrounding creativity can be succinctly captured with an expressive formal logic. We do not claim here that we have invented and specified this logic, but we have taken appreciable steps toward doing so, by way of our modeling framework, $\mathcal{CEC}$, upon which $\mathscr{M}$ is based. We turn now to $\mathcal{CEC}$ and $\mathscr{M}$.

14.3 The Music Calculus

While considerable work has been devoted to modeling music at various levels, from the raw signal-processing stage to representing hierarchical structures, modeling of the cognitive, social, and doxastic dimensions of music has not been carried out.

We provide a small glimpse of the foundations of our approach to constructing the music calculus $\mathscr{M}$; these foundations give at least a provisional account of these three dimensions. Why do we need such a formalism? As we begin to examine the act of musical conducting in a bit more detail, we begin to see why:

> Consider a simple situation in which there is a composer c, a performer p, a listener listener, and a conductor h. The composition, or score, in question is *score*. The performance of *score* by p is *performance*. Composer c creates *score* with the intention of inducing a single emotional effect $effect_1$ in the listener of the piece, listener. Performer p has a belief that the composer intends the music to draw out $effect_1$ in listener, but performer p might want his performance to have effect $effect_2$ on listener. The conductor h might in turn have beliefs regarding what the composer and the performer intend, and c might have their own intentions for the performance. Each participant in such a scenario can have further iterative beliefs: for example, the conductor believing what the performer believes the composer intended the performance should be. The conductor should also have an understanding of emotional effects and their inter-relations. For example, h should know that a melancholic effect is incompatible with a joyous effect. Such knowledge of effects should allow the conductor to dynamically alter a performance to elicit compatible effects. . . .

Obviously, even this simple, informal analysis reveals that cognitive, social, and doxastic factors are quite real, and quite central. Our music calculus, designed to allow formal capture of such factors, is based on the *cognitive event calculus* ($\mathcal{CEC}$), which we review briefly now.

The $\mathcal{CEC}$ is a first-order modal logic. The formal syntax of the $\mathcal{CEC}$ is shown in Fig. 14.1. The syntax specifies sorts S, signature of the function and predicate symbols f, syntax of terms t, and the syntax of sentences ϕ. We refrain from specifying a formal semantics for the calculus as we feel that the possible-worlds approach, though popular, falls short of the *tripartite analysis of knowledge* (Pappas [20]), according to which *knowledge* is a *belief* that is true and justified. The standard possible-worlds semantics for epistemic logics skips over the justification criterion for knowledge.[1] Instead of giving here a full formal semantics for our calculus based in a formalization of justification, we specify a set of inference rules that capture our informal "justification-based" semantics (Fig. 14.2).

We denote that agent a knows ϕ at time t by $\mathbf{K}(a, t, \phi)$. The operators **B**, **P**, **D**, and **I** can be understood to align with belief, perception, desire, and intention, respectively. The formula $\mathbf{S}(a, b, t, \phi)$ captures declarative communication of ϕ from agent a to agent b at time t. Common-knowledge of ϕ in the system is denoted by $\mathbf{C}(t, \phi)$. Common-knowledge of some proposition ϕ holds when every agent knows ϕ, and every agent knows that every agent knows ϕ, and so on *ad indefinitum*. The Moment sort is used for representing timepoints. We assume that timepoints are isomorphic with $\mathbb{N}$; and function symbols (or functors) $+, -$, relations $>, <, \geq, \leq$ are available.

The $\mathcal{CEC}$ includes the signature of the classic *event calculus* (EC) (see Mueller's [18]), and the axioms of EC are assumed to be common knowledge in the system [2]. The EC is a first-order calculus that lets one reason about events that occur in

[1] The possible worlds approach, at least in its standard form, also suffers from allowing logically omniscient agents: agents which know all logically valid sentences.

Syntax

$$S ::= \begin{array}{l}\textsf{Object} \mid \textsf{Agent} \mid \textsf{Self} \sqsubset \textsf{Agent} \mid \textsf{ActionType} \mid \textsf{Action} \sqsubseteq \textsf{Event} \mid \\ \textsf{Moment} \mid \textsf{Boolean} \mid \textsf{Fluent} \mid \textsf{Numeric}\end{array}$$

$$f ::= \begin{array}{l} \mathit{action} : \textsf{Agent} \times \textsf{ActionType} \rightarrow \textsf{Action} \\ \mathit{initially} : \textsf{Fluent} \rightarrow \textsf{Boolean} \\ \mathit{holds} : \textsf{Fluent} \times \textsf{Moment} \rightarrow \textsf{Boolean} \\ \mathit{happens} : \textsf{Event} \times \textsf{Moment} \rightarrow \textsf{Boolean} \\ \mathit{clipped} : \textsf{Moment} \times \textsf{Fluent} \times \textsf{Moment} \rightarrow \mathit{Boolean} \\ \mathit{initiates} : \textsf{Event} \times \textsf{Fluent} \times \textsf{Moment} \rightarrow \textsf{Boolean} \\ \mathit{terminates} : \textsf{Event} \times \textsf{Fluent} \times \textsf{Moment} \rightarrow \textsf{Boolean} \\ \mathit{prior} : \textsf{Moment} \times \textsf{Moment} \rightarrow \textsf{Boolean} \\ \mathit{interval} : \textsf{Moment} \times \textsf{Boolean} \\ * : \textsf{Agent} \rightarrow \textsf{Self} \\ \mathit{payoff} : \textsf{Agent} \times \textsf{ActionType} \times \textsf{Moment} \rightarrow \textsf{Numeric} \end{array}$$

$$t ::= x : S \mid c : S \mid f(t_1, \ldots, t_n)$$

$$\phi ::= \begin{array}{l} p : \textsf{Boolean} \mid \neg\phi \mid \phi \land \psi \mid \phi \lor \psi \mid \phi \rightarrow \psi \mid \phi \leftrightarrow \psi \mid \forall x : S.\ \phi \mid \exists x : S.\ \phi \\ \mathbf{P}(a,t,\phi) \mid \mathbf{K}(a,t,\phi) \mid \mathbf{C}(t,\phi) \mid \mathbf{S}(a,b,t,\phi) \mid \mathbf{S}(a,t,\phi) \\ \mathbf{B}(a,t,\phi) \mid \mathbf{D}(a,t,\mathit{holds}(f,t')) \mid \mathbf{I}(a,t,\mathit{happens}(\mathit{action}(a^*,\alpha),t')) \\ \mathbf{O}(a,t,\phi,\mathit{happens}(\mathit{action}(a^*,\alpha),t')) \end{array}$$

Fig. 14.1 Cognitive event calculus syntax

time and their effects on fluents. The $\mathcal{CEC}$ is versatile: it provides a formal account of mendacity [11], the false-belief task (modeled by Arkoudas and Bringsjord in [2]), and the mirror test for self-consciousness, described in [9]. The latter can be consulted to read more about the calculus.

Our preliminary music calculus has at its core an EC-based hierarchical representation of the syntax and semantics of music. To our knowledge, this work represents the first attempt at modeling the hierarchical structure of music in the event calculus, or any augmentation thereof.

While the syntactic hierarchical structure of music has been commented upon in [21, 22], there has been precious little study of the compositional or hierarchical semantics in music. Our calculus is intended to remedy this. Our representation also draws upon observations that music exhibits syntactic structure similar to that found in natural language. The alphabet of music, for us, consists of events representing idealized notes combining information about the pitch, time, duration, and timbre of the note. This is exactly similar to the CHARM representation described in [22]. The CHARM system allows much leeway in how such events can be combined together

Rules of Inference

$$\frac{}{\mathbf{C}(t,\mathbf{P}(a,t,\phi)\rightarrow\mathbf{K}(a,t,\phi))}\;[R_1] \qquad \frac{}{\mathbf{C}(t,\mathbf{K}(a,t,\phi)\rightarrow\mathbf{B}(a,t,\phi))}\;[R_2]$$

$$\frac{\mathbf{C}(t,\phi)\; t\leq t_1 \ldots t\leq t_n}{\mathbf{K}(a_1,t_1,\ldots\mathbf{K}(a_n,t_n,\phi)\ldots)}\;[R_3] \qquad \frac{\mathbf{K}(a,t,\phi)}{\phi}\;[R_4]$$

$$\frac{t_1\leq t_3, t_2\leq t_3}{\mathbf{C}(t,\mathbf{K}(a,t_1,\phi_1\rightarrow\phi_2)\rightarrow(\mathbf{K}(a,t_2,\phi_1)\rightarrow\mathbf{K}(a,t_3,\phi_2)))}\;[R_5]$$

$$\frac{t_1\leq t_3, t_2\leq t_3}{\mathbf{C}(t,\mathbf{B}(a,t_1,\phi_1\rightarrow\phi_2)\rightarrow(\mathbf{B}(a,t_2,\phi_1)\rightarrow\mathbf{B}(a,t_3,\phi_2)))}\;[R_6]$$

$$\frac{t_1\leq t_3, t_2\leq t_3}{\mathbf{C}(t,\mathbf{C}(t_1,\phi_1\rightarrow\phi_2)\rightarrow(\mathbf{C}(t_2,\phi_1)\rightarrow\mathbf{C}(t_3,\phi_2)))}\;[R_7]$$

$$\frac{}{\mathbf{C}(t,\forall x.\ \phi\rightarrow\phi[x\mapsto t])}\;[R_8] \qquad \frac{}{\mathbf{C}(t,\phi_1\leftrightarrow\phi_2\rightarrow\neg\phi_2\rightarrow\neg\phi_1)}\;[R_9]$$

$$\frac{}{\mathbf{C}(t,[\phi_1\wedge\ldots\wedge\phi_n\rightarrow\phi]\rightarrow[\phi_1\rightarrow\ldots\rightarrow\phi_n\rightarrow\psi])}\;[R_{10}]$$

$$\frac{\mathbf{B}(a,t,\phi)\;\;\mathbf{B}(a,t,\phi\rightarrow\psi)}{\mathbf{B}(a,t,\psi)}\;[R_{11a}] \qquad \frac{\mathbf{B}(a,t,\phi)\;\;\mathbf{B}(a,t,\psi)}{\mathbf{B}(a,t,\psi\wedge\phi)}\;[R_{11b}]$$

$$\frac{\mathbf{S}(s,h,t,\phi)}{\mathbf{B}(h,t,\mathbf{B}(s,t,\phi))}\;[R_{12}] \qquad \frac{\mathbf{I}(a,t,happens(action(a^*,\alpha),t'))}{\mathbf{P}(a,t,happens(action(a^*,\alpha),t))}\;[R_{13}]$$

$$\frac{\begin{array}{l}\mathbf{B}(a,t,\phi)\;\;\mathbf{B}(a,t,\mathbf{O}(a^*,t,\phi,happens(action(a^*,\alpha),t')))\\ \mathbf{O}(a,t,\phi,happens(action(a^*,\alpha),t'))\end{array}}{\mathbf{K}(a,t,\mathbf{I}(a^*,t,happens(action(a^*,\alpha),t')))}\;[R_{14}]$$

$$\frac{\phi\leftrightarrow\psi}{\mathbf{O}(a,t,\phi,\gamma)\leftrightarrow\mathbf{O}(a,t,\psi,\gamma)}\;[R_{15}]$$

Fig. 14.2 Cognitive event calculus inference rules

to form hierarchical structures. We impose some constraints that stipulate that such structures must correspond to some abstract syntax:

1. Events in music must have some *syntax* with which they can combine with other events in music;
2. events in music must have *semantics* or meaning which interact with the meaning of other events to produce a composite meaning for the whole musical piece.

To this end, we use a representation inspired by the Combinatory Categorial Grammar approach to modeling meaning in natural and formal languages. (See [28] for a quick introduction to the CCG formalism.) Informally, each word in a language is assigned an expression in the typed lambda calculus. The types also specify one of two possible directions in which the lambda function can take arguments. The types allow certain parses of sentences to be ruled out. The meaning of a piece of text is one of the many functional reductions that can be carried out.

The following example illustrates this. The word 'John' has syntactic type *NP*, that is, noun phrase, and has semantic value $john$; similarly, 'Mary' has syntactic type *NP* and semantic value $mary$. The word 'loves' is a bit more complex. It has syntactic type $(S/NP)\backslash NP$, which means that the word 'loves' combines with an NP on the left to give a phrase with type (S/NP); it then combines with an NP on the right to give a phrase of type S, which is of course a complete sentence. The word 'loves' has a lambda function as its semantic value; this function indicates the operations we just described. The following is a parse tree for "John loves Mary", which results in an analysis that gives us $loves(john, mary)$ as the meaning of the whole sentence at the bottom of the parse.

$$\dfrac{\dfrac{\dfrac{John}{NP : john} \quad \dfrac{loves}{(S/NP)\backslash NP : \lambda xy.loves(x, y)}}{S/NP : \lambda y.loves(john, y)} \quad \dfrac{Mary}{NP : mary}}{S : loves(john, mary)}$$

We observe that the CCG formalism can be adapted to music in order to enable us to handle semantically rich theories of music that can go beyond shallow syntactic forms to the deep meaning of musical pieces. While Steedman [23] presents a CCG grammar for music, the grammar lacks any formal semantics; this is understandable, given that the motivation behind Steedman's project was not to give a cognitive account of music. Our semantics, specified using the Cognitive Event Calculus, has intensional operators and fluents, both of which would eventually be necessary for modeling all aspects of music and cognition, from composition and generation to performance and listening.

Fig. 14.3 shows a part of the signature of our preliminary music calculus. We have self-explanatory sorts for representing different aspects of the musical universe. Note has its usual standard interpretation. A Score is a sequence of notes formed using the function symbol *add*. A MusicParticle is a note played at a particular moment and can be considered the simplest MusicPhrase. Simpler MusicPhrases combine in myriad ways to form complex MusicPhrases; this is represented using *combine*. The rendition of a Score using a Recommendation in a *performance* results in a MusicPhrase. The music phrases have meanings Meaning which form a subset of the lambda expressions; the meanings combine using *reduce*. The phrases have abstract types Type; the types combine using *apply*. Allowed combinations of music phrases are represented using *allowed*. Recommendations by the conductor

S: =Recommendation $\subseteq$ Action | Person $\subseteq$ Agent | *Level* | Label

f: = *MaxLevel*: Person $\times$ Level $\rightarrow$ Boolean
Level: Person $\times$ Level $\rightarrow$ Boolean
WithinOne: Level $\times$ Level $\rightarrow$ Boolean
LessThan: Level $\times$ Level $\rightarrow$ Boolean
a, b, c, d, e, f, g: Level
Tutti: Label $\rightarrow$ Boolean
low, *high*: Label

Fig. 14.3 Signature of the Music Calculus

are represented using objects from the sort Recommendation. Simple machinery for representing affects is achieved using the sort Affect and the predicate symbol *feels*. We model affects as a subsort of fluents in the event calculus. The way the meaning of a music phrase produces an affect is to be captured by *translates*.

With this syntactic machinery we can account for different agents interpreting a piece of music differently. What might be the meaning of a musical piece? It definitely includes affects produced in a listener. In addition to affects, the meaning can include objective properties of the music, such as its tempo, which the current version of Handle can process.

The General Problem of Conducting: The general problem of conducting can be stated as follows: Given a score *score* and the composer's intention that the listener listener should feel affect a, is there a music phrase p which is the result of performing *score* with the conductor's recommendation r such that the meaning of the phrase p translates into affect a in listener?

$$\begin{aligned}
&\mathbf{I}(\mathsf{h}, t, \mathit{feels}(\mathsf{listener}, a)) \Rightarrow \\
&\exists p : \mathsf{MusicPhrase} r : \mathsf{Recommendation}. \\
&(\mathbf{B}(\mathsf{h}, t, \mathit{performance}(r, \mathit{score}) = p \wedge \mathit{translates}(\mathit{meaning}(p), a))).
\end{aligned}$$

What axioms do we need to enable the conductor to determine his actions? At a minimum, we need a formula specifying combination of the music particles into music phrases. Axiom $\mathcal{A}_1$ states that two music phrases can combine if and only if their syntactic types let them combine. If they combine, the combined phrase has syntax and semantics dictated by the original pieces.

$$\boxed{\mathcal{A}_1}\quad \begin{aligned}
&\forall m_1\, m_2 : \mathsf{MusicPhrase} \\
&\mathit{allowed}(m_1, m_2) \\
&\Leftrightarrow \\
&\exists m : \mathsf{MusicPhrase}.\ \mathit{combine}(m_1, m_2) = m \\
&\mathit{type}(m) = \mathit{apply}(\mathit{type}(m_1), \mathit{type}(m_2)) \wedge \\
&\mathit{meaning}(m) = \mathit{reduce}(\mathit{meaning}(m_1), \mathit{meaning}(m_2))
\end{aligned}$$

We need knowledge capturing how the meaning of music translates into affects in agents. Before formalizing this, we need an axiom stating that musical meanings produce affects. Axiom $\mathcal{A}_2$ states that if a piece of music has some meaning, there is an event that causes an affectual response. Here *start* is a defined function symbol giving us the start of a music phrase.

$$\boxed{\mathcal{A}_2}\quad \begin{aligned}
&\forall m : \mathsf{MusicPhrase}\ \exists\, e : \mathsf{Event}\ a : \mathsf{Affect}\ t : \mathsf{Moment} \\
&\mathit{initiates}(e, a, t) \wedge \mathit{translates}(\mathit{meaning}(m), a) \wedge t > \mathit{start}(m)
\end{aligned}$$

Axiom $\mathcal{A}_3$ states a basic property of affects: affects have to be instantiated or associated with agents.

$$\boxed{\mathcal{A}_3} \quad \forall a : \mathsf{Affect}\ \exists\ p : \mathsf{Agent}.\mathit{feels}(p, a)$$

The *translates* predicate is supposed to capture the translation or production of affects in agents via the semantic properties of music. Upon some reflection, the reader may suspect that we have swept under this predicate symbol the hard-to-formally-model processes that operate in the production of affects. We expect that, when axiomatized, determining whether $\mathit{translates}(m, a)$ holds could be as hard as general-purpose deductive reasoning. Let the axioms governing *translates* be Γ. The problem of conducting can be now stated as finding an r such that:

$$\begin{aligned}
&\{\mathcal{A}_1, \mathcal{A}_2, \mathcal{A}_3, \ldots\} \cup \Gamma \vdash \\
&\mathbf{I}(\mathsf{h}, t, \mathit{feels}(\mathsf{listener}, a)) \Rightarrow \\
&\exists p : \mathsf{MusicPhrase}\ r : \mathsf{Recommendation}. \\
&(\mathbf{B}(\mathsf{h}, t, \mathit{performance}(r, \mathit{score}) = p \wedge \mathit{translates}(\mathit{meaning}(p), a)))
\end{aligned}$$

14.4 The Handle Trajectory

Handle was originally a rather humble logico-mathematical component of a music-creating entity known as CAIRA [for more information on CAIRA and "humble" Handle, see [15]]. CAIRA is powered in significant part by *human* muscial creativity, and, as explained above, our goal is a standalone AI that demonstrates across-the-board muscial creativity$_T$. The first step in the engineering devoted to creating such as AI was reported by by Ellis and Bringsjord; see Fig. 14.4 for the simple, initial architecture of Handle, conceived as an artificial conductor. As promised, we now describe three versions of Handle, beginning with the conductor version.

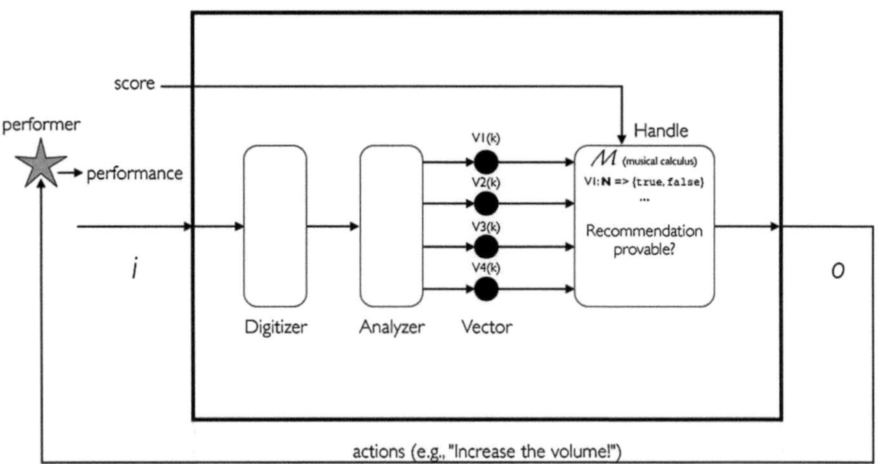

Fig. 14.4 Primitive Handle-as-conductor architecture

14.5 Three Versions of Handle

14.5.1 Handle as Conductor

Handle$_{cond}$ is both a microcosmic version of the logic-based parts of CAIRA and a standalone *creative*$_{T+}$ machine conductor. Can a computing machine "understand" music and reason from that understanding to the direction of a great conductor, itself issued in real time so as to improve the performance in question? While we are confident the answer is Yes, the only way to provide via engineering an existence proof of this affirmative answer is to start humbly, and gradually scale up. Accordingly, Handle$_{cond}$ was created around a single pianist playing a short piece, and specifically to investigate the understanding and "conducting" of this playing. A screenshot of Handle$_{cond}$ in action is shown in Fig. 14.5.

Handle$_{cond}$ has two major components; the first is an audio analysis module running within MATLAB that controls low-level audio and signal processing routines on incoming live or recorded audio. This module passes information, using a predetermined protocol and format, to the musical calculus system, which runs via Common Lisp. Handle$_{cond}$ is capable of computing the tempo of live or recorded audio using the system described in [14]. The result of this computation is then passed on to the reasoning system, which in turn determines whether the song is being played at a tempo appropriate for the audience and context. Figure 14.5 shows Handle$_{cond}$ responding to a performance of the Prelude in C major from Book 1 of Bach's *The Well-Tempered Clavier* by asking for it to be replayed at a slightly faster tempo.

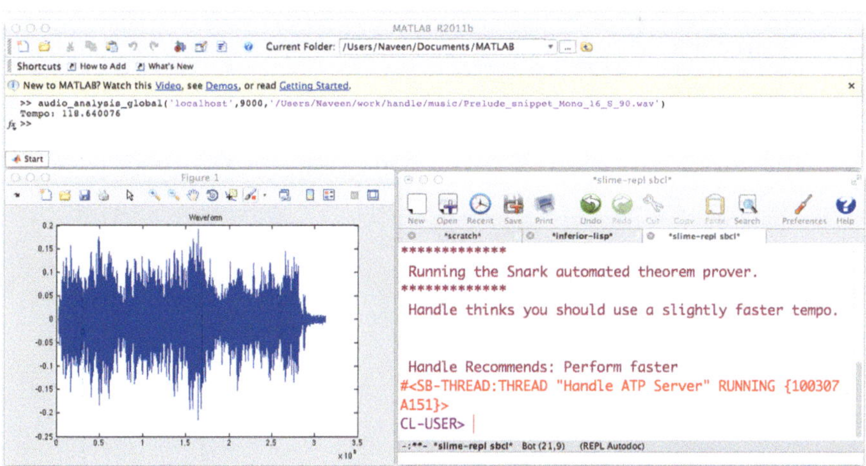

Fig. 14.5 Sample output from Handle$_{cond}$

14.5.2 Handle as Jazz Muscician

Whereas Handle$_{cond}$ is designed to work with score-based, structured music, and to deviate only minimally from a known form, Handle$_{jazz}$ is designed to explore the field of unstructured music in the form of improvised, free "fusion jazz," an area explored by other systems such as OMax.[2]

As formerly, the architecture of Handle$_{jazz}$ consists of two major components, one running in MATLAB and the other in Common Lisp. The most notable change is that the MATLAB component is the controlling client, rather than the Lisp system. The MATLAB client interfaces with FILTER and MaxMSP via a simple set of OpenAudio commands sent periodically via the network, which provide information on the current tension level of the three players as integers in the range 1–7. These values are used to look up the current 'perceived state' of the performance for FILTER, and appropriate information is returned. If, however, the state had not been encountered, a best guess is made, derived from a weighted nearest-neighbor heuristic, and the state is marked for formal calculation using an implementation of the music calculus in the Lisp-based SNARK automated theorem prover. Due to the length of time required to perform such operations, all logic calculations are currently done off-line between performances. An overview of the instantiation of the Handle architecture corresponding to the work reported on in the present section is given in Fig. 14.6.

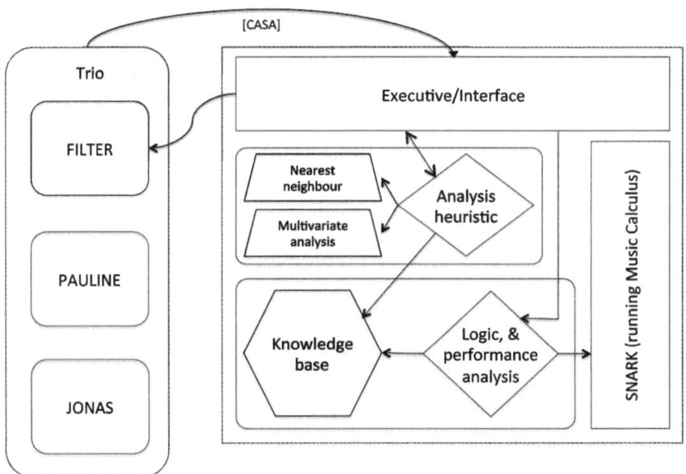

Fig. 14.6 Instantiation of Handle$_{jazz}$ architecture

[2] OMax generates interactive improvisational music in collaboration with a human performer, via statistical and learning methods, in a manner similar to how FILTER operates [3]. However, OMax differs from CAIRA as a whole in that it does not have a component that is a counterpart to Handle$_{jazz}$, which uses logic-based reasoning to influence how the improvised music is generated.

Table 14.1 Ensemble states as determined by $Tension_B$

Ensemble states	Musician A	Musician B	CAIRA C
Solo A	A > B+1	B+1< A	C+1< A *
Solo B	A+1< B	B> A+1	C+1< B *
Solo C	0 < A < 5	0< B < 5	??
Low Level Tutti	0< A < 5	0< B < 5	??
High Level Tutti	A> 4	B> 4	C> 4 *
Ending	A = 0	B = 0	C = 0 *
Uncertain	None of the above		??

A, B, and C represent the calculated state of Musicians A, B, and CAIRA, respectively. CAIRA may disregard the recommendation and respond differently. CAIRA must decide which state to prefer. States are ordered hierarchically such that overlap goes to the higher-ranking state [4]

Handle$_{jazz}$, as the logical reasoning portion of the larger CAIRA system, accepts inputs from three audio sources: two of these correspond to two human players and the third is the output from FILTER. Computational Audio Scene Analysis (CASA) is used on all three audio streams at various points during the piece to extract data on various features of each player's performance. These features—$tension_B$, $tension_{MIR}$, $dynamic$, $tempo$, $valence$ and $activity$—are mapped to an integer scale in the range 0–6 [4]; using combinations of 3-tuples of these features, information on the current state of the performance is derived. States used in this work and their definitions may be found in Table 14.1: possible states are solos for players A, B and C, low- and high-tension $tutti$, ending and an "uncertain" state. The latter state indicates there is insufficient information for Handle$_{jazz}$ to supply meaningful data. When CASA information is supplied to Handle$_{jazz}$, it returns an integer value representing current performance state of the piece, using the definitions in Table 14.1. Based on this value, FILTER then adjusts its own parameters to take the recommendation into account in its performance. In the work described, only $tension_B$ (as defined in [4]) is used for state calculations.

As is the case for human players, Handle$_{jazz}$ has gaps in its knowledge when asked for its opinion on a state it has not encountered. In such cases it attempts an "intuitive leap," using a predetermined heuristic to calculate a "best-guess" response. If such a guess is made, it is stored in the knowledge-base together with a marker indicating it *was* a guess. At the end of the performance, all states where a guess existed are passed to SNARK for formal proof, and the resultant values are stored permanently in the knowledge-base. Figures 14.7 and 14.8 show output from Handle$_{jazz}$ operating over a short piece of music and the beginning of the resulting SNARK proof resulting from an unknown state, respectively.

Handle$_{jazz}$ operates primarily by logical deduction of state using axioms derived from predetermined rules. The rules given in Table 14.1 are used to construct premises in first-order logic by which one may prove each state. Numerical values received via the CASA system are mapped to a corresponding player-level value (0 = A, 1 = B,

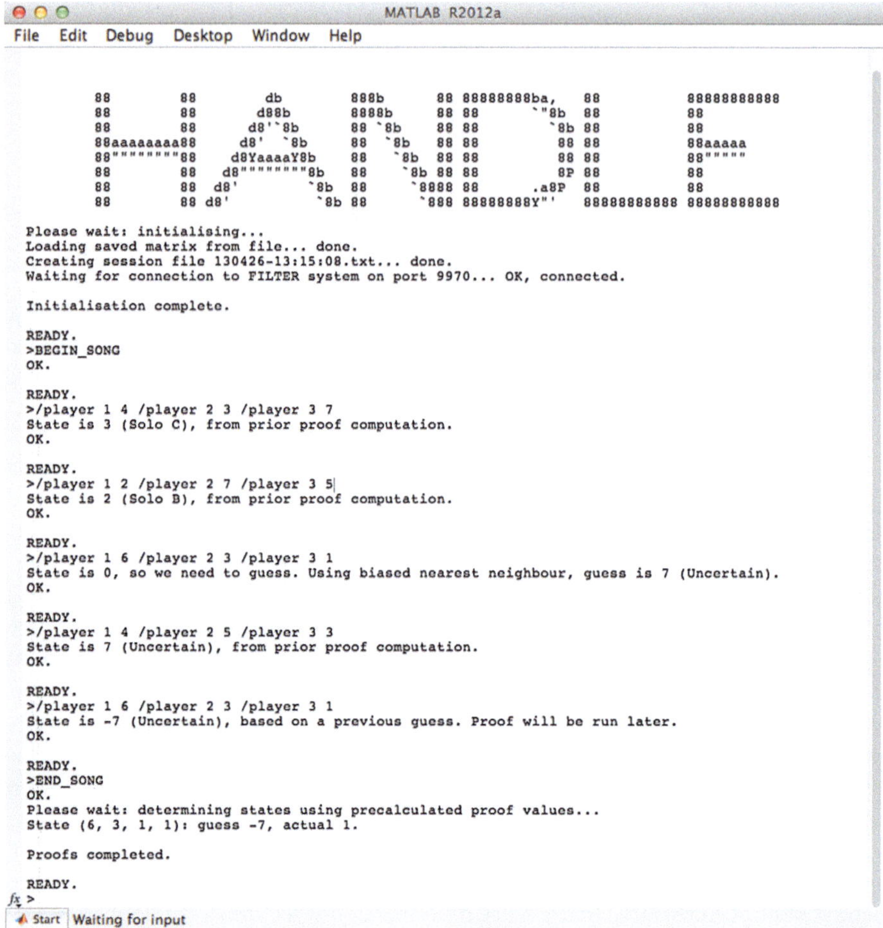

Fig. 14.7 Sample Output from Handle$_{jazz}$

etc.); these are used for comparisons during theorem-proving, replacing the ordinary arithmetic operators which SNARK cannot reason over.

The following are some key examples of premises in FOL used to define the music states [27], some of which are visible as SNARK axioms in Fig. 14.8.

1. A person has a solo iff they have the highest level and there are no other players within one level.

$$\forall p_1 : \textsf{Person}\ Solo(p_1) \leftrightarrow \exists l_1 : \textsf{Level}\ MaxLevel(p_1, l_1) \land \neg\left(\begin{matrix} \exists p2 : \textsf{Person} \\ \exists l_2 : \textsf{Level} \end{matrix} \left[\begin{matrix} Level(p_2, l_2) \land WithinOne(l_1, l_2) \\ \land\ (p_1 \neq p_2) \end{matrix}\right]\right)$$

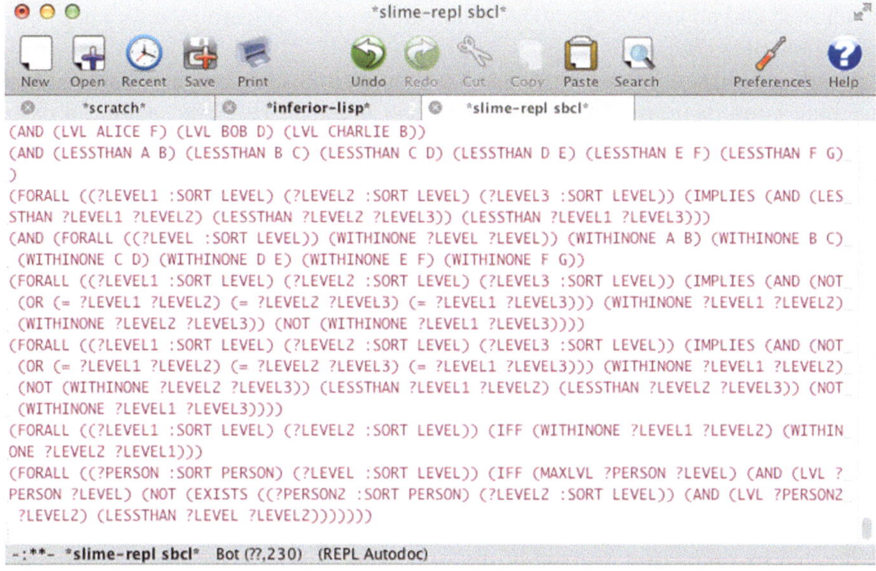

Fig. 14.8 Sample Output from SNARK

2. There is a low-level *tutti* iff every player has a level greater than zero and less than five, and no player has a solo.

$$Tutti(low) \leftrightarrow \forall p : \mathsf{Person}\ \exists l : \mathsf{Level} \begin{pmatrix} Level(p,l) \wedge LessThan(l,f) \\ \wedge\ LessThan(a,l)) \wedge \neg Solo(p) \end{pmatrix}$$

3. There is a high-level *tutti* iff every player has a level greater than four.

$$Tutti(high) \leftrightarrow \forall p : \mathsf{Person}\ \exists l : \mathsf{Level}(Level(p,l) \wedge LessThan(e,l))$$

4. It is the end iff all players are at level zero.

$$\mathsf{End} \leftrightarrow \forall p : \mathsf{Person}\ (Level(p,a))$$

By determining states logically in this fashion, the agent is guided to play in a fashion that is expected on a social level. However, given that the state determines certain weightings within the FILTER machine-learning aspect, the agent remains able to improvise freely within the expected range, as a response to the other players; in effect, this is the same as a human player reasoning about what the other players are doing and what they intend, and reacting accordingly. Similarly, when the agent is first initialized, it has no knowledge about what combinations of inputs relate to which state and no knowledge about which are required to learn by reasoning about new inputs over time.

Handle$_{jazz}$ uses axioms provided and computed in SNARK to derive state values matching combinations of 3-tuples of musical features. However, again, in cases where precomputed information is unavailable Handle is designed to make an "intuitive leap," by using one of a number of possible heuristics to guide its judgment.

We note at this point that for the given three-player scenario we may represent each state as a color corresponding to the most extreme coordinates for a feature 3-tuple. For example, a solo for player A corresponds to red (6, 0, 0 $\rightarrow$ 255, 0, 0), Solo B is 0, 6, 0, which encodes to green, and Solo C is 0, 0, 6, blue. Low-tension *tutti* is represented by grey (calculated by finding the centroid of the state), high-tension *tutti* is white, end is black, and uncertainty is determined by the coordinate values themselves. After computing all state combinations, such a representation appears as in Fig. 14.9.

Our approach for resolving an unknown state at runtime is to use a *nearest-neighbor* algorithm. Given the (safe) assumption that the states for all 8 corners of the state cube are absolutely known ab initio, the value of any unknown state may be derived by taking the average of the sum of the distance vector between it and each (fully known) corner. As more and more states are derived and the knowledge-base becomes progressively more complete, however, we would prefer to use more accurate information; that is, state values which are more proximate to the unknown state. These 8 cubes, nearest to the unknown state in diagonal distance, are determined using a simple vector approach and the mean of their values returned. Thus, given an unknown state α with coordinates $(\alpha_x, \alpha_y, \alpha_z)$ and $\langle x \rangle$, $\langle y \rangle$, $\langle z \rangle$ vectors of the co-ordinates of all known states, the 8 nearest neighbors of α can be found as follows:

$$
\begin{aligned}
index_1 &= \text{find}(x[i] \geq \alpha_x \;\&\; y[i] \geq \alpha_y \;\&\; z[i] \geq \alpha_z \\
index_2 &= \text{find}(x[i] \geq \alpha_x \;\&\; y[i] \leq \alpha_y \;\&\; z[i] \geq \alpha_z \\
index_3 &= \text{find}(x[i] \geq \alpha_x \;\&\; y[i] \leq \alpha_y \;\&\; z[i] \leq \alpha_z
\end{aligned}
$$

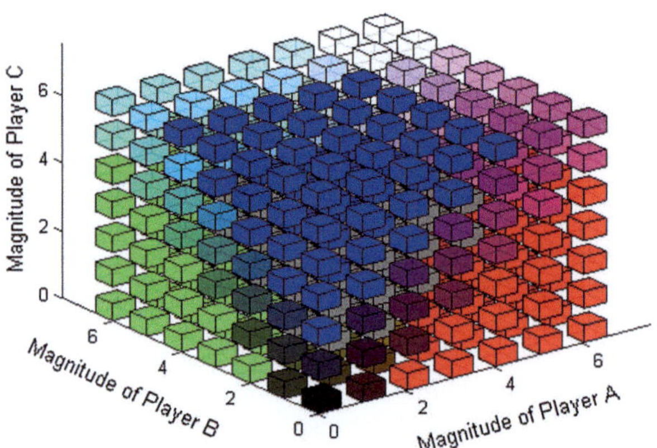

Fig. 14.9 Chromatic representation of complete 3D state cube for $tension_B$

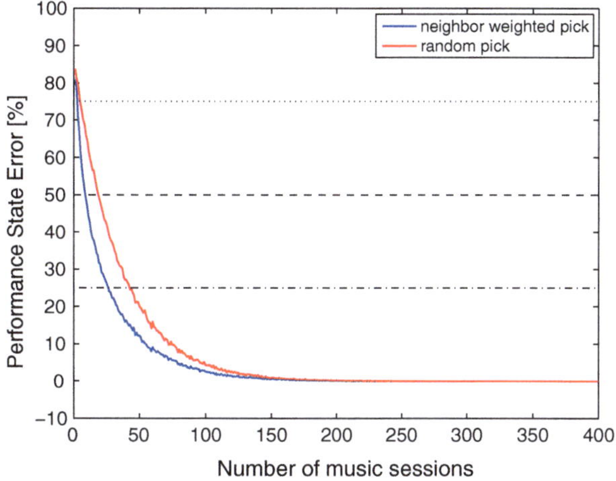

Fig. 14.10 Nearest-Neighbor Error versus Random Guess

$$index_4 = \text{find}(x[i] \geq \alpha_x \ \& \ y[i] \leq \alpha_y \ \& \ z[i] \leq \alpha_z$$
$$index_5 = \text{find}(x[i] \leq \alpha_x \ \& \ y[i] \geq \alpha_y \ \& \ z[i] \geq \alpha_z$$
$$index_6 = \text{find}(x[i] \leq \alpha_x \ \& \ y[i] \geq \alpha_y \ \& \ z[i] \leq \alpha_z$$
$$index_7 = \text{find}(x[i] \leq \alpha_x \ \& \ y[i] \leq \alpha_y \ \& \ z[i] \geq \alpha_z$$
$$index_8 = \text{find}(x[i] \leq \alpha_x \ \& \ y[i] \leq \alpha_y \ \& \ z[i] \leq \alpha_z$$

While admittedly simplistic, this approach nonetheless conforms to our definition of 'intuition.' In order to determine the efficacy of this approach, a simulation was run of over 400 sessions to determine the percentage error of the nearest-neighbor approach by picking 10 random-player inputs each session. It can be seen that, as new states were accumulated in the knowledge-base, the accuracy of the nearest neighbor method improves (Fig. 14.10). A comparator is also provided, based on an approach which guesses randomly on uncomputed states.

We acknowledge that we may stand accused of not properly addressing, or exploring, the concept of 'intuition' in Handle$_{jazz}$. We do not deny this. However, in defense, we would reply by stating that 'intuition,' as it is commonly understood, requires a much larger dataset than the one we were able to use, one that in all likelihood contains a certain amount of conflicting data, together with an appropriate means of accessing and using those data. Handle$_{jazz}$ currently functions using only one musical feature, $tension_B$, thereby yielding a dataset containing a paltry $7 \times 7 \times 7 = 343$ states; as a consequence, 'intuitive leaps' are inevitably barely distinguishable from mere arithmetic computation. But, should we ever have the additional information, the problem of how we manage it will inevitably arise.

One of the most interesting effects to emerge from explorations with Handle$_{jazz}$ is the concept of the state cube, shown above in Fig. 14.9. Originally developed by Valerio based on work by Ellis, it appeared to have considerable implications for computation using multiple musical features, even leading to a method of mapping

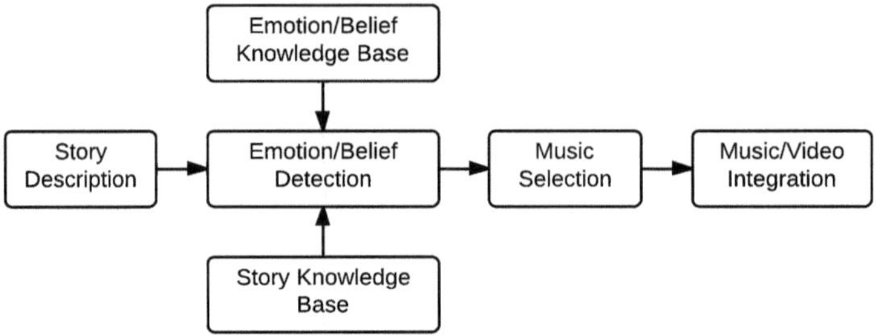

Fig. 14.11 Process for creating music from narrative

for the use of visual stimuli in the composition of music. (For full information, see [27].) Most importantly for Handle$_{jazz}$, however, it offers a resource to manage and easily explore the large data-space resulting from the many possible combinations of musical features captured hitherto, as well as any we hope to capture in the future.

14.5.3 Toward Handle as Film Composer

Handle$_{film}$ reflects a turn toward a new component of the creative process: composing music for film. Our starting point in this endeavor parallels the starting point of film music as a whole. When silent films first started using music, there was a pianist or organist who would play music along with the video. This music was often improvised, but for those who were not as comfortable with improvisation there were books of music printed with basic instructions indicating what music was appropriate for many different types of scenes; initially, we take a similar approach [24]. Taking a story in the form of video, we represent it in a form that Handle$_{film}$ can reason over and gather knowledge about. Then, using this knowledge, Handle$_{film}$ can select a piece of music from a predefined database of music that contains both a list of the pieces available and a description of when each piece is most suitable. Once a piece of music is selected, it is then matched to the video. The entire high-level process can be seen in Fig. 14.11.

14.5.3.1 Story Foundation

In 1944, Fritz Heider and Marianne Simmel carried out a seminal and now-famous experiment to study human perception of the behavior of others. The experiment used videos portraying abstract shapes—triangles and circles to be exact—interacting in an ambiguous but highly suggestive manner [16]. Heider and Simmel found that most people watching the video assign personalities to the shapes and a storyline to the video to explain the interactions between these shapes; they also found that people invent *similar* personalities and storylines. Handle$_{film}$ is currently based on stories we invent for a type of action seen in this original experiment. These stories are relatively short, but are still interesting this makes them a good starting point

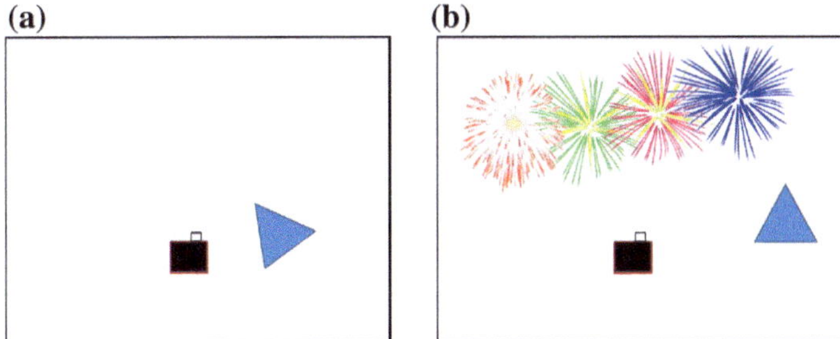

Fig. 14.12 Two Frames from a Heider-Simmel-based video (**a**) Triangle "Examining" Box (**b**) Triangle "Celebrating"

for story representation and film composing. The fact that the stories are natively in a video format also helps with aligning music to the story. In addition, since there is substantial consensus regarding what the stories are "about," we can partially measure how successful our music generation is by analyzing whether the music enforces the perceived storyline or deprecates it. Lastly, by using the Heider-Simmel stories we open up the opportunity for collaboration with other research teams (see [25]). Figure 14.12 shows two frames from a Heider-Simmel based story we have created.

14.5.3.2 Story Representation

In order for Handle$_{film}$ to generate music to match a given story, we need a representation scheme for stories that is in a form Handle$_{film}$ can use to reason about the story, and to make decisions about what music is most appropriate. Unsurprisingly, we have opted to use formal logic, starting with a proper fragment of the music calculus discussed above; viz., the *event calculus* (EC), elegantly presented, in connection with stories (specifically with the field of *story understanding*) in [18]. EC allows us to represent a variety of simple stories and then reason about the events in them, and about some basic emotional constructs related to those events. This information can then be used to determine what kind of music is best suited for the story. Figure 14.13 shows the event-calculus representation of the story in Fig. 14.12.

As the stories and our descriptions of them become more complex, EC is no longer sufficient to represent all the narratological input that Handle$_{film}$ needs in order to compose music to match the relevant story. For example, EC cannot properly represent the beliefs and knowledge of the agents in the story. To represent these facets of the story, we must supplement the EC by using the $\mathcal{CEC}$ described previously (also see [9]). Figure 14.13 shows our first steps in representing the Triangle's beliefs and emotions via the following simple conditionals:

$$\mathbf{P}(Triangle, t, Fireworks) \Rightarrow \mathbf{B}(Triangle, t, Celebration)$$
$$\mathbf{B}(Triangle, t, Celebration) \Rightarrow Happy(Triangle)$$

```
(defparameter *CIRCInitiates*
  (list '(forall ((?e :sort event) (?f :sort fluent) (?t :sort integer))
              (iff (Initiates ?e ?f ?t)
                   (or (exists ((?s :sort square) (?a :sort agent))
                           (and (= ?e (Touches ?a ?s))
                                (= ?f (ActivationTime ?s ($$sum ?t 3)))))
                         (exists ((?s :sort square) (?a :sort agent))
                           (and (= ?e (Touches ?a ?s))
                                (= ?f (Beeping ?s))))
                         (and (= ?e (StartFireworks))
                              (= ?f (Fireworks))))))))

(defparameter *CIRCTerminates*
  (list '(forall ((?e :sort event) (?f :sort fluent) (?t :sort integer))
              (iff (Terminates ?e ?f ?t)
                   (exists ((?s :sort square))
                           (and (= ?e (StartFireworks))
                                (= ?f (Beeping ?s))))))))

(defparameter *CIRCReleases*
  (list '(forall ((?e :sort event) (?f :sort fluent) (?t :sort integer))
              (not (Releases ?e ?f ?t)))))

(defparameter *CIRCHappens*
  (list '(forall ((?e :sort event) (?t :sort integer))
              (iff (Happens ?e ?t)
                   (or (and (= ?e (Touches Triangle Square))
                            (= ?t 8))
                       (and (= ?e (StartFireworks))
                            (HoldsAt (ActivationTime Square ?t) ?t)))))))

(defparameter *StateConstraints*
  (list '(forall ((?s :sort square) (?t :sort integer) (?t1 :sort integer) (?t2 :sort integer))
              (implies (and (HoldsAt (ActivationTime ?s ?t1) ?t)
                            (HoldsAt (ActivationTime ?s ?t2) ?t))
                       (= ?t1 ?t2)))))

(defparameter *Observations*
  (list '(forall ((?t :sort integer))
              (not (HoldsAt (ActivationTime Square ?t) 0)))
        '(forall ((?f :sort fluent) (?t :sort integer))
              (not (ReleasedAt ?f ?t)))
        '(not (HoldsAt (Beeping Square) 0))
        '(not (HoldsAt (Fireworks) 0))))

(defparameter *Emotions*
  (list '(forall ((?t :sort integer)) (iff (HoldsAt (Beeping Square) ?t)
                                           (HoldsAt (Frightened Triangle) ?t)))
        '(forall ((?t :sort integer)) (iff (HoldsAt (Believes Triangle Celebration) ?t)
                                           (HoldsAt (Happy Triangle) ?t)))))

(defparameter *Beliefs*
  (list '(forall ((?t :sort integer))
              (iff (HoldsAt (Fireworks) ?t)
                   (HoldsAt (Believes Triangle Celebration) ?t)))))

(defparameter *Queries* '(((HoldsAt (Frightened Triangle) ?t) "frightening")
                          ((HoldsAt (Happy Triangle) ?t) "happy")
                          ((HoldsAt (Believes Triangle Celebration) ?t) "celebration")))
```

Fig. 14.13 Event calculus representation of the fireworks scene

These conditionals give us a general model of the Triangle's "psyche," allowing us to determine what the Triangle's reactions are to events in the story, and to then generate music based on these reactions using the method described in the next section. Currently, the process of taking the story as we would understand it and representing it in the forms described above is done manually, but Handle$_{film}$ will later be able to perform this task on its own. This autonomy will require an ambitious level of story understanding, but such capabilities are likely required for creativity$_T$ in the realm of film composing.

14.5.3.3 Music Generation

Once the story has been formally represented, we can use a theorem prover to determine what emotions or beliefs are present at any given timepoint by simply requesting a series of proofs. One challenge in this method is that if we ask the theorem prover to prove that an emotion is present at a given timepoint when it is not, the theorem prover may run forever. However, since our stories are relatively simple, a proof is usually found quickly if one exists we can confidently declare that an emotion is absent if it is not found after a reasonable time period, so we still obtain good results. As we scale up to more complex stories, this technique may not be valid unless we are willing to wait a long time for every proof, but for now it achieves our desired goal. Once the emotions, beliefs, or other factors important to the music are determined for every timepoint, we query a knowledge-base using these descriptive factors to determine what song is most appropriate at that time. Currently, our knowledge-base is in accord with *Motion Picture Moods for Pianists and Organists*, a book of songs for silent films like one mentioned previously [23]. For the fireworks scene, Handle knows that the Triangle will perceive fireworks going off and can therefore use the formulae presented above to infer that the Triangle will believe there is a celebration and that the Triangle will be happy. Handle then queries the knowledge-base for music that best matches the "happy" and "celebration" descriptors and returns Giacomo Meyerbeer's *Coronation March*, the first line of which can be seen in Fig. 14.14.

As we move toward modern-day film music, the interaction between the story and the music becomes much more complex, and therefore our method for generating music will have to become correspondingly intelligent. For example, while most silent music is parallel in nature—meaning that it directly correlates with what is happening in the story—it is just as common for modern-day film music to be counterpunctual—meaning that it contradicts what is happening in the story, often to emphasize some specific aspect. To cope with such techniques, we need to expand our representation of the story to include not only what is happening in the story, but what the story is trying to convey. We could then reason about what music best matches the purpose of the story rather then just its contents. Our hope is that a "reversal" of the story-generation approach in [7] will serve as a basis for this deeper form of story understanding.

Fig. 14.14 Music selected by Handle for the fireworks scene

14.6 Concluding Remarks

We suspect that some learned readers of a theoretical bent will be at least somewhat disppointed that we haven't engaged foundational questions regarding the nature of music, and that we therefore have taken no positions on such questions. Instead, yes, we have reported on our goal to mechanize "middle-standard" creativity in the realm of music. Nonetheless, our commitment to the music calculus, truth be told, does in fact reflect some rather deep positions (on the part, at least, of Bringsjord), ones affirmed in the context of the excellent work of relevant theoreticians. We end by encapsulating two such positions here.

One: Those familiar with Occidental music know that it is clearly sometimes *propositional* in nature. Handel (not Handle) clearly intended to at a minimum bolster belief, in his audience, in certain propositions, via *Messiah*. And there are countless other such examples. But *must* music be propositional in nature? As Cross [13] eloquently points out, No. Music can be participatory and interactive, with the declarative dimension eliminated, or at least allowed to approach zero. In this context, it's important to realize that our $\mathscr{M}$, despite the fact that it is a logical system, is quite well-suited to modeling dynamic, multi-agent interaction in cases where the interaction in question is free of propositional content. To put the point starkly, the event calculus within the music calculus can be used to model information flow that is entirely non-semantic and inanimate.

Second position: There are other formal frameworks, if not outright formal theories, designed to capture large parts of music and music cognition. Above, we cited, and briefly discussed [26], but of note also is the framework of Murray-Rust & Smaill [19], the framework provided by Allwood et al. [1], and the insightful modification of Allwood et al.'s framework carried out by Cross [13]. Our music calculus, in this context, has a distinguishing mark, one that we seek to refine and enhance into the future: viz., $\mathscr{M}$ promotes the explicit representation and reasoning of iterated doxastic attitudes (and, for that matter, other intensional attitudes) to the level of full explicitness. Creative human conductors, jazz muscicians, and film composers all, we believe, are truly creative only when they premeditatedly reason, and indeed reason well, over the beliefs that they have about the beliefs that listeners/viewers are likely to have when these listeners/viewers experience the music in question. Accordingly, unlike other frameworks, ours allows for formulae in which explicit and iterated intensional operators are allowed.

Acknowledgments This project is made possible by generous sponsorship from both the NSF (grant no. 1002851, to Braasch, Bringsjord, and Oliveros), and the John Templeton Foundation, to Bringsjord and Govindarajulu. The authors would additionally like to thank the anonymous referees of this paper for their insights.

References

1. Allwood, J., Nivre, J., Ahlsé, E.: On the semantics and pragmatics of linguistic feedback. J. Semant. **9**(1), 1–26 (1992)
2. Arkoudas, K., Bringsjord, S.: Toward formalizing common-sense psychology: an analysis of the false-belief task. PRICAI 2008: Trends in Artificial Intelligence pp. 17–29 (2008)
3. Assayag, G., Bloch, G., Chemillier, M.: Omax-ofon. Sound Music Comput. (SMC) **2006** (2006)
4. Braasch, J.: The μ-cos$m$ project: an introspective platform to study intelligent agents in the context of music ensemble improvisation (2012) (to appear).
5. Bringsjord, S.: Declarative/Logic-based cognitive modeling. In: Sun, R. (ed.) The Handbook of Computational Psychology, pp. 127–169. Cambridge University Press, Cambridge (2008). http://kryten.mm.rpi.edu/sb_lccm_ab-toc_031607.pdf
6. Bringsjord, S.: The logicist manifesto: at long last let logic-based AI become a field unto itself. J. Appl. Logic **6**(4), 502–525 (2008). http://kryten.mm.rpi.edu/SB_LAI_Manifesto_091808.pdf
7. Bringsjord, S., Ferrucci, D.: Artificial Intelligence and Literary Creativity: Inside the Mind of Brutus, A Storytelling Machine. Lawrence Erlbaum, Mahwah (2000)
8. Bringsjord, S., Ferrucci, D., Bello, P.: Creativity, the turing test, and the (better) lovelace test. Mind. Mach. **11**, 3–27 (2001)
9. Bringsjord, S., Govindarajulu, N.S.: Toward a modern geography of minds, machines and math. SAPERE Philosophy and Theory of Artificial Intelligence (forthcoming)
10. Bringsjord, S., Licato, J.: Psychometric artificial general intelligence: the Piaget-MacGuyver room. In: Wang, P., Goertzel, B. (eds.) Foundations of Artificial General Intelligence, pp. 25–47. Atlantis Press, Amsterdam, The Netherlands (2012). http://kryten.mm.rpi.edu/Bringsjord_Licato_PAGI_071512.pdf. This url is to a preprint only
11. Clark, M.: Cognitive illusions and the lying machine. Ph.D. thesis, PhD thesis, Rensselaer Polytechnic Institute (RPI) (2008)
12. Cope, D.: Computer Models of Muscial Creativity. MIT Press, Cambridge, MA (2005)
13. Cross, I.: "Does not compute?" Music as real-time communicative interaction. AI Soc. **28**, 415–430 (2013)
14. Ellis, D., Poliner, G.: Identifying cover songs' with chroma features and dynamic programming beat tracking. In: Proceedings of International Conference on Acoustical, Speech, and Signal Processing ICASSP-07, vol. 4, pp. 1429–1432. IEEE (April 2007). A MATLAB library is available at: http://labrosa.ee.columbia.edu/projects/coversongs/
15. Ellis, S., Govindarajulu, N.S., Bringsjord, S., Haig, A., Kuebler, C., Taylor, J., Braasch, J., Oliveros, P., Nort, D.V.: Creativity and conducting: handle in the CAIRA project. In: Proceedings of the Workshop "Computational Creativity, Concept Invention, and General Intelligence" (2012)
16. Heider, F., Simmel, M.: An experimental study of apparent behaviour. Am. J. Psychol. (1944)
17. Inhelder, B., Piaget, J.: The Growth of Logical Thinking from Childhood to Adolescence. Basic Books, New York, NY (1958)
18. Mueller, E.: Commonsense Reasoning. Morgan Kaufmann (2006)
19. Murray-Rust, D., Smaill, A.: Towards a model of musical interaction and communication. Artif. Intell. **175**, 1697–1721 (2011)
20. Pappas, G., Swain, M.: Essays on Knowledge and Justification. Cornell University Press (1978)
21. Patel, A., et al.: Language, music, syntax and the brain. Nat. Neurosci. **6**(7), 674–681 (2003)
22. Pearce, M.: Notes on CHARM—a specification for the representation of musical knowledge (2002)
23. Rapee, E. (ed.): Motion Picture Moods for Pianists and Organists. Arno Press Inc, Arno (1970)
24. Redner, G.: Deleuze and Film Music : Building a Methodological Bridge Between Film Theory and Music, chap. Introduction, Intellect Ltd (2010)
25. Roemmele, M., Archer-McClellan, H., Gordon, A.S.: Triangle charades: a data-collection game for recognizing actions in motion trajectories (2014)

26. Steedman, M.: The Blues and the abstract truth: music and mental models. In: Garnham, A., Oakhill, J. (eds.) Mental Models in Cognitive Science: Essays in Honour of Phil Johnson-Laird. Psychology Press, East Sussex (1996)
27. Valerio, J.: Heuristic for Multi-Variable Music State Recommendations. Master's thesis, Rensselaer Polytechnic Institute, Troy (2013)
28. Zettlemoyer, L.S., Collins, M.: Learning to map sentences to logical form: structured classification with probabilistic categorial grammars. In: Proceedings of the 21st Conference on Uncertainty in Artificial Intelligence, vol. 5, pp. 658–666 (2005)

Chapter 15
Computational Creativity and Music

David Cope

Abstract In this chapter, I first define the word 'creativity' for the purposes of my research and then compare this definition to others found in the literature. Following this, I present two simple examples of creativity involving games—words and chess—in which I invite readers to participate. I then describe two musical situations and subsequent computer programs that, following the discussion surrounding the examples, creatively provide output as tentative proof of their success. Finally, I describe ways this software can be improved to output more elegant results.

15.1 Definition

Defining creativity, especially in the context of computer emulation of such can prove incredibly difficult. To wit,

> It should be obvious, but it is sometimes forgotten, that musical creativity cannot be defined without reference to the quality of the music it produces. If a greater degree of creativity does not result in a better piece of music, what is the meaning of creativity? And with that one might feel compelled to abandon the topic forthwith, because judgements about what constitutes good music are notoriously contentious ([5], p. 25).

One can deflect a precise definition in the form of,

> Computational creativity is the term used to describe the subfield of Artificial Intelligence Research where we study how to build software that exhibits behaviours deemed creative in people ([2], p. 3).

but at the risk of finding oneself more or less up the creek when it comes to programming. As is typical in her writing, Margaret Boden comes closest to defining the subject of computational creativity,

> Creativity isn't magical. It's an aspect of normal human intelligence, not a special faculty granted to a tiny elite. There are three forms: combinational, exploratory, and transformational. All three can be modeled by AI—in some cases, with impressive results. AI techniques

D. Cope (✉)
University of California, Santa Crus, CA
e-mail: howell@ucsc.edu

T.R. Besold et al. (eds.), *Computational Creativity Research: Towards Creative Machines*,
Atlantis Thinking Machines 7, DOI 10.2991/978-94-6239-085-0_15

> underlie various types of computer art. Whether computers can "really" be creative isn't a scientific question but a philosophical one, to which there's no clear answer. But we do have the beginnings of a scientific understanding of creativity ([1], p. 23).

Many other referenced definitions appear in the first chapter of *Computer Models of Musical Creativity* I refer here to as CMMC [4]. There, readers will find a plethora of other definitions and academic sources of further definitions for a ore complete review of the subject.

In this chapter, however, I will define 'creativity' as "the association of two ideas heretofore *not* considered related but now revealed as logically connected." This definition is somewhat more readable than the one I use in CMMC as "The initialization of connections between two or more multifaceted things, ideas, or phenomenon hitherto not otherwise considered actively connected," ([4], p. 11). Both of these definitions are computable in the sense that one can produce code directly from them that can be effectively used to generate potentially creative results. Buy potentially, here, I mean that one must of course agree with my definition of creativity if one is to appreciate the output of the code as creative.

Unfortunately, my definition for creativity, as all the others I've seen, still leaves significant questions unanswered. For example, many would include cooking, exercising, even skydiving as potentially creative activities. Certainly my definition does not exclude these things. But, one might therefore ask whether creativity represents an approach to *anything* that involves unexpected associations between two things not previously considered related.

Of course, the same could be said for *who* or *what* can be creative—my two cats, for example, since I've certainly seen them act in very creative ways. And I've also seen rocks roam around under wind power at the Racetrack near Death Valley in what seem like creative ways. And who's to say rocks are any less intelligent than cats? After all, many rocks are more complex than animals lower on the food chain.

Therefore, I underscore my definition (above) by making sure it is understood that *any activity* or *thing* doing that activity can be creative if it 'associates two ideas heretofore not considered related but now revealed as logically connected.' In a sense, then, my job is finished. Computers certainly fall under the above stated criteria.

For this chapter, however, I will stick to humans and to technical devices of human construction. Even here, though, I've known individuals for whom what I consider creativity comes so easily I wonder if it can be called creativity. After all, consistently producing surprises makes one suspicious it represents the true nature of something as special as creativity. In contrast, I've known others who've made only one or two creative discoveries in their entire lives, discoveries that have made significant impacts on others. Should such things as these numbers matter? And is creativity simply something voted on in silent elections by how much it impresses the society in which it occurs? If so, then the objective becomes subjective to the point of making true determination impossible.

And, of course, I've not included self-awareness—whether one must be consciously aware of choices being made—or cognition—whether one person will hear the results as creative and another not, based on previous experience. These notions, too, obviously play significant roles in determining whether something is creative.

Not finished taking potshots at my definition, I would also ask whether creativity can be pigeonholed as a single observable type. For example, few would doubt that John Cage was creative. At the same time, how can we consider the more subtle musical techniques created by Debussy and Schoenberg, for example, in the same category? One—Cage—clearly revolutionized the entire definition of the meaning of the word 'music,' while the other two revolutionized within the canon of music itself. And, of course, there are yet further refinements that separate one work of music from another and one part of one work from another part, and many refinements exist between these extremes that require creativity or otherwise remain stagnant.

Luckily, my definition includes these differences. At the same time, however, I offer the following three categories to help refine what I hope will become a clearer notion of creativity. These three areas reflect the previous distinctions: Global, Stylistic, and Constrained. Cage falls clearly in the Global category, as do many other experimental composers who attempt to expand the traditional definitions of music in their work. Debussy and Schoenberg fall into the Style category, since they clearly mark expansions of previous music but without straining standard definitions. As for Constrained, one would place most composers who follow whatever rules of their time and yet attempt to expand them in their own ways as those rules allow. I note here that these three distinctions can be applied to other areas of creativity besides music.

My reason for explicitly defining the above categories is to ensure that readers will find my examples of creativity clear rather than vague, and reasonably executable by computer programs. To clarify my definition here, then, I propose to investigate the Constrained category of creativity within traditional concert music. The computer programs I develop, therefore, do not attempt to redefine anything or expand on a particular style of music, but use given rules to demonstrate how those rules can be used creatively to produce new-sounding music.

15.2 Games

Before presenting musical examples, I often find it useful to explore metaphors in other fields—most often games for me—in that any remaining concerns regarding definitions can be clarified.

One simple example of creativity aside from the arts is a game I often play with family and friends. This game involves choosing three letters, any three letters, and finding words—discounting plurals and proper names—that contain those three letters grouped in the order presented. 'GST,' for example, can be found together in 'angst,' or 'RST' from 'overstuffed.' Obviously playing this game without having a computer or dictionary available can become quite challenging, especially if you're pulling letters out of thin air. A simple brute force approach, while possible, could

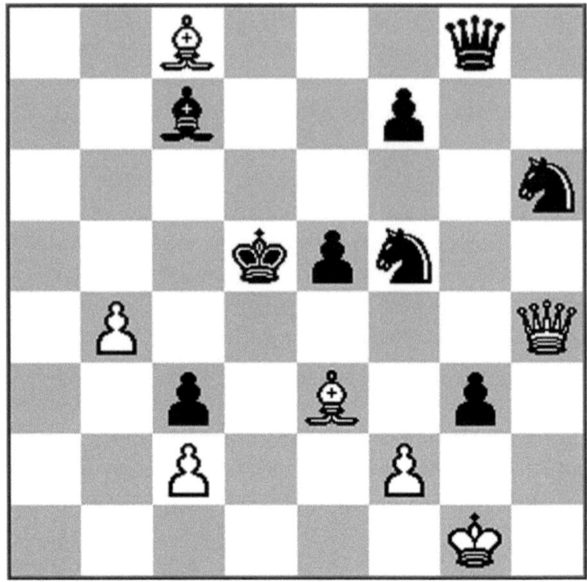

Fig. 15.1 White to mate in two moves

take a very long time even for a single random three-letter combination if, that is, the combination actually exists in a word. Clearly, some kind creative process must be involved to solve some of these combinations.

For example, one type of creativity employed that obviously uses association involves phonetically sounding out loud the three-letter combination in ways that make associations more apparent. Thus, 'NGL,' three consecutive letters one would not expect to find quickly in a word, almost immediately brings to mind 'angle,' 'strangle,' or 'entangle.' Here are a few other examples that might strain your own abilities at this game:

RFGONO PIF TYP LDO RBL XTA MBL RRI BKE XYZ.

Those groupings that contain three consonants pose the biggest problems, but either way I don't believe it's possible to succeed for very long without sounds, association of sounds with words, and ultimately, creativity. This type of game, while simple and with few rules, certainly belongs to the Constrained category of creativity as described in the previous section.

As a somewhat more elaborate Constrained category of creative game playing, I present a straightforward example in chess (see Fig. 15.1).

Like most games, chess has definite and highly limited rules that cover the board, movements of pieces, initial setup, and completion. At every step of the game, one has a limited choice of possible moves. Of course, I contend here that making the right move, the one that ultimately wins the game or at least contributes to winning the game in two moves, requires something more than brute force, and that something is, arguably, creativity. Even in situations like that shown in Fig. 15.1, where White makes the first move and must checkmate Black after Black's next move, little is clear except what pieces White can logically move (likely the Queen or one of the two Bishops). Note that Black here already leads White by a Knight and Pawn (depending on whether those counting equate Bishops with Knights), roughly accumulating to six points.

I encourage readers to attempt to solve the puzzle on their own before reading the following explanation. My point is that, even with only a relatively few likely choices each of which can be worked through to discover its potentials, the resulting checkmate requires creative thinking that many would find difficult or impossible. Without examining all potential moves here—my count has White with thirty-five in all, many of these obviously irrelevant—most inexperienced players will still find this a difficult problem to solve. Good chess players typically sample all possibilities, looking ahead to Black's potential responses which are far more than White's thirty five, to see if by some quirk a checkmate is hidden somewhere in them. With chess etiquette being what it is, all this must take place in the player's head and not on the board itself or on paper.

For those not successful in discovering the solution, I present Queen to E-4 (letters mark A-G left to right, and numbers 1–8 from bottom to top). Clearly a gambit, this gives Black nine more points to add to the four already in hand if Black takes White's Queen. This last move, however, leads to an immediate checkmate for White (Bishop at C-6–B-7). If Black decides against taking White's Queen, its only option is D-6, which leads to checkmate by White's bishop E-3–C-5. None of these moves are illegal, all are known by both parties, yet most players of White won't give the Queen's gambit a second thought since it immediately produces a point differential that most reasonable players will think impossible to surmount. In chess, position is everything. Most players know this, but it's difficult to consider when facing such staggering odds.

This chess example, like the word game given previously, works within rules but extends reasonable expectations by requiring unusual, even obtuse, associations. Assumptions that XYZ belongs in a word or that sacrificing a piece to fall thirteen points behind will win the game, most assuredly represents 'association of two ideas heretofore not considered related but now revealed as logically connected.'

15.3 Music

While many might doubt the serious potentials for creativity in playing games like letter combinations and chess, very few doubt its presence in the process of composing music. But, since music has rules, composing poses the same kinds of problems exampled by such game playing.

In the following sections, I initially discuss melodic creativity based on human-composed models, and then describe simple computer-composed single lines using those models. In the section following that—more complex due to its harmonic nature—I center on triads and voice leading along with description of a computer program and its output. Finally, I present a sample of creativity in counterpoint, once again demonstrating both human and computer-generated music.

15.3.1 Melody

Most composers sketch as an integral part of their creative processes, often using associations on their way to developing variations and new ideas.

Figure 15.2 provides an example of such a process.

Beethoven is noteworthy for keeping notebooks of sketches of melodic ideas that have, since his death, been published and referenced for all to see. In Fig. 15.2, we see roughly the same theme in two different guises composed two years apart. The projected symphony in 2(a) was never completed, but by cannibalizing it, Beethoven reveals he thought well enough of the theme to include it in a slightly different guise in a later published work. Is (b) then more creative than (a) according to Beethoven?

Note that while transposed to G-major and metrically offset, Fig. 15.2b's opening measures are identical to (a) for the first five notes. The music then varies in quite interesting ways both in rhythm and pitch and ends with a hint of G-minor with the Bb nine notes from the end (2b). The range, when appropriately transposed, remains the same with the final note the same as well. Within those frameworks, however, Beethoven has not just embellished his original idea, but also neatly crafted a surprisingly creative variation.

Figure 15.3 provides an even more interesting example of sketching with (a) being the first and (b) the eleventh sketch with, obviously, nine continuously improving (at least in Beethoven's opinion) ones in between.

These examples show a much more drastic range of changes than those in Fig. 15.2, with the more or less final version—excepting the second and fourth measures—varying significantly from the initial one. Unlike Fig. 15.2, which simply leaps from one version to the other, this example slowly—through the missing sketches—

Fig. 15.2 **a** Opening bars of finale theme by Beethoven of projected symphony from 1795; **b** Rondo from Sonata Op. 49, No. 1 from 1797 by Beethoven

Fig. 15.3 First (**a**) and eleventh (**b**) sketches from Beethoven's Piano concerto in D (unfinished) 1815 from [3], pp. 346 and 352

evolves from simple origins to a quite chromatic (at least for its time) elegant form. I have attempted on numerous occasions to fill in the missing sketches myself without consulting the actual Beethoven ones, and found my own attempts difficult to smoothly transition from (a) to (b). I invite readers to make this attempt themselves. There is no better way to understand creativity than to produce examples yourself and compare them with someone who has proven an expert in the field, like Beethoven. The full set of sketches appears in CMMC ([4], p. 82) for reference.

Interestingly, Figs. 15.2 and 15.3 resemble the game situations of letter/words and chess respectively. Figure 15.2 leaps from a smaller subset to a larger one like letters to words, and Fig. 15.3 demonstrates a step-by-step approach to developing a resolution to a problem like chess.

As an example of how a computer program might create new melodies following the rules presented, imagine a group of pitches that contains, say, all the iterations of the same pitch each including its previous and following pitches. For example, the second-space A group in Fig. 15.3 would contain two instances (one from each theme). The first instance would also have a record of fourth space E as preceding pitch and fourth line D as following pitch, and the second instance of second-space A would have fourth space E as preceding pitch (the same as in the previous instance), and fifth line F-natural as following pitch. It would then not be difficult to imagine similar groups of pitches representing all the pitches in the two themes.

The composing part of this simple program would first remove a beginning pitch from the appropriate group (in this case either from the ledger line A above the staff group) or the fifth line F-sharp group, the beginning pitches of the two examples. For our example, let's choose the A above the staff from the first example. Then the program would save that choice and select the next pitch's group based on the following pitch of that choice (in this case G above the staff's group). The choices now would be many since that G occurs six times—F-sharp, B, F-sharp, F-sharp, A, and E being the following pitches—with F-sharp the most probable choice given we choose randomly and there are three F-sharps present. The program continues in like manner until it arrives at a pitch with no following pitch, which signals the end of the melody. Note that at this point, previous pitches have had no impact on the decision-making process, something I will soon describe.

With enough data in each group—enough choices—recombinations like this can occur following the basic models from the database. By using the term 'recombination' here, I'm referring to a Markov chaining process that uses probabilities as just described. Unfortunately, this process can produce excessively long or almost immediately short outputs, neither of which really imitates the melodies shown. Therefore, the program needs more rules that will, on the one hand, provide more than next-pitch probabilities to give shape and logic to these melodies and on the other hand ensure that the output won't simply repeat one of the melodies exactly.

As a simple example of how this might occur, imagine that the program's groups are labeled with single digit numbers, each containing a significant number of options within it. The models from which the program creates new output might then have this as input:

1 2 3 4 5 6 7 8
1 4 3 5 8 6 7 2
3 2 1 4 5 7 6 8

The numbers tell us is that the orders provided here are acceptable. The number 1 can be followed by 2 or 4, with 4 a more likely choice by a probability of two to one. The number 2 can be followed by 3, 1, or nothing—a cadence in music—each with equal chance. Thus, correctly recombining the models could produce the following outputs 3 4 3 5 8 6 7 2 and 1 2 3 5 6 7 6 8, and so on. Both of these are correct as far as they go.

What I've just described is often called a First Order Markov forward-chaining process. First order, because only one member (the previous one) of a chain of items is necessary to predict the next. A Second Order Markov forward-chaining process requires two numbers to gauge the next, and so forth. In this manner, almost any algorithmic process can be described as a Markov chain. As example and using the previous models, 1 and 2 can only be followed by 3, 1 and 4, can be followed by 3 or 5 with equal probability. This process certainly helps provide more context to the choices made. Unfortunately, second order chaining does not help with ending an output with reasonable length phrases. Furthermore, the higher the order number used, the more likely the output will repeat exactly one of the inputs.

To make the process more constrained and contextual, therefore, I now add orders of backward-chaining Markov processes used in conjunction with the forward-chaining ones. This means that the just-described process will also include the objects that follow the to-be-chosen one. Based on the previous models as input, 1 can be followed by 2 or 4 with 4 having a two to one better chance of being chosen as previously mentioned. However, backward chaining suggests that the 3 group determines the opposite; that 2 and 4 are equally probable given that 3 is the most probable of the following numbers.

To make this clearer, the number sets of 123, 143, and 145 give the probability that 1 will be followed by 2 one third of the time or 4 two thirds of the time, while taking the third number and working backward gives the number sets of 321, 341, and 541, clearly showing that 4 is the more likely choice and thus balancing the forward-chaining process. While subtle, this kind of contextual framework produces

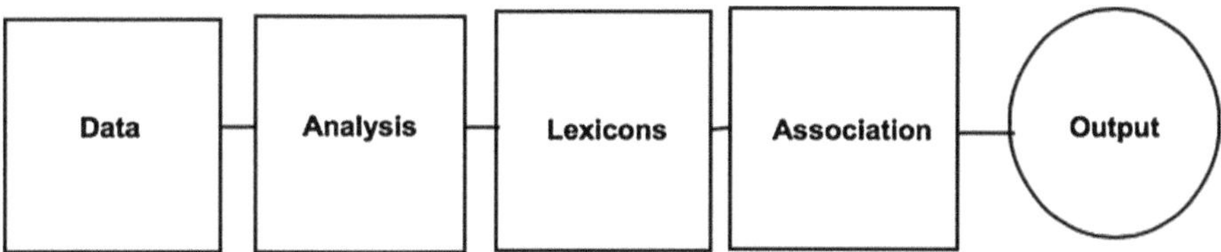

Fig. 15.4 A data-driven creativity program

more sensitive and creative possibilities that likely will end in reasonable time lengths and on reasonable pitches.

The resulting program from the description just given looks something like that shown in Fig. 15.4.

Figure 15.4 begins with as much data as possible presented in a form that allows logical segmentation in the form previously described. The Analysis portion then stores these segments in their appropriate groups. The association process follows the forward and backward chaining Markov reorganization processes producing original output. As should be clear, this design is simple, data-driven, and highly contextual in both its forward and backward directions.

Figure 15.5 presents eight examples representing potential creativity, four of which by humans and four by a version of the program just described.

Note that each melody begins identically and none contain chromatic pitches. The durations of the computer-composed examples were similarly derived from durations provided by the extended human-composed ones. I won't indicate which are human composed and which by computer program, but leave that up to readers to decide. Note that the database, and thus the models for each, consisted of Mozart melodies in Classical style simply because the rules are so apparent and the style obvious. The ones completed by humans (four different people not including myself) may seem obvious to some readers. However, I caution those making serious guesses here, that what may seem obvious may not be so at all.

For example, note that melodies (c), (d), (g), and (h) each contain two imitative phrases of equal length, that (a), (f), and (h) contain leaps of a sixth or more, all but (d) and (g) end with downward motion, most contain only half, quarter, and eighth notes, and that (e) and (h) have the widest ranges. Thus, the examples present significant diversity that makes it difficult to determine which demonstrate human and which computer origins.

I would agree with those that insist that such short and simple examples do not provide a perfect test upon which to base judgment as to human versus computational creativity. At the same time, short and simple melodies such as those presented here can focus our attention on the nitty-gritty of the problem. While creativity can and does exist at all levels of music composition, the smallest decisions may likely be the most telling.

Fig. 15.5 Eight completions of the same four note beginning, four by human composers (not the author) and four by the simple computer program described in the text

15.3.2 Harmony

Harmony in Classical-style music follows very defined vertical and horizontal rules. In brief, these rules consist of triads and stepwise motion of horizontal lines. There are, of course, exceptions to these rules such as certain seventh chords as harmonies

Fig. 15.6 Two phrases of diatonic music for use as a database

and voice-leading leaps, especially in the lowest line. However, as the examples in Fig. 15.6 demonstrate, there is a clarity of progression that suggests function, direction, key, cadence, and so on, all supported by the basic rules stated above. All the chords here are triads or partial seventh chords (partial since seventh chords require four pitches with only three pitches allowed here).

Voice leading, one of the most important factors in tonal music, requires that each line (here designated from top to bottom as soprano, alto, and tenor) move as little as possible, with the lowest line able to leap if necessary. Using half-steps as the smallest interval here, the soprano in Fig. 15.6a follows 0, −1, 1, −1, −2, −2, 0, 0. Note that no interval moves more than 2 (a major second, not a leap) and the number of intervals is naturally one less than the number of pitches. The same is true of the alto voice 2, −2, 0, 0 −2, −1, −2, 2, while the bass has one interval larger than a second (the −4 in measure 3), legal since the tenor is the lowest voice. Figure 15.6b follows the same principles with but one exception (the 5 in the alto voice in measure 4).

In my book *Computer Models of Musical Creativity* [4], I describe a computer program called Emily Howell that I maintain is creative. It accomplishes this feat via a human collaborator using a carrot and stick approach. I invite readers to explore not only my description of Emily's processes in CMMC, but also the code available on my website. As readers can tell from my previous references in this chapter, however, the program I will now describe requires that it operate creatively on its own.

Here, I begin by establishing three basic principles to which this program must adhere. The first, and likely the most important, is simplicity. While some of my software has become more complicated over the years, it never began that way. The second principle involves something I maintain in all my work, that it be data-driven. This means, simply, that the program requires as many appropriate examples for analysis as possible. Finally, the third principle requires that each example in the database be coherent in the sense that its context follows the same rule set. So, to make myself clear, this program must be

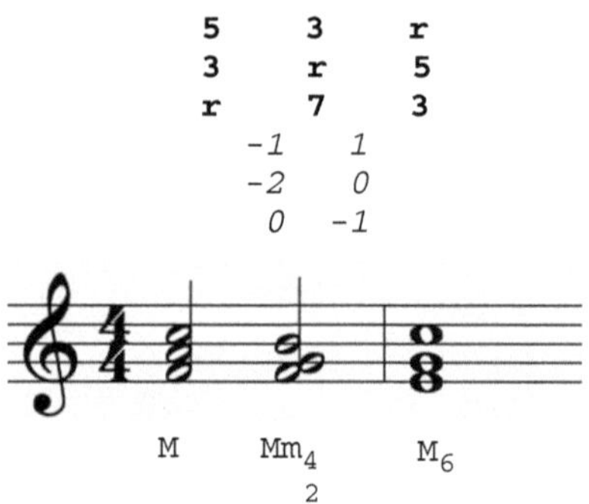

Fig. 15.7 Second, third, and fourth chords of Fig. 15.6a with analysis

1) Simple,
2) data-driven,
3) and contextual.

With these three principles in mind, I believe that creativity can be achieved in a computer program.

Since the music in Fig. 15.6 represent good examples, I will use them as a database for a data-driven computer program acting creatively. Both of these examples are tonal, diatonic (in C-major without any chromatic notes), and use standard voice-leading principles as previously mentioned and detailed. Thus, the music in Fig. 15.6 is consistent and will supply all the rules for the program that I will now describe.

Figure 15.7 now presents an analysis of the second, third, and fourth chords of Fig. 15.6a for close inspection.

The nomenclature in Fig. 15.7 is—from top to bottom—chord members ('r' equals root) in boldface, and below those three lines, interval motion (0 equaling no motion and minus sign indicating downward motion) in italics. Underneath the music itself, M refers to Major, m minor, with the combination of the two used for seventh chords, the first letter the nature of the bottom triad, and the second letter the nature of the interval of the seventh. The numbers next to these letters indicate inversions, something that need not concern us here, but I have included for those interested (figured bass shorthand is actually redundant due to the uppermost description of chord members). For those familiar with music theory, I've excluded function in this presentation to avoid confusing readers without that knowledge.

Linking Fig. 15.7 with a computer program analysis process could be accomplished by stating this data by voice in the following manner:

$$(((5\,\mathrm{M}) - 1(3\,\mathrm{Mm}))\quad((3\,\mathrm{Mm})\ 1\ (\mathrm{r}\,\mathrm{M})))$$

read as the fifth of a major triad (M) moving downward by minor second to the third of a major/minor seventh chord (Mm), which then (the second parenthetical grouping) moves up a minor second to the root of a major triad. This rule can then be used for producing new output. In other words, whenever a pitch represents the

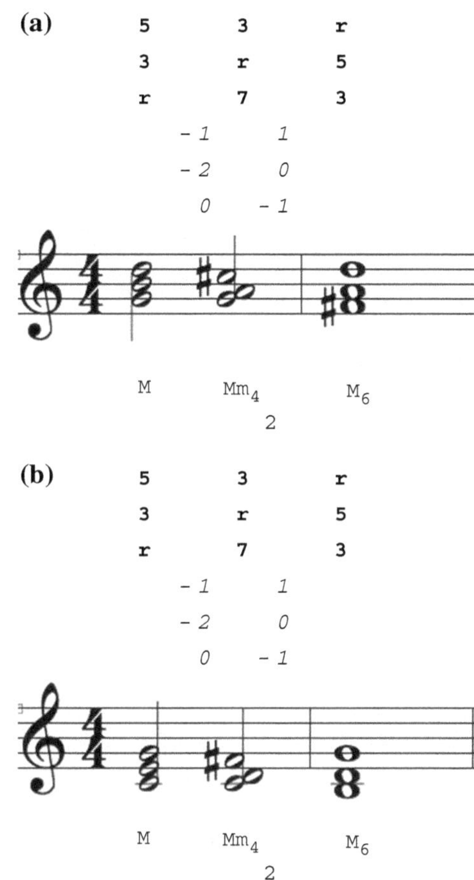

Fig. 15.8 Computer-created extrapolations of a chromatic pitch from Fig. 15.7

fifth of a major triad it can correctly move downward a minor second to the third of a Mm seventh chord given the roles of the other voices conform as well. The alto and tenor voices can be stated in exactly the same ways with the resultant rule confirmed by the last three chords in Fig. 15.6b.

With this rule in mind, a computer program can and will—without instruction to the contrary—use it to create both of the examples in Fig. 15.8.

These two examples follow exactly the rule derived from the voice leading in the database, but in doing so create chromatic pitches not present in those two phrases as exemplified in the new phrase shown in Fig. 15.9.

Fig. 15.9 Output using materials from Fig. 15.6 and rules shown in Fig. 15.7

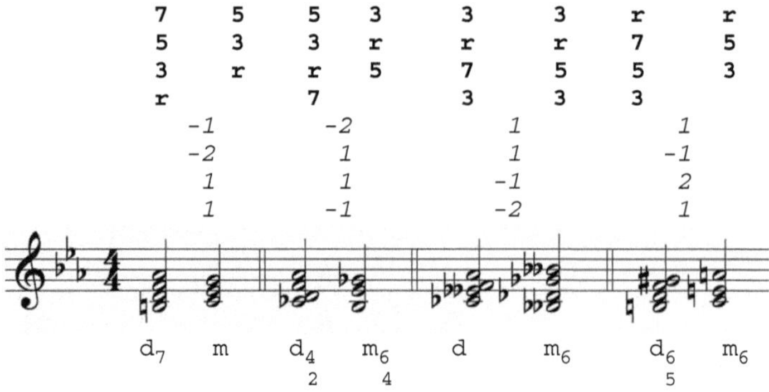

Fig. 15.10 The same-sounding diminished seventh chord resolving four different but correct ways

In this new phrase, the chromatic F-sharp in measure one creates an 'association of two ideas heretofore not considered related but now revealed as logically connected.' An average student in a music theory class would never think of turning in an assignment like Fig. 15.9. A highly creative student, however, would, since it follows the rules the models follow. One might argue that the new output in Fig. 15.9 was created by a machine program having no idea what it was doing. I would argue that it was I who created Fig. 15.9 first, knowing full well what I was doing, and then I programmed my computer following exactly the process just described to imitate the process. Since my computer program had only the information I had when composing its output and we both followed the same rules, are we not both creative?

For those interested, the second chord in Fig. 15.9 is called a secondary dominant of the dominant chord in C-major, a simple kind of chromatic motion that was most likely one of the first to invade diatonic harmony in the late Renaissance or early Baroque periods of music history. Whether such chords occurred by the method just described or not, no one knows. But I contend that it's certainly one reasonable hypothesis.

Figure 15.10 presents something more advanced than secondary dominants, a diminished seventh chord which by virtue of its symmetrical nature and voice leading allows the root to exist as any of the pitches present in the chord.

Without going into details here, each of the four examples sounds the same but are differently spelled and thus allow the voice leading from a database of examples to provide it with several options for resolution. Figure 15.11 shows how this is possible.

In the case of the output in Fig. 15.11b, the final example of the diminished seventh chord reaffirms the second measure chord excepting the latter B-natural is respelled as a C-flat. The respelling allows for the modulation to E-flat.

Again, I must emphasize here that I created this output first, and then used the program to output the same thing based on the data in the database.

As a final example of machine musical creativity, I here present an even more complex example involving chromatic pitches in both the database and in the output, but with the output involving a different and creative result. In Fig. 15.12, the

Fig. 15.11 The same-sounding diminished seventh chord resolving to the C-minor triad or as a modulatory diminished seventh chord of E-flat in final measures of (**a**) and (**b**)

first measure is a simple motion typical of the diatonic data in previous examples. The second measure represents an Italian augmented sixth chord in C-major resolving appropriately to a major triad. The third measure presents the same Italian augmented sixth chord respelled as a Major-minor seventh chord of the far-related key of C-major—C-sharp major or D-flat major. All this based on a very simple initial two phrases in a database, neither of which modulates. Thus, new output (not shown here due to space restrictions) will accomplish something none of the input could, but based entirely on the rules derived from that input. And, once again, I state that I created the output before allowing the simple computer program to accomplish the same thing.

The program that produced the above output is so simple that I have not included a visual schematic. It consists of just three components. The first of these is the database itself, represented in numbers. The second component is an analysis

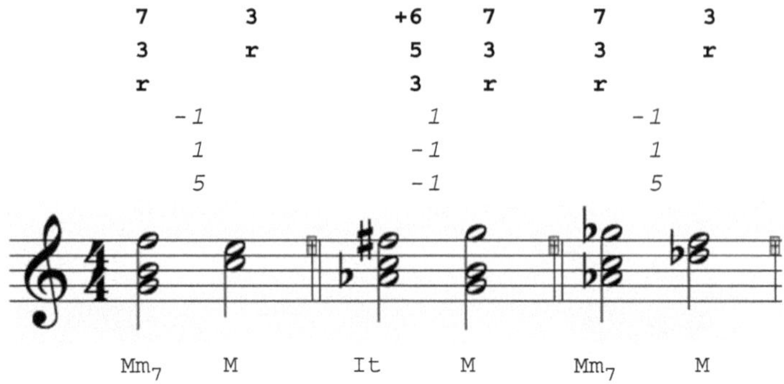

Fig. 15.12 The same-sounding major-minor seventh chord resolving first to a major triad missing fifth, then as an Italian augmented sixth chord to a M triad missing fifth, and then a respelled Italian augmented sixth chord to a different M

sub-program that extracts the rules in the manner described earlier in this chapter. And the last component uses those rules to produce output. With the examples presented here, any good programmer could code such a program in a very short time.

15.3.3 Counterpoint

Even given these arguments for machine creativity, however, I here present more complex examples, one created by a human and the other by a computer not unlike that just described but having greater degrees of analytical and compositional constraints than those previously described.

Composing Baroque fugues requires at least as many rules as chess and language, probably more so considering that not only does the form itself have rules regarding entrance of the theme and keys, but that these rules must fit within the strict conditions of triadic harmonies and tonal voice-leading involving consonance and dissonance as well as contrary and oblique voice motions. Rules once again, therefore, are particularly important ingredients to the creative engineering of a good result. Interestingly, Bach ends his exposition as required on a tonic B-minor triad at the end of his exposition (measure 21 of Fig. 15.13a), while the Emmy example ends its exposition on a B-major triad in the middle of its measure 15 (Fig. 15.13b). Is this creativity, or simply a mistake by the program? Possibly it's an interesting correlation with Bach's immediate (measure 22) reference to B-major as well, since the rules used for creating Fig. 15.13b included the Bach in Fig. 15.13a. Given the forward and backward chaining of the Markovian processes described previously, the latter makes perfectly good sense.

All of the examples in the figures of this chapter can be heard from files presented on my website at https://arts.ucsc.edu/faculty/cope/new.html. My inclusion of clearly more elegant examples here may seem unfair to those who would like to

Fig. 15.13 **a** Fugue 48 from The Well-Tempered Clavier by J.S. Bach; **b** Fugue 48 from The Well-programmed Clavier by Experiments in Musical Intelligence (Emmy)

discover more about the processes involved in creating both. For those that wish this, I suggest treading several dozens of books on Bach, or the books on Emmy listed on my website where the MIDI files reside. I have also identified the two sources of the fugue expositions so as to not encourage gamesmanship as to which is which when none is intended. For those believing that the Bach here is far more creative and intelligent music, I have no disagreement. I do believe, however, that to not consider the Emmy example as creative, at least to a small degree would be foolish.

15.4 Conclusions

This brings me to my final thought; how do we recognize one thing as creative and another not? The answer for me is twofold. First, creativity must be surprising. By this I mean it must be unusual, unique, unexpected, or previously considered unrelated. Second, these surprising results must reveal something heretofore hidden, produce an insight not yet considered, and most importantly associate two or more ideas not previously seeming related. In short, it must associate two ideas not considered logical but now revealed as imaginatively connected. And the examples in this chapter, I believe, do just that.

From this point of view, the magic elixir of creativity is a split infinitive that 'boldly goes where no one has gone before,' a gambit least expected that wins a game, or a sudden chromatic modulation that only makes sense when contextually understood. All of these made possible by the intricate interrelationships between what has and what will occur; surprise followed by revelation. And, lest I forget, all of these being choices from already extant possibilities that generate the rules used.

References

1. Boden, M.A.: Computer models of creativity. AI Mag. **30**(3), (2014)
2. Colton, S.: The painting fool: stories from building an automated painter. In: McCormack, J., d'Inverno, M. (eds.) In: Computers and Creativity, p. 3. Springer, Berlin (1998)
3. Cook, Nicholas: Beethoven's unfinished piano concerto: a case of double vision? J. Am. Musicol. Soc. **42**(2), 338–373 (1989)
4. Cope, David: Computer Models of Musical Creativity. MIT Press, Cambridge (2005)
5. Merker, B.H.: Layered constraints on the multiple creativities of music. In: Irène, Wiggins, G.A. (eds.) Musical Creativity: Multidisciplinary Research in Theory and Practice, p. 25. Psychology Press, New York (2006)

Chapter 16
A Culinary Computational Creativity System

Florian Pinel, Lav R. Varshney and Debarun Bhattacharjya

Abstract Compared to artifacts in expressive or performance domains, work products resulting from scientific creativity (including culinary recipes) seem much more conducive to data-driven assessment. If such products are viewed as an assembly of constituents that follow certain association principles, one could apply computationally intensive techniques to generate many possible combinations and use automated assessors to evaluate each of them. Assembly work plans for the selected novel products could subsequently be inferred from existing records. In this chapter, we report on our efforts to build a computational creativity system for culinary recipes. After gathering data and creating a knowledge base of recipes and ingredients, the system generates ingredient combinations that satisfy user inputs such as the choice of key ingredient, desired dish, and cuisine. Once a combination has been selected with the help of novelty and quality evaluators, the system further recommends ingredient proportions using a distributional conformance method and generates recipe steps using a subgraph composition algorithm. The time durations or efforts of atomic steps are estimated by solving an inverse problem from data on complete recipes. The example of culinary recipes could be generalized and applied to other scientific domains; manufacturing products and business processes could potentially follow a similar recipe for success.

F. Pinel (✉) · D. Bhattacharjya
IBM Thomas J. Watson Research Center, Yorktown Heights, New York, USA
e-mail: pinel@us.ibm.com

D. Bhattacharjya
e-mail: debarunb@us.ibm.com

L.R. Varshney
Coordinated Science Laboratory, Department of Electrical and Computer Engineering, University of Illinois at Urbana-Champaign, Urbana, USA
e-mail: varshney@illinois.edu

T.R. Besold et al. (eds.), *Computational Creativity Research: Towards Creative Machines*, Atlantis Thinking Machines 7, DOI 10.2991/978-94-6239-085-0_16

16.1 Introduction

Traditional work in computational creativity has typically focused on domains such as visual arts, writing, and music [1–5]. In characterizing human creativity, however, it has been found that there are at least three distinct groupings of creative domains: *expressive creativity* (visual arts, writing, humor); *performance creativity* (dance, drama, music); and *scientific creativity* (invention, science, culinary) such that abilities within one member of a category transfer over to other members of the category [6]. Given the potential business impacts of creativity in science and invention [7, 8], we focus our efforts here on scientific creativity, rather than expressive or performance creativity. In particular, we take culinary creativity as our specific domain; the basic techniques developed here could also be useful for other kinds of scientific creativity, especially in those settings where artifacts are to be constructed from components.

We adopt a definition of creativity used in human creativity research [9]: *Creativity is the generation of a product that is judged to be novel and also to be appropriate, useful, or valuable by a suitably knowledgeable social group.* By defining principled descriptions of artifacts as compositions of parts, we compute novelty using information-theoretic measures that have been validated with psychology experiments across domains [10]. For appropriateness and value, we draw on fundamental chemical properties of flavor and neurobiological properties of flavor perception to define measures [11, 12]. In both the novelty and quality dimensions, the scientific nature of the domain lends itself to data-driven assessment. The ability to assess enables the selective step of creativity, which is arguably just as important as the generative one [9, 13].

Notwithstanding, the central focus of this chapter is on the generative algorithms we use for creating novel and flavorful culinary recipes. Note that unlike Morris et al. [14], who also looked at culinary creativity, generating ingredient proportions and recipe steps (with timings) rather than simply the list of ingredients is critical for our work. Our system produces a complete work plan for physically creating a dish.

A block diagram for our computational creativity system is presented in Fig. 16.1. Data from a domain knowledge database is categorized and fed to the work product designer, which interacts with the work product assessor to suggest one or more work products, and with the work planner to generate the corresponding work plan. Applied to the culinary domain, the work product designer outputs combinations of ingredients and their proportions—the recipes—and the work planner produces preparation instructions.

The domain knowledge database consists of a collection of existing work products and their constituents, including information pertaining to their properties, quality assessments, styles, and cultural associations. The knowledge categorizer is responsible for classifying these products and constituents, defining an ontology that helps users specify the profile of the novel products they wish to create. This is described in Sect. 16.2.

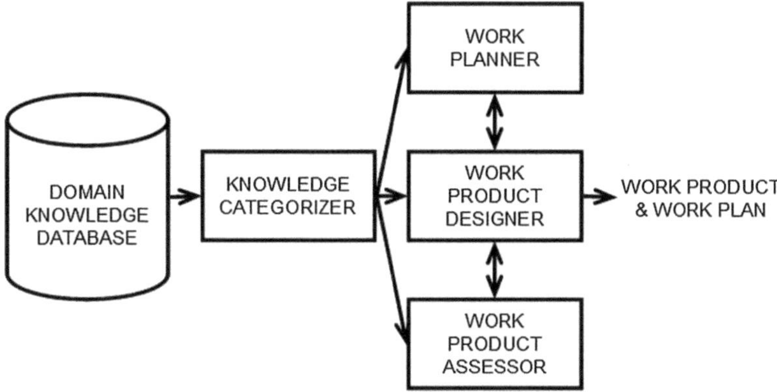

Fig. 16.1 Block diagram of computational creativity system that produces a work product and a work plan

The designer is the lead component of the system, though its task depends heavily on the other modules. Once system users select target categories of work products and constituents, it generates combinations of constituents that satisfy these inputs. As the number of possible combinations rapidly exceeds what current computers can handle, it becomes necessary to introduce additional filtering criteria. The generation of ingredient combinations is explained in Sect. 16.3. Once the combinations of constituents are selected, their proportions must be determined. Section 16.4 describes the generation of ingredient proportions through a distributional conformance method based on the learned properties of the dish to be prepared.

The assessor examines the combinations produced by the designer, and evaluates them along several dimensions. Certain dimensions are specific to the creative domain of interest, in particular metrics measuring the quality of the proposed artifacts—the flavor pleasantness of a recipe in the culinary arts, or the longevity of a fragrance in perfumery. Others are generic enough to be used in various contexts, such as novelty [15] or cost. Although we briefly review the assessor in Sect. 16.3, it is described in much more detail elsewhere [16]. The designer and the planner are the central focus of this chapter.

Next, the work planner determines how a specific work product, selected by the system user based on the outputs of the designer and assessor, can be executed. The nature of the work plan varies with the creative field. For material work products, it typically entails constituent proportions and manufacturing instructions. Section 16.5 discusses the generation of recipe instructions. There is also a need to estimate the time durations or efforts of the various atomic recipe steps. Section 16.6 explains how this can be done by solving an inverse problem to deal with the fact that original recipe data typically only lists timings for larger "molecular" recipe steps. Finally, we conclude the chapter in Sect. 16.7.

16.2 Assembling a Culinary Knowledge Database

We propose a representational model for culinary computational creativity that captures all the aspects of domain knowledge necessary to design, assess, and plan recipes. In this model, shown in Fig. 16.2, the basic unit is the recipe, which is then decomposed into all the elements one usually finds in a cookbook: an ingredient list, and a sequence of steps with their inputs, outputs, and properties. Another set of entities (dish, cuisine, ingredient type, and ingredient pairing) are used by the knowledge categorizer to define the ontology of ingredients and recipes.

However, additional elements are needed to enable the data-driven assessment of new ideas. Through chemical analysis, ingredients are broken down into flavor compounds. The compounds are in turn characterized by odor descriptors and pleasantness evaluations, all of which are used by the assessor to compute quality metrics. Similarly, ingredient proportion generation makes critical use of the ingredient nutrition facts, such as carbohydrate content, as we will see later.

Peer-produced online repositories make knowledge in several creative domains accessible to be learned. By extracting data from the Wikia recipe repository [17], we can populate the greater portion of the representation model gravitating around the recipe entity. The human-readable text, though less structured than recipes in published cookbooks, can be parsed using natural language processing. Statistical parsing with domain-specific tokens is able to identify the correct task, tools, ingredients, and tips from a recipe instruction with sufficient accuracy. Other databases, whether public like the USDA National Nutrient Database [18], or commercial like the Volatile Compounds in Food database [19], provide additional ingredient properties.

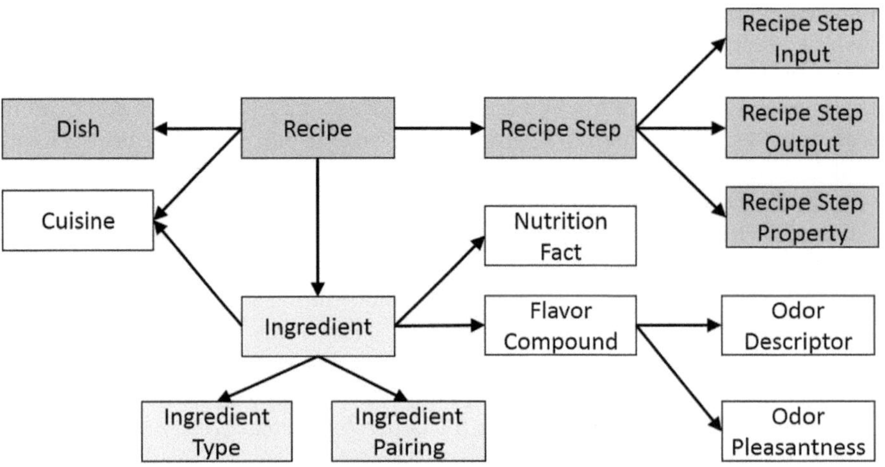

Fig. 16.2 Knowledge representation for culinary recipes and ingredients

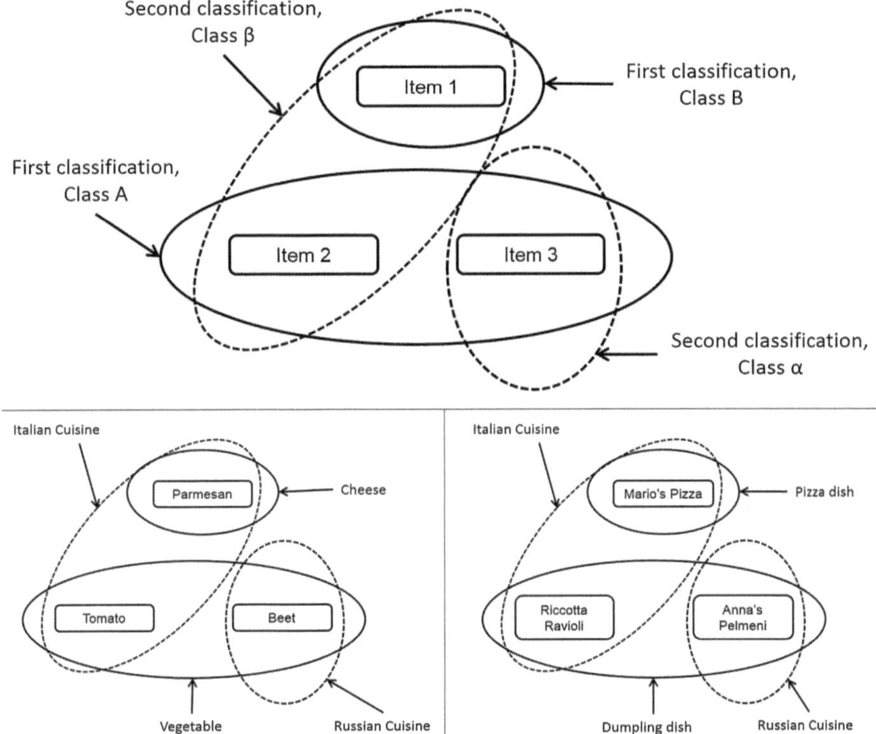

Fig. 16.3 Classifying constituents in artifacts; *Top* General model with items; *Bottom Left* Ingredients; *Bottom Right* Recipes

The knowledge categorizer serves two purposes. First, it classifies both the work products and the constituents, possibly using more than one classification, as illustrated in Fig. 16.3. Recipes are categorized into dishes (pizza, quiche, pasta…) and cuisines (American, Italian…), and ingredients into ingredient types (meat, vegetables, fruits…) and also by the cuisines in which they are prevalent. The online repositories used to populate the knowledge database provide a starting point for this categorization effort. Wikia recipes are tagged with ethnic cuisine categories, and the dish represented by each recipe is inferred from the recipe name. By parsing the Wikia and Wikipedia [20] pages on the various cuisines and ingredients, one can further associate ingredients and dishes with cuisines. The pairwise compatibility of ingredients is studied by analyzing how often any given pair of ingredients is found together in the corpus of recipes. Some classifications, such as the ingredient types, may be difficult to derive solely from the information contained in existing repositories. One approach consists in clustering the ingredients based on their nutrient or flavor compound composition. Domain experts can also refine the classification with a crowdsourcing tool.

Ingredient Type	Frequency	Min	Max
Cheese	98%	1	2
Egg product	98%	1	1
Dairy	91.8%	1	2
Seasoning / spice	83.7%	1	3
Vegetable	83.7%	1	2
Oil / fat	42.9%	1	1
Herb	32.7%	1	2
Dough	30.6%	1	1

Fig. 16.4 Example of an ingredient bill for a quiche recipe

Categories can overlap or be disjoint. A given recipe usually represents a single, well-defined dish, but may be shared by the cuisines of multiple neighboring countries. In this case, it might make sense to establish a hierarchy of categories, in order to calculate a similarity distance between them.

The second purpose of the knowledge categorizer is to provide a bill of constituents for the work product categories. An example of an ingredient bill is provided in Fig. 16.4. Using an association rule algorithm and performing a statistical analysis of existing recipes, the categorizer determines what combination of ingredient types is required to execute a given dish, as well as the frequency of each ingredient type and the minimum and maximum numbers of ingredients of that type commonly used in the recipe.

The ingredient bill is not immutable. A single dish can lead to several possible bills. A pie can represent a savory dish or a dessert, both of which command very different lists of ingredient types. At a more fine-grained level, a pie can be made with store-bought dough, or the dough can be made from scratch, which results in more variations that should be captured by the associative rules. Furthermore, the bills are refined using online learning methods. As users create recipes with the system, their inputs constitute new data points that are fed back to the knowledge categorizer. Only successful ingredient bills are used for feedback.

Creation, in our view, is the process of decomposing work products into their constituents as depicted in the data model, categorizing the entities, and then recomposing and reconstituting new work product ideas. This is different from simple modification of existing work products that are often modularly designed, because no modular structure is maintained from existing work products. Since data-driven idea generation can only use features that are derivable from the underlying data, the knowledge database is the cornerstone of our computational creativity approach. Although this limits the creative universe to combinations of previously seen components, completely new pairs of components certainly can and do arise.

16.3 Generating Ingredient Combinations

Although the system we have developed could operate autonomously, it can achieve greater impact by initiating a dialog in which both user and machine contribute ideas [21]—even in purely human contexts, interacting groups are more creative than individuals. Since the first step in creativity is to find a problem [9], we created a novel interactive interface design for mixed initiative, with the human and the computer system taking turns in the creative conversation. In order to create a new recipe, users are asked to select a base ingredient, dish, and cuisine. More than one of each selection is allowed, leading to fusion cuisines or "frankenfoods" such as ramen burgers or churro ice cream sandwiches. The system offers assistance in the process: by searching the knowledge database, the system suggests cuisines and dishes that use a given ingredient most often. Once the choices have been made, one or more ingredient bills representing the selected dish are presented to the user, who may tweak the ingredient type composition before the system begins designing the new recipe.

A typical ingredient bill calls for 10–15 ingredients, and can be satisfied by trillions or quadrillions of ingredient combinations. The main role of the combinatorial algorithm is to narrow down the possible recipes to those combinations that comply with the user inputs, pulling information from the knowledge database as a guideline.

Figure 16.5 illustrates this process. On the left side, the ingredient bill for the selected dish lists how many ingredients of each type are needed to create the recipe, and the system generates all the possible combinations accordingly. Since the knowledge database categorizes ingredients by cuisine and associates them with others based on pairwise compatibility, the base ingredient and cuisine picked by the user

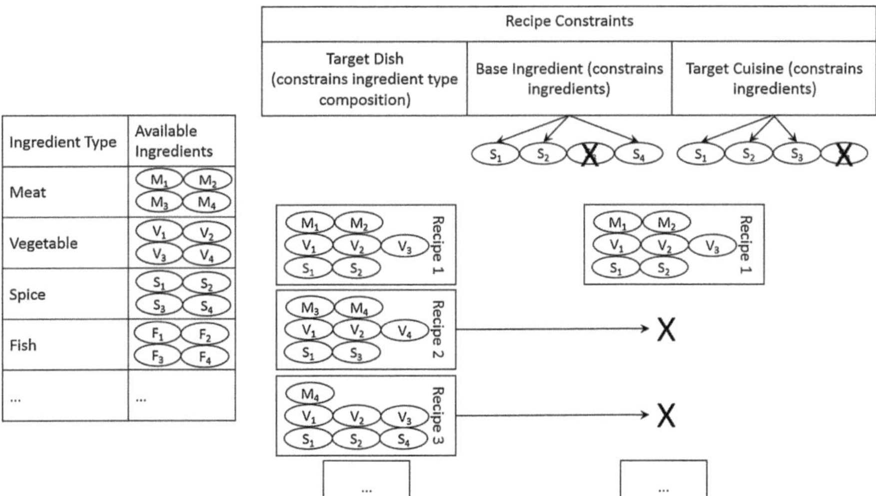

Fig. 16.5 Generating ingredient combinations

further constrain the combinations: ingredients that either do not pair well with the base ingredient, or are not found in the target cuisine, are eliminated from the possibilities.

Additional filters could be applied. Most dietary restrictions (vegetarian, low-fat, gluten-free, Kosher) can be mapped to a list of ingredients to avoid, sometimes by checking the ingredient types or their nutrition facts. In another example, a profile, either built explicitly by the users or inferred from their online history, may indicate preferred foods and help produce personalized recipes.

In most cases, the above approaches are not sufficient to narrow down the combinations to a number that can be handled by today's computers in real time. The work product assessor can alleviate this issue by retaining only the recipes that rank the highest with respect to one or more evaluations, such as novelty or estimated pleasantness. Ultimately, if too many combinations subsist, the recipes displayed to the user may be chosen randomly. Generating random subsets of possibilities has the added benefit of ensuring that every run of the system, even with the same inputs, will produce slightly different results, thus avoiding repetition and keeping the system creative.

Recall that the work product assessor ranks ideas according to metrics defined using basic ideas from information theory, psychology, and chemistry to predict the novelty and flavorfulness of newly-created ingredient combinations. This approach does not require training data on full recipes/dishes and is distinct from a supervised learning approach [14]; the benefit is that the system is more likely to create novel and untrained ingredient combinations. Data sources from culinary traditions, chemoinformatics, and hedonic psychophysics are, however, required.

16.4 Calculating Ingredient Proportions

Once the list of ingredients for the new recipe has been generated by the system, the next step is to calculate the ingredient proportions in the new recipe. Since our system is almost entirely data-driven, striving to generate a creative recipe using a corpus of recipes, we use the corpus itself to help guide these calculations. Our goal is to try to match the proportions in the new recipe with certain distributional characteristics of existing recipes, for instance, the distribution of ingredients and of certain nutritional coefficients. This bears strong resemblance to methods for texture synthesis [22, 23].

When measuring ingredients for a new recipe, chefs rely on several principles, depending on the nature of the dish being created. Desserts and pastries (sweet and savory) require precise measurements of ingredients to achieve a nutrient composition that in turn creates a specific texture. In the example of a custard, the amount of egg protein (and its nature, depending on whether one uses yolks or whites) as compared to the amount of liquid determines the firmness of the cooked mixture. Minerals are also required, so as to obtain a coherent gel instead of curdled eggs floating in liquid [24, p. 94]. In cake recipes, the impact of the proportions of various batter ingredients on the final baked product has been studied at length [25]; the effects can be traced back to the nutrients that the ingredients contain. A formula for ice cream [26] constrains the proportions of five groups of nutrients (fat, milk solids that are not fat, sugar, other solids, and water) within certain ranges, regardless of the actual ingredient types. It is shown that irrespective of whether the fructose comes from corn syrup, strawberry, or carrot—as long as one follows the formula and the general instructions to make such a frozen dessert, one will obtain a properly balanced ice cream. Following the nutrient ratios of existing recipes is clearly a key factor in determining ingredient proportions for a new recipe.

It is important to note that dishes that do not fall in the above category should be prepared using a different template. For instance, adding one more carrot or one more cup of stock to a stew will not compromise the structure of the dish; measurements of this sort are often approximated by chefs. In these cases, the relative proportions of the various ingredient types such as the meat-to-vegetable or the sauce-to-meat ratios are more crucial.

Our system uses the notion of balancing ingredients to determine the proportions. The essence of the idea is that ingredient proportions are determined so that the proportions of each nutrient (protein, fat, carbohydrate, etc.) and each ingredient type (meat, vegetable, herb, etc.) of the new recipe appropriately conform to the distributions of those proportions in the existing recipes of the same dish. A major advantage of our method is that it is data-driven, i.e. the required information is acquired entirely from the data. It should be noted that additional expertise and ontological information could easily be used in conjunction with our method.

We now describe the mathematical model behind our system's ingredient proportions algorithm. We assume that the dish and the ingredient bill for the new recipe have already been determined. Suppose that all existing recipes for the dish in the database have been identified. We introduce the notation in Table 16.1.

Since our method determines the ingredient proportions, $\mathbf{x}$ represents the vector of decision variables. These are determined as the output of an optimization

Table 16.1 Mathematical notation for the ingredient proportion model

Notation	Description
M	Number of ingredients in new recipe
P	Number of nutrients in new recipe
Q	Number of ingredient types in new recipe
c_{mp}	Percentage of pth nutrient in mth ingredient
t_{mq}	Indicator which is 1 only if mth ingredient is of qth ingredient type
$\mu_p^{nutrient}$	Mean percentage of pth nutrient in existing recipes
$\sigma_p^{nutrient}$	Standard deviation of percentage of pth nutrient in existing recipes
$\mu_q^{ing\ type}$	Mean percentage of qth ingredient type in existing recipes
$\sigma_q^{ing\ type}$	Standard deviation of percentage of qth ingredient type in existing recipes
$\mathbf{x} = \{x_m : m = 1, \ldots, M\}$	Percentage of ingredients in the new recipe

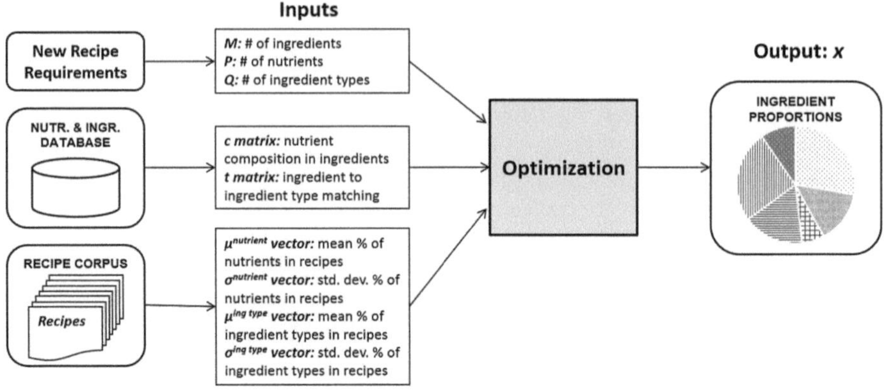

Fig. 16.6 Calculating ingredient proportions

formulation, the inputs to which are provided by the new recipe requirements, the corpus of recipes, and nutritional and ingredient type information from the knowledge database. Figure 16.6 shows the various inputs described above as well as their sources, depicting how they feed into the optimization module. The optimization attempts to balance the nutrient and ingredient type composition of the target recipe.

Let us first consider the nutrient composition of the new recipe. Since there are M ingredients in the new recipe and the mth ingredient has proportion c_{mp} of the pth nutrient, the total amount of the pth nutrient in the new recipe can be obtained by summing over all ingredients: $\sum_{m=1}^{M} x_m c_{mp}$. We propose the following cost function for nutrient balancing:

$$C^{nutrient} = \frac{1}{P-1} \sum_{p=1}^{P} \left[\left(1 - \frac{\sigma_p^{nutrient}}{\sum_{p=1}^{P} \sigma_p^{nutrient}} \right) \left(\sum_{m=1}^{M} x_m c_{mp} - \mu_p^{nutrient} \right)^2 \right]$$

When this nutrient balancing cost is minimized, proportions where the nutrient composition in the new recipe deviates from the mean nutrient composition are penalized. The weight/importance of any particular nutrient is inversely proportional to its standard deviation. The rationale behind this approach is that the composition of the new dish should conform more closely to existing dishes for those aspects that exhibit little variation in the database (and are therefore considered more tightly constrained from a structural formula point of view).

In a similar vein, we use the following cost function to conform to the distribution of ingredient types in the existing recipes:

$$C^{ing\ type} = \frac{1}{Q-1} \sum_{q=1}^{Q} \left[\left(1 - \frac{\sigma_q^{ing\ type}}{\sum_{q=1}^{Q} \sigma_q^{ing\ type}} \right) \left(\sum_{m=1}^{M} x_m t_{mq} - \mu_q^{ing\ type} \right)^2 \right]$$

The balancing cost is expressed as a weighted sum of the balancing costs for nutrients and ingredient types:

$$C^{balance} = wC^{nutrient} + (1 - w)\, C^{ing\ type}$$

Since the computational creativity system has already identified the ingredient bill, the proportions algorithm should ideally (wherever possible) avoid zero proportions, as this would effectively remove certain ingredients. To account for this issue, we add a regularization term to the cost function, for instance:

$$C^{reg} = \frac{1}{M}\sum_{m=1}^{M} x_m^2$$

The total cost can in turn be a weighted sum of the balancing cost and the regularization cost, of the form:

$$C^{cost} = \lambda C^{balance} + (1 - \lambda)\, C^{reg}$$

Any of a number of well-known techniques could be used to calculate the ingredient proportions $\mathbf{x}$ that minimize the total cost. The optimization problem can be formulated as a quadratic programming problem; when the Hessian matrix is positive definite, the problem is a convex optimization problem and a local solution to the problem is also a global solution.

There are several extensions that can be made to this optimization formulation to address other aspects of creating recipes. An issue that may be particularly important revolves around the quantities of specific ingredients. For instance, there may be certain dishes that revolve around a particular ingredient or set of ingredients. These ingredients must be present, albeit in low amounts if and when desirable, for the dish to preserve its identity. A moussaka without layers of fried eggplant, or an aioli without a noticeable garlic taste may not be recognizable dishes. A potential way to tackle this in the optimization is to set lower bound constraints on proportions of certain key ingredients. As another example, there are often ingredients such as herbs and seasoning/spices which are typically used in smaller proportions. Sometimes recipes do not even mention their exact quantities, and therefore it may be necessary for the system to treat proportions for these ingredients in a different way. This could be tackled with the help of upper bound constraints, or by using a different regularization expression for these low-proportion ingredients.

Finally, there are other extensions that should be easy to incorporate in the optimization formulation. The target dish could be required to conform to other aspects of dishes (beyond nutrients and ingredient types), such as the composition of chemical compounds like capsaicin (which is responsible for hot pepper pungency), the balancing of a subset of the five taste sensations (sweet, salty, sour, bitter, umami), etc. If there happens to be incomplete nutrient information in the database, then ingredient amount analysis may be able to compensate. For example, the knowledge database may not include capsaicin concentration, but data on the average amount of pepper used in existing recipes serves as a substitute. Similarly, in the absence of data regarding basic tastes such as bitterness or sourness, the recipe database would likely indicate that lime and grapefruit are used in smaller amounts than navel oranges. Even in the absence of the expertise of a qualified chef, the data-driven approach that we have described can create effective and balanced recipes.

16.5 Generating Recipe Instructions

In the knowledge database, recipe instructions have been parsed into sequences of tasks that connect the ingredients to the final product. These sequences can be viewed as a directed acyclic graph, as shown in the example on Fig. 16.7. Vertices represent either ingredients or tasks, and the edges capture their ordering. Unlike in a cookbook where the writing imposes that all the actions follow a single thread, a graph displays many steps being completed in parallel: by analyzing the inputs and outputs of each step, we only retain the edges that are strictly necessary.

At the highest level, tasks are described by a verb that represents a category of similar cooking actions. Figure 16.8 presents an ontology of preparation methods, compiled by Michael Laiskonis, Creative Director at the Institute of Culinary Education [27]. The ontology can be completed with synonymous verbs using a lexical database such as WordNet [28].

The recipes created by our system are entirely novel and do not result from simple ingredient substitutions in existing recipes. The large number of possible dishes and the many variations within the same dish also make it impossible to design templates that could simply be customized to match a given ingredient combination. For these reasons, we choose to proceed in two steps.

In the first step, we examine each ingredient individually, and create one or more plausible sequences of tasks it could be involved in. Each sequence constitutes a subgraph that starts with one ingredient and ends with the last recipe task, which produces the final product. For each ingredient, we start by extracting all the subgraphs for the

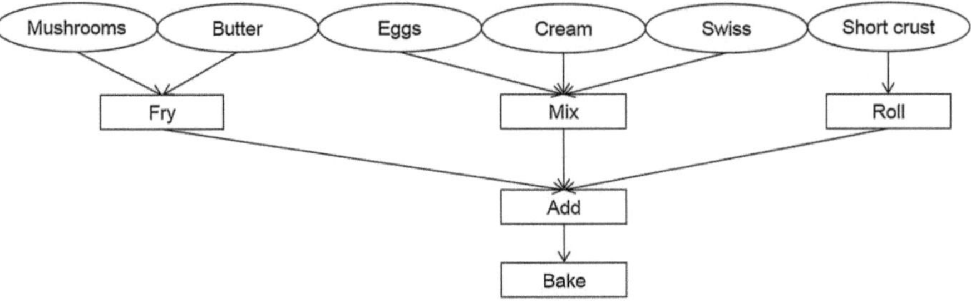

Fig. 16.7 Example of an acyclic graph representing recipe instructions for a simple mushroom quiche

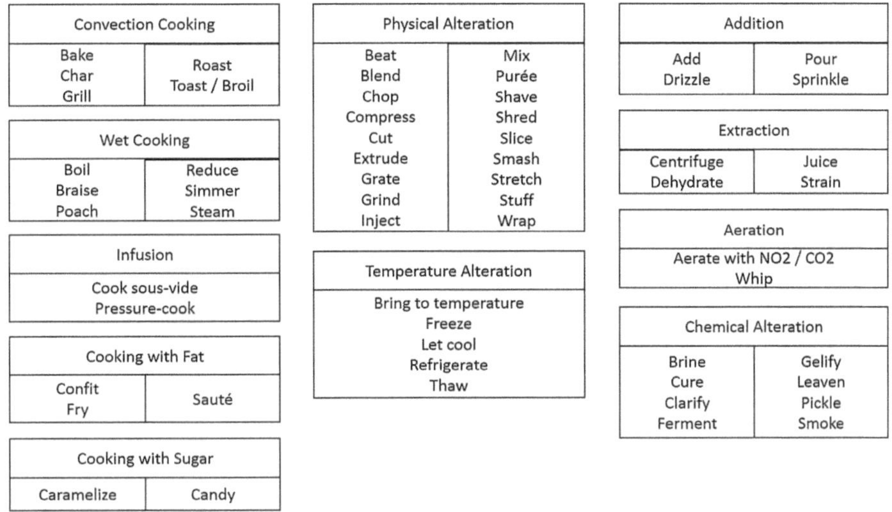

Fig. 16.8 An ontology of preparation methods

same dish and the same ingredient or ingredient type from the knowledge database. In case no subgraph can be found for the ingredient or ingredient type, the search can be extended to similar ingredient types. In the example of a salmon quiche, if the repository does not contain any existing recipe for seafood quiche, we assume that fish can be handled as meat, and consider the meat subgraphs of the quiche recipe graphs.

The subgraphs are simplified following several preset rules. Certain tasks bring no essential information to the high-level preparation process ("preheat", "transfer"), and are eliminated. Others become redundant once the subgraph has been pulled out of the whole graph (such as two consecutive "add" or "mix" verbs), and are merged.

The resulting subgraphs for a given ingredient are then clustered, and each cluster is assigned a suitability score. Clustering uses a weighted graph edit distance [29] and a bisecting k-means algorithm, where k can be chosen so that the largest distance in any cluster remains below a certain threshold. As a result, each cluster contains subgraphs that all handle an ingredient roughly the same way from start to finish. The score of each cluster is based on the number of subgraphs in the cluster, weighted by the relevance of each subgraph to the recipe ingredient: subgraphs that start with the same ingredient receive a higher weight than the ones that just deal with the same ingredient type, which in turn weigh more than the subgraphs that use a substituted ingredient type. A higher score indicates a cluster that describes a very common action sequence, and/or closely matches the ingredient used in the new recipe.

With the subgraph clusters now sorted by decreasing suitability score, we select the cluster with the highest rank as the preferred method for handling the ingredient. Clusters with lower scores will be used as backups if a merged graph cannot be obtained otherwise. It is important to note that since the suitability score calculation is

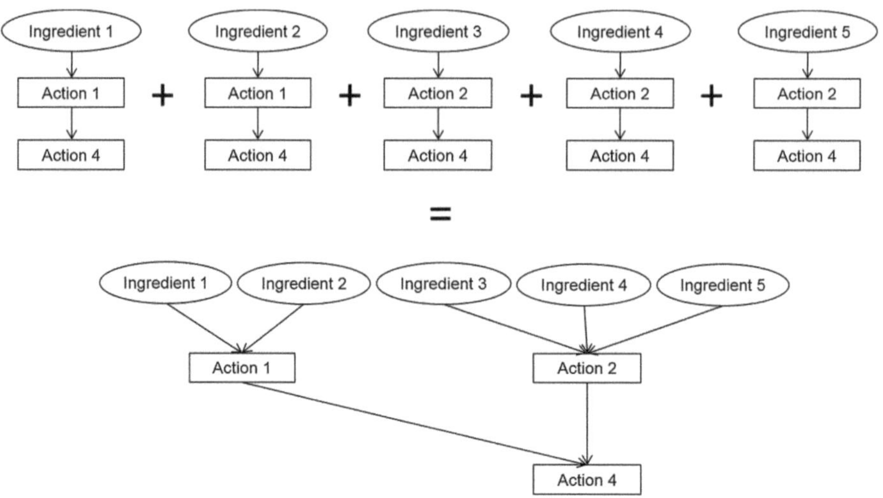

Fig. 16.9 Merging ingredient subgraphs to create a work plan

ingredient-specific, two ingredients of the same type will not necessarily be assigned the same subgraph.

In the second step, represented in Fig. 16.9, we proceed with merging all the selected ingredient subgraph clusters together. Our approach is a variation on the minimum common supergraph computation [30], with differences in where the merging occurs, what is being merged, and in what order.

Our task ontology contains a limited number of verbs that can accept more than one input—mainly "add" and "mix". Therefore, these are the only vertices where the merging can occur. Another constraint inherent to recipes is that the graph must end with one single output (even if the last instruction indicates serving several dish elements together).

As we are dealing with clusters of subgraphs, a representative subgraph must be picked within each cluster. The selection can take into account the weight of each subgraph and its distance to the other members of the clusters.

The recipe graph is built by merging the representative ingredient subgraphs one at a time, with the minimum common supergraph being obtained by choosing the highest possible merge point in the action sequence. In the example in Fig. 16.9, the subgraphs for ingredients 1 and 2 are merged on action 1 rather than action 4. There may be cases when a subgraph cannot be merged, because of a lack of a suitable merge point. One could conceivably insert an "add" action anywhere in a subgraph, since one can always combine two ingredient mixtures together, even though the result is not guaranteed. Another, more preferable option, is to reconsider the choice of the representative ingredient subgraph, or of the ingredient subgraph itself.

As representative ingredient subgraphs are reconsidered, the order of the ingredients being merged becomes important. Indeed, not all ingredient types are determinant to shape the recipe outline. Herbs and spices can be added to a dish at many different stages without compromising the result; oil is generally used for frying, and its position in the graph is mainly determined by which other ingredients need to be fried. A distinction between primary (such as proteins) and secondary (such as seasoning) ingredients emerges as a useful classification that can be added to the knowledge database.

The final graph produces high-level recipe instructions. By looking back at the ingredient subgraph clusters that were retained, it is possible to make a second pass to refine the specifics of each task, from the oven temperature to how vegetables should be chopped.

16.6 Estimating Recipe Step Durations

To enable work planning, it is important to estimate how long a given recipe will take to prepare. Since we are constructing completely new recipes, we must estimate the durations of individual recipe steps and then sum them together to find an estimate of the complete recipe's duration. Unfortunately, duration measurement data is typically available for complete recipes rather than for individual steps, e.g. we may have data on how long it takes to prepare a crostini, but not how long it takes to peel and cube lychees. This section describes an approach for "unmixing" durations of individual steps from data pertaining to complete recipes.

Before proceeding, we provide the following definitions:

- *Atom*: a work task that cannot be broken into smaller constituents,
- *Molecule*: a structured collection of atomic tasks that are linked together as a larger task, e.g. through a directed acyclic graph,
- *Equivalence Class of Atoms*: a set of atomic tasks that are thought to require the same amount of time,
- *Molecule Catalog*: a predefined list of all possible molecules, their constituent atoms, and their durations,
- *Incomplete Atom Catalog*: a predefined list of all possible atoms and which molecules contain them, and
- *Complete Atom Catalog*: a predefined list of all possible atoms and their durations.

The core idea is that if we have estimates of the duration of a large number of complete recipes (molecules), with overlapping sets of individual steps (atoms), we can use inference algorithms to find the effort of the individual steps by solving an inverse problem. The basic approach is depicted diagrammatically in Fig. 16.10 and proceeds as follows:

1. Each molecule in a catalog is broken down into its constituent atoms,
2. The resulting incomplete atom catalog is categorized into equivalence classes,
3. A measurement operator is constructed from the listing of which atom is part of which molecule, and a measurement vector is constructed from the durations of the molecules,

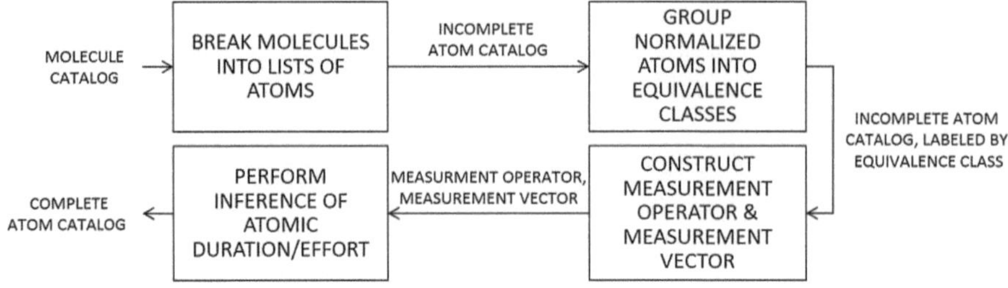

Fig. 16.10 Method for estimating the time durations of atomic steps in a recipe

4. If the measurement operator is sufficiently invertible, proceed to Step 5; otherwise return to Step 2 and categorize into coarser equivalence classes,
5. The measurement operator and the measurement vector are used as inputs to an inference algorithm that solves the inverse problem to find the durations of each atom.

Since we want to determine how much time it takes to perform granularized atomic tasks when we only have access to the duration of molecules composed of several atomic tasks all mixed together, several difficulties must be overcome. First, we need to ensure the inverse problem to be solved is not "undercomplete" (more equations than unknowns) by grouping atomic tasks into equivalence classes whose members will be treated as having the same duration as any other member of the class. Second, we need to normalize/rescale duration to take into account disparities in quantities, tools, etc. Finally, we need to develop an inference algorithm to find the unknown atomic durations from the known durations of molecules and the known memberships of atoms in molecules.

Time is an extensive quantity and essentially adds up linearly. Letting y be the total time of the molecular piece of work, and x_a, x_b, x_c be the times for three atomic steps, we assume that $y = x_a + x_b + x_c$. We also assume that there is a finite set of N possible recipe steps from which these steps are chosen. We can therefore use indicator variables a_i to write a generalized expression for the sum, where $a_i = 0$ for absent steps and $a_i = 1$ for present steps:

$$\begin{bmatrix} a_1 \cdots a_N \end{bmatrix} \begin{bmatrix} x_1 \\ \vdots \\ x_N \end{bmatrix} = y,$$

where the $\{a_i\}$ and y are known whereas the $\{x_i\}$ are unknown and to be inferred. We see that using data on just a single molecule, the inference problem is undercomplete. Now consider a catalog with many molecules M, with the number of possible recipes steps denoted as N:

$$\begin{bmatrix} a_{11} & \cdots & a_{1N} \\ \vdots & \ddots & \vdots \\ a_{M1} & \cdots & a_{MN} \end{bmatrix} \begin{bmatrix} x_1 \\ \vdots \\ x_N \end{bmatrix} = \begin{bmatrix} y_1 \\ \vdots \\ y_M \end{bmatrix},$$

where A is a sparse binary matrix.

As part of data preprocessing, it is important to normalize durations according to weight, measure, tool/instrument, etc. by rescaling. One way to implement this normalization in the linear setting is to take the A matrix as weighted, with the weights as described, rather than interpreting the matrix as binary.

We wish to solve the inference problem $A\mathbf{x} = \mathbf{y}$ for $\mathbf{x}$, resulting in an estimate $\hat{\mathbf{x}}$. Since data may be inaccurate, insufficient and inconsistent, one possible algorithm to use is the Lanczos inverse [31]:

$$\hat{\mathbf{x}} = (A^T A)^{-1} A^T \mathbf{y}.$$

If there is some prior knowledge about the statistical nature of $\mathbf{x}$, another possible algorithm to use is message-passing Bayesian inference [32, 33].

If all atoms in the several molecules were unique, then it would be impossible to perform the inference with any degree of validity. To ensure this does not happen, we should group normalized atoms from disparate molecules into equivalence classes. This can be performed using an unsupervised clustering algorithm, e.g. based on features from the ontology. Note that grouping is not restricted to steps that operate on single ingredients.

In this clustering, there is a tradeoff between κ, the estimated poorness of inverse problem solution obtainable (in the linear setting, this can be measured using the condition number of the measurement matrix A), and σ, the internal coherence in the equivalence classes, which must be balanced to obtain best overall performance. One way to trade condition number and coherence is to define a hierarchy for equivalence class formation, e.g. using tree-structured k-means clustering: the hierarchy can be defined jointly by both ingredient (red apple $<$ apple $<$ fruit $<$ produce) and action (brunoise $<$ dice $<$ cut). Then we may proceed up the hierarchy decreasing σ until κ is sufficiently small.

Thus we have a basic way to estimate the durations of recipes, using data from other recipes.

There are several extensions that could be made to this basic recipe step duration estimation procedure. Doing certain atomic steps together as a molecular step may take less time than doing them separately as atomic steps, which introduces a nonlinearity in the additivity of work times. Also, a molecular step may be able to exploit certain parallelization possibilities to do work faster than split up into atomic steps, which also introduces a nonlinearity in the additivity of work times [34]. These nonlinearities should be handled when considering how atoms lead to molecules. Furthermore, although here we assume that each atomic step has a precise fixed duration x_i, the duration could be modeled as a stochastic quantity governed by a probability measure $X_i(\omega)$. In such a model, it would be of interest to estimate probability measures rather than merely scalar values.

16.7 Conclusion

In this chapter, we have described algorithms that form the foundation of a computational creativity system that can automatically or semi-automatically discover, design, and plan culinary recipes that are flavorful, novel, and perhaps also healthy. We described an architecture for a computational creativity system in which to embed these algorithms, comprising a designer, an assessor, and a planner, all fed by a domain knowledge database.

Recipes created by the computational creativity system, such as a Caymanian Plantain Dessert, have been rated as more creative than existing recipes in online repositories by expert judges [35]. Moreover, professional chefs at various hotels, restaurants, and culinary schools have indicated that the system helps them explore new vistas in food. We foresee further innovations in the future that will continue to enhance the functionality of our culinary computational creativity system.

Although we took a particular creative application domain—culinary recipe design and planning —as an example, the system architecture, approaches and algorithms developed in facing the challenges should be applicable across creative scientific domains [8]. Indeed, culinary recipes, when viewed as being constructed from their constituents into proportions and structured plans, may not be so different from travel itineraries, feature sets for new products, or even business processes that are characterized by components, combining rules, and plans.

Acknowledgments The authors thank the Institute of Culinary Education for their support assembling the culinary knowledge database and testing the recipes produced by the system.

References

1. Boden, M.A.: The turing test and artistic creativity. Kybernetes **39**(3), 409–413 (2010)
2. Cardoso, A., Veale, T., Wiggins, G.A.: Converging on the divergent: the history (and future) of the International joint workshops in computational creativity. A I Mag. **30**(3), 15–22 (2009)
3. Cope, D.: Experiments in Musical Intelligence. A-R Editions, Madison, Wisconsin (1996)
4. McCorduck, P.: Aaron's Code. W. H. Freeman, New York (1990)
5. Wiggins, G.A.: Searching for computational creativity. New Gener. Comput. **24**(3), 209–222 (2006)
6. Carson, S.H., Peterson, J.B., Higgins, D.M.: Reliability, validity, and factor structure of the creative achievement questionnaire. Creat. Res. J. **17**(1), 37–50 (2005)
7. IBM: Capitalizing on complexity: Insights from the global chief executive officer study (2010)
8. Jagmohan, A., Li, Y., Shao, N., Sheopuri, A., Wang, D., Varshney, L.R., Huang, P.: Exploring application domains for computational creativity. In: Proceedings of International Conference on Computational Creativity (ICCC 2014) (2014)

9. Sawyer, R.K.: Explaining Creativity: The Science of Human Innovation. Oxford University Press, Oxford (2012)
10. Baldi, P., Itti, L.: Of bits and wows: a Bayesian theory of surprise with applications to attention. Neural Netw. **23**(5), 649–666 (2010)
11. Ahn, Y.Y., Ahnert, S.E., Bagrow, J.P., Barabási, A.L.: Flavor network and the principles of food pairing. Sci. Rep. **1**, 196 (2011)
12. Shepherd, G.M.: Neurogastronomy: How the Brain Creates Flavor and Why It Matters. Columbia University Press, New York (2012)
13. Colton, S., Wiggins, G.A.: Computational creativity: the final frontier? In: De Raedt, L., Bessiere, C., Dubois, D. (eds.) ECAI 2012: 20th European Conference on Artificial Intelligence, pp. 21–26. IOS Press BV, Amsterdam (2012)
14. Morris, R.G., Burton, S.H., Bodily, P.M., Ventura, D.: Soup over beans of pure joy: culinary ruminations of an artificial chef. In: Proceedings of International Conference Computational Creativity (ICCC 2012), pp. 119–125 (2012)
15. Varshney, L.R.: Surprise in computational creativity and machine science. Decis. Anal. Today **32**(1), 25–28 (2013)
16. Varshney, L.R., Pinel, F., Varshney, K.R., Bhattacharjya, D., Schoergendorfer, A., Chee, Y.M.: A big data approach to computational creativity. arXiv preprint arXiv:1311.1213 (2013)
17. Wikia Recipes. http://recipes.wikia.com/
18. USDA National Nutrient Database for Standard Reference. http://ndb.nal.usda.gov/. U.S. Department of Agriculture, Agricultural Research Service (2013)
19. Volatile Compounds in Food. http://vcf.tno.nl/
20. Wikipedia. http://www.wikipedia.org/
21. Smith, G., Whitehead, J., Mateas, M.: Computers as design collaborators: interacting with mixed-initiative tools. In: Proceedings of the ACM Creating Cognitive Worshop Semi-Automatic Creating (SAC 2011) (2011)
22. Efros, A.A., Leung, T.K.: Texture synthesis by non-parametric sampling. In: Proceedings of 7th IEEE International Conference Computer Vision, pp. 1033–1038 (1999)
23. Heeger, D.J., Bergen, J.R.: Pyramid-based texture analysis/synthesis. In: Proceedings of 22nd Annual Conference Computer Graphics Interactive Techniques (SIGGRAPH'95), pp. 229–238 (1995)
24. McGee, H.: On Food and Cooking: The Science and Lore of the Kitchen. Scribner, New York (2004)
25. Mizukoshi, M.: Model studies of cake baking. VI. Effects of cake ingredients and cake formula on shear modulus of cake. Cereal Chem. **62**(4), 247–251 (1985)
26. Liddell, C., Weir, R.: Frozen Desserts: The Definitive Guide to Making Ice Creams, Ices, Sorbets, Gelati, and Other Frozen Delights. St. Martin's Griffin, New York (1996)
27. Institute of Culinary Education. http://www.ice.edu/
28. WordNet: A lexical database for English. http://wordnet.princeton.edu/. Princeton University (2010)
29. Bunke, H.: Graph matching: theoretical foundations, algorithms, and applications. Proc. Vis. Interface **2000**, 82–88 (2000)
30. Bunke, H., Jiang, X., Kandel, A.: On the minimum common supergraph of two graphs. Computing **65**(1), 13–25 (2000)
31. Jackson, D.D.: Interpretation of inaccurate, insufficient and inconsistent data. Geophys. J. R. Astron. Soc. **28**(2), 97–109 (1972)
32. Fletcher, A.K., Rangan, S., Varshney, L.R., Bhargava, A.: Neural reconstruction with approximate message passing (NeuRAMP). In: Shawe-Taylor, J., Zemel, R., Bartlett, P., Pereira, F., Weinberger, K. (eds.) Advances in Neural Information Processing Systems 24, pp. 2555–2563. MIT Press, Cambridge (2011)
33. Rangan, S.: Generalized approximate message passing for estimation with random linear mixing. In: Proceedings of 2011 IEEE International Symposium on Information Theory, pp. 2168–2172 (2011)

34. Oppenheim, D.V., Varshney, L.R., Chee, Y.M.: Work as a service. In: Bouguettaya, A., Sheng, Q.Z., Daniel, F. (eds.) Advanced Web Services, pp. 409–430. Springer, Berlin (2014)
35. Davis, A.P.: Digital gastronomy: when an IBM algorithm cooks, things get complicated-and tasty. Wired **20**(10) (2013)

Chapter 17
Interactive Meta-Reasoning: Towards a CAD-Like Environment for Designing Game-Playing Agents

Ashok K. Goel and Spencer Rugaber

Abstract We posit that experimentation is a central element of the creative process. The question then becomes how can we support experimentation in creative tasks? We take inspiration from the success of computer aided design (CAD) environments that enable designers to construct, evaluate and revise models of engineering systems. Design of game-playing software agents is another creative task. By analogy, we present a CAD-like environment that enables designers to construct, evaluate and revise models of game-playing agents. However, unlike engineering systems, intelligent agents may learn from experience. In particular, intelligent agents may use meta-reasoning over their own models to redesign themselves. Thus, we envision a CAD-like environment in which the human designer and the intelligent software agent cooperate to perform interactive meta-reasoning to redesign the agent. In this article, we describe three elements of this vision: (2) an agent modeling language called TMKL2, (2) an interactive environment called GAIA for experimenting with the models of game-playing software agents, and (3) GAIA's module called REM that performs meta-reasoning for self-adaptation in game-playing software agents. We illustrate these concepts for the task of design of software agents that play variants of Freeciv, a turn-based strategy game.

17.1 Background, Motivations and Goals

Generate and Test is a common general-purpose method in artificial intelligence (AI) [79]. Given a problem, in the Generate and Test method, an intelligent agent first generates a candidate solution and then evaluates the solution. If the solution is acceptable to the agent, then the problem is solved; If not, then the agent may generate another solution for evaluation. The agent may repeat this until an acceptable solution

A.K. Goel (✉) · S. Rugaber
Design & Intelligence Laboratory, School of Interactive Computing,
Georgia Institute of Technology, Atlanta, USA
e-mail: goel@cc.gatech.edu

S. Rugaber
e-mail: spencer@cc.gatech.edu

T.R. Besold et al. (eds.), *Computational Creativity Research: Towards Creative Machines*,
Atlantis Thinking Machines 7, DOI 10.2991/978-94-6239-085-0_17

is found (if one exists). Thus, the Generate and Test method is iterative. The balance between generation and testing in the method depends on the knowledge available to the agent. At one end of the knowledge spectrum, if the agent has complete and correct knowledge of the problem and the problem world, then the agent may generate a provably optimal solution to the problem and little testing is needed. At the other extreme, if the agent has little knowledge of the problem and the problem world, it may generate a candidate solution randomly, and then evaluate the solution, for example, by applying an objective function to the solution or by trying out the solution in the world, with the world acting as its own "objective function". The logic school of AI is closer to the former end of the knowledge spectrum; the evolutionary computing school is closer to the latter.

If Generate and Test is a basic process of intelligence, it is even more fundamental to creativity. Many problems that intelligent agents encounter routinely are well defined. For example, in planning, the agent may have knowledge about the initial state and the goal state, as well as the actions available in the world; Further, the agent already may have solved the problem, or similar problems, in past. Thus, a robot planning a navigation route to drive me from my home to my office may require some intelligence, but, under normal conditions, not much creativity. Most problems requiring creativity, on the other hand, are ill defined and open ended. From understanding a story to comprehending a movie, from painting a portrait to composing a symphony, from inventing a technological system to discovering a scientific phenomenon, for creative tasks, the agent typically has neither sufficient knowledge of the problem and the problem world to generate an acceptable solution initially, nor an objective function to efficiently evaluate a candidate solution. Thus, *for creative tasks, the agent must experiment:* it must generate candidate solutions, evaluate the solutions, use the knowledge gained from the testing to generate better solutions, and repeat the cycle until it has acquired sufficient knowledge of the problem and the problem world that it can generate an acceptable solution. Again evaluation may take several forms, but *the central element in the creative process is experimentation.*

Given that experimentation is a central element of the creative process, a core question for research on computational creativity is how can we support humans in experimentation? Note that the design of the experiments must be such that each experimental run produces new information about the problem or the problem world so that the agent can try to produce a better candidate solution in the next iteration of Generate and Test. To answer this question, we take inspiration from the enormous success of computer-aided design (CAD) as a scientific and technological enterprise [9, 77]. Design in general is a creative task because novel design problems typically are open-ended and ill defined. CAD environments are successful in part because they enable designers to experiment with candidate solutions to design problems. When a designer specifies a candidate solution to a problem, CAD tools can help the designer, for example, in constructing geometric models and performing numerical simulations for evaluating the solution. The geometric modeling and numerical simulations produce new information that the designer can take into account in refining or revising the candidate solution.

The question for research on computational creativity then becomes how can we develop CAD-like interactive environments for enabling experimentation for other creative tasks? Thus, our research laboratory has been studying the nature of experimentation in creative tasks such as technological invention and scientific discovery, including thought experiments, simulated experiments, and real experiments. For example, we have studied how scientists construct explanatory models of complex phenomena through thought experimentation [20, 34, 35]. We have also developed interactive tools for supporting simulated experiments in both task domains. For example, we have developed interactive tools that enable science students to construct conceptual models of complex phenomena, evaluate the models through simulation, and revise the models based on the results of the simulation, repeating the cycle until the student has constructed an acceptable model [33, 39, 76].

In this article, we view another creative task domain through this lens of experimentation as a central element of the creative process: design of game-playing intelligent agents. Human often play interactive games against one or more autonomous game-playing software agents. In general, most humans are more likely to enjoy playing interactive games against intelligent software agents instead of trivial agents. Thus, an important question in design of interactive games is how to design intelligent game-playing agents. This is a difficult task: in its most general form, it appears equivalent to the general task of designing AI agents! The complexity and creativity of the task depends on the nature and scale of the game. In this article, we focus on turn-based strategy games such as Freeciv.[1] Multi-player turn-based strategy games such as Freeciv are dynamic, only partially observable, non-deterministic, and often have huge state spaces. Thus, design of software agents that can play such turn-based strategy games is a very complex and creative task.

In this article, we describe a CAD-like interactive environment for designing game-playing software agents called GAIA that enables experimentation with an agent's design. GAIA provides the designer with a visual editor for constructing a conceptual teleological model of the agent in terms of its tasks, methods and knowledge (TMK) in a high-level agent modeling language called TMKL2. GAIA also provides the designer the ability to simulate the agent: GAIA automatically translates the TMK model into executable code, interacts with game server, and runs the simulation. The designer may use GAIA to not only monitor the results of the simulation, but also to inspect the agent's knowledge and reasoning during the simulation. The designer may then modify the teleological model of agent; GAIA provides a persistence mechanism for managing agent models. The designer may repeat this cycle of experimental modeling and simulation until the designer is satisfied with the simulation results.

While the above process of experimentation in the design of game-playing software agents is parallel to the experimentation processes in CAD, it is important to note that there are also important differences between the tasks of engineering design and agent design. A fundamental difference between the two is that unlike most engineering systems, an intelligent software agent can *learn* from its experiences

[1] http://freeciv.wikia.com/.

in the world! More specifically in our context of experimentation in creativity, just as the human designer may use the results of simulating the agent's design in the game world to modify the teleological model of the agent, at least in principle, the agent too may use the simulation results to modify its own model and thus redesign itself. This is *meta-reasoning* [17], and in particular *meta-reasoning for self-adaptation* [32].

An agent's potential ability to redesign itself raises both challenges and opportunities for developing a CAD-like environment for designing game-playing software agents. The challenge is empowering the agent to experiment with its design and to redesign itself as needed. The opportunity is that if the agent can reflect on its own design, then, at least in principle, the human designer and the software agent could cooperate in the redesign process. We call this *interactive meta-reasoning*. We envision a CAD-like environment in which a human designer begins the process of interactive meta-reasoning by constructing an initial conceptual teleological model of the agent's design. The designer may then run a simulation in the game world to evaluate the design. Once the simulation results are available, the software agent may reflect on its simulated behavior and on the knowledge and reasoning that led to the behavior, and refine or revise its knowledge and/or reasoning. The human designer may intervene at any point in the redesign process, for example, when the redesign task is beyond the capacity of the agent, until the designer and the agent collaboratively reach an agent design acceptable to the designer.

In this article, we focus on three core elements of the above vision. Firstly, we describe the TMKL2 agent modeling language and GAIA's visual editor for enabling the construction of a TMK model of a game-playing agent in the language such that the model is both human and machine understandable. Secondly, we describe GAIA's architecture for interactive meta-reasoning in which the designer can construct, evaluate and revise TMK models of the game-playing agent. Thirdly, we describe GAIA's process of meta-reasoning for self-adaptation in game-playing software agents. We also briefly relate this work with similar work on agent modeling, meta-reasoning, game-playing, design thinking, and computational creativity.

17.2 Freeciv, A Multi-player Turn-Based Strategy Game

Our work takes place in the context of a multi-player turn-based strategy game called Freeciv. Freeciv is an open-source variant of a class of *Civilization* games with similar properties. A human player can play Freeciv against one or more opponents, some or all of whom can be software agents. The aim in these games is to control and grow a civilization while competing for limited resources against other players' civilizations. The major activities in this endeavor are exploration of the randomly initialized game world, resource acquisition and development, and warfare, which may at times be either offensive or defensive in nature. Winning the game is achieved most directly by destroying the civilizations of all opponents, but can also be accomplished

Fig. 17.1 A screenshot of the Freeciv client interface

through peaceful means such as scientific advancement resulting in the building of an interstellar spaceship. It can take a few hours to play a complete game.

Figure 17.1 illustrates a screenshot of the Freeciv client. The largest pane displays a map of a player's cities and units (warriors, workers, etc.). The pane in the upper left corner displays a smaller overview map of the entire game world, with which the player can navigate. The icons below the small map display the allocation of player resources among taxes, luxuries, and scientific research. Below these icons is a button that the player can click to end his/her current turn. The large white text box on the bottom of the screen is the information window, containing all messages sent from the Freeciv server to the Freeciv client. The tool bar at the top has game options and actions enabling the player to, for example, save the game, develop research technologies, and create new playing units.

On one hand, Freeciv is so large, complex, and dynamic that no simple, rigid strategy is likely to succeed in winning the game. On the other, Freeciv provides a controlled environment in which we can experiment with different game rules, different opponents (human, built-in, or of our own devising), and different strategies (developing a city, building an army, collecting resources, defending a city, attacking an opponent, etc.). More importantly from our perspective, the problem of designing a software agent for playing Freeciv is open ended and ill defined, and the emergent behavior of the game is almost always new and often unexpected. In other words, designing a software agent for playing Freeciv is a creative task.

17.3 TMKL2: A Language for Agent Modeling For Meta-Reasoning

The first element of our approach to addressing experimentation in designing software agents that can play turn-based strategy games is an agent modeling language called TMKL2. The TMKL2 language enables construction of TMK models of software agents. For the purposes of this discussion, a TMK model of an agent is a teleological model that explicitly represents the agent's tasks (T), the agent's methods for accomplishing the tasks (M), and the agent's knowledge of its environment that enables its methods to accomplish its tasks and for it to live in the environment (K). A task describes the computational goal of producing a specific result in the world. A method is a mechanism that produces a result in a specified manner: A method decomposes a task into subtasks, specifies the ordering among the subtasks and is represented as a finite state machine. The knowledge element in a TMKL2 agent models knowledge about the concepts and relations in the world in which the agents reside as well as concepts and relations in the agent's architecture.

In GAIA, TMKL2 models of game-playing software agents are realized through an interpreter. When used to model Freeciv agents, the interpreter is capable of executing a model in conjunction with Freeciv's client to play the game. A main feature of TMKL2 is that it directly connects the goals of an agent to the mechanisms for realizing those goals. The original TMK language (TMKL) was implemented as a Lisp mini-language; that is, both the model itself and the agent were written in the Lisp language. Hence, the connection between the goals, as expressed in the model, and the agent amounted to a few additional data structures. In TMKL2, models are expressed as Java objects, with the composite software agent needing to communicate with the Freeciv server. Consequently, the connection between the goals and the mechanisms is less direct, being realized by a run-time environment. This indirection however has the benefit of making TMKL2 more independent of the target language in which the agent is written.

The TMKL2 language has several features that are important to the subsequent discussion:

- TMKL2 was designed to model software agents expressly for the purpose of supporting agent adaptation.
- An agent may interact with other, external software that has not been modeled. For example, the other software may be part of an operating system that offers access to devices for interacting with the external world. Also, a TMKL2 model may describe only a part of the agent. In either case, TMKL2 assumes that the

division between the modeled and external parts is clearly defined, and, in GAIA, it takes the form of an API to the external code.
- There is an explicit distinction between the run-time behavior of the agent and adaptation-time redesign of the agent. Adaptation takes place on a TMK model of the agent, and the model is then interpreted to effect agent behavior.
- TMKL2, as described here, is domain independent. That is, there is nothing about it that is specific to games, and thus it is applicable to a wider variety of software agents.
- TMKL2 has three subnotations: Goals, Mechanisms and Environment, corresponding to the tasks, methods and knowledge portions of the language, respectively.

17.3.1 Goals

The first subnotation of the TMKL2 language describes the agent's goals. A Goal expresses a reason that the agent does what it does, in terms of its intended externally observable effects on the world. Goals may be parameterized enabling the agent to target specific elements of its Environment, such as, for example, a specific city. A Goal is expressed via a pair of logical expressions describing the precondition for Goal accomplishment (called its Given condition) and the expected effect of Goal accomplishment on the agent's Environment (its Makes condition). The final element of a Goal specification is an indication of the means by which the Goal is to be accomplished. This takes the form of a Mechanism invocation. That is, Goals are directly tied to the means by which they are to be achieved.

17.3.2 Mechanisms

The Mechanism portion of a TMK model in the TMKL2 language is responsible for describing how an agent accomplishes its Goals. There are two kinds of Mechanisms: Operations and Organizers, which are each defined in terms of two logical expressions describing their precondition for execution (Requires condition) and their effect (Provides condition).

An Operation is a parameterized invocation of a computational resource provided to the software agent via an API to external software such as the Freeciv server. That is, each Operation corresponds to an action in the Freeciv world.

An Organizer is a Mechanism defined as a finite state machine comprising States and Transitions. Start, failure and success States are all explicitly indicated. States, in turn, define subGoals, enabling hierarchical refinement of an agent's specification. Transitions may be conditional (dependent on a Data Condition) with respect to the agent's current perception of the world, as expressed in its Environment. An Organizer composes Operations into Mechanisms for achieving Goals.

17.3.3 Environment

A TMK model of an agent in the TMKL2 language includes a description of the agent's knowledge of the world. In particular, the agent's Environment comprises a set of typed Instances and Triples (3-tuples) relating the Instances to each other. In order to describe Instances and Triples, TMKL2 provides two modeling constructs, Concepts and Relations.

A Concept is a description of a set of similar Instances. It is defined in terms of a set of typed Properties. Moreover, Concepts are organized in an inheritance hierarchy promoting compositionality and reuse. There is a built-in concept called Concept. When a Model is constructed, Instances of a Concept are automatically added to it for each defined Concept, enabling reflection by the agent over its own definition.

Relations in TMKL2 allow the modeling of associations among Instances. In particular, an Instance of one Concept can be related to Instance of another via a Triple.

17.3.4 Example Model

Figure 17.2 illustrates part of the TMK model of an agent, called *Alice*, capable of playing a simplified version of Freeciv. In the simplified Freeciv, a city is either producing a warrior or collecting gold pieces on any given turn, and Alice can win by collecting 185 pieces of gold. The figure illustrates GAIA's visual syntax for the TMKL2 language. The partial model includes Alice's top-level Goals and Organizers.

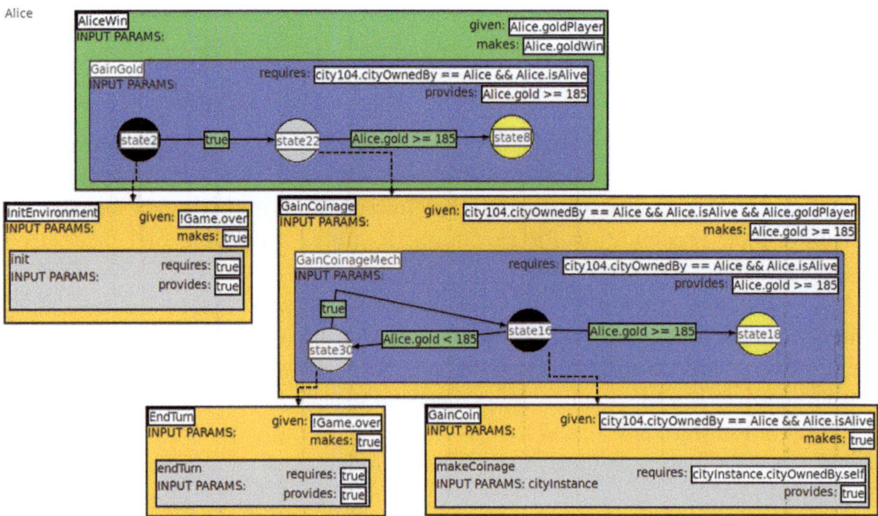

Fig. 17.2 TMKL2 Model of a Freeciv agent

In particular, the top rectangle of the diagram denotes Alice's top-level Goal (that she has collected sufficient gold pieces). Contained within this rectangle is another, depicting an Organizer comprising three States—an initial State, a subGoal and a final State.

The subGoal is shown as the rightmost of the two rectangles on the second level. Its Organizer, in turn, has two subGoals—one that continually mints more gold until enough has been produced and the other determining when to end the game. The bottom two rectangles represent Operations, responsible for interacting with the Freeciv server.

Complementing the Goals and Mechanisms shown in the figure is Alice's Environment (not shown). Example Concepts represented by Instances in the Environment include City, Tile, Player, and Unit.

17.3.5 TMKL2 Semantics

A TMKL2 program connects the Goals of a software agent to the Mechanisms by which the Goals are accomplished. The program is *declarative* in the sense that all behavior is defined in terms of logical expressions (Given, Makes, Requires, Provides). Consequently, one semantic interpretation of a TMKL2 program is that it describes the behavior that a software agent must exhibit in order for it to accomplish a set of top-level Goals.

TMKL2 programs are not just descriptive, however: They can be used to actually control the modeled agent. To convey how this is accomplished requires a description of the *operational semantics* of the language. Operationally, a TMKL2 Model can be viewed as a hierarchy of finite state machines (FSMs) controlling communication with the external software with which the agent interacts and with the Environment. Superior state machines correspond to superior Goals. FSMs corresponding to Goals without any subGoals are called *leaf* FSMs.

All state machines execute synchronously; that is, at any given time, each machine is in a specific State. At the next virtual clock tick, all pending Data Conditions for active leaf machines are evaluated, and the outgoing Transitions evaluating to true are traversed, resulting in entry into new States. Upon entry into a State, the corresponding Mechanism is interpreted. Mechanism interpretation ultimately resolves into Operation invocations and updates to the Environment. After all invocations have been processed, the Environment is updated to reflect any changes to the agent's run-time data structures made by the invocations. Interpretation terminates if the Organizer for the top-level Goal enters either a success or failure State.

17.4 GAIA: An Interactive Environment for Meta-Reasoning for Agent Design and Redesign

GAIA is an interactive environment for constructing, evaluating and revising TMK models of game-playing software agents in the TMKL2 language. In this section, we give a high-level description of its architecture and major subsystems, including REM, the knowledge-based reasoner used for self-adaptation of software agents. We also describe its run-time interface to the world in which the agent executes. GAIA is written in the Java programming language and was built using the Eclipse[2] software development environment.

17.4.1 GAIA Architecture

Figure 17.3 illustrates the conceptual architecture of GAIA. In the center left of the figure is SAGi, the GAIA user interface. Through the SAGi interface, a designer can enter and edit TMK agent models in the TMKL2 language, submit the mod-

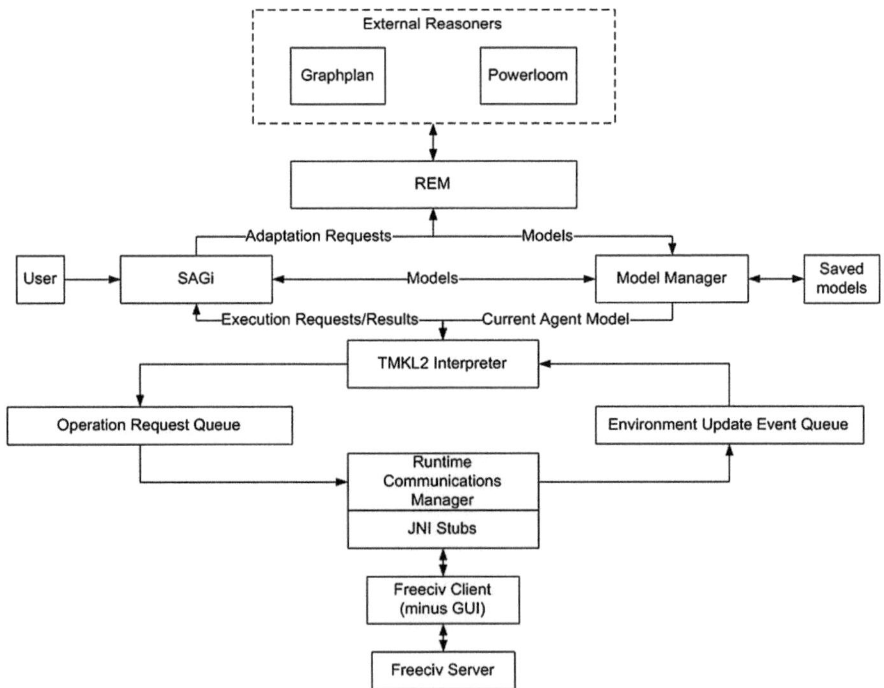

Fig. 17.3 The architecture of GAIA

[2] http://www.eclipse.org/.

els to the TMKL2 Interpreter (shown in the center of the figure) for execution in the game world, monitor the behavior of the agent in the simulation, and inspect the agent's knowledge at each stage of the simulation. In-memory representation of TMK agent models takes the form of Java objects that are interpreted by the TMKL2 Interpreter. Also part of the GAIA architecture is the Model Manager (shown in center right of the figure) that is responsible for encapsulating access to agent models and persisting them to permanent storage. Given that the task of designing a game-playing agent entails experimentation, the human designer is likely to have many model versions existing at any point in time. The Model Manager Models saving models for later access. REM, illustrated at the top center of Fig. 17.3, is the reasoning module responsible for automatic adaptation of TMK models based on the results of a simulation. REM, besides having its own inferencing capabilities is designed to allow use of external reasoners, such as planners and situated learners. Finally, the Runtime Communications Manager, depicted at the bottom of Fig. 17.3, interacts with the Freeciv client and server to carry out the simulation in the Freeciv game world.

17.4.2 SAGi

SAGi provides the human designer with a palette of icons representing TMK model elements of different types. The designer can construct agent models of the kind illustrated in Fig. 17.2 by dragging and dropping the icons. SAGi provides a property panel for each specific kind of element to set values of the element's attributes. SAGi also provides the means by which model interpretation and execution is initiated, paused, and stopped.

17.4.3 TMKL2 Interpreter and Interface to the Game World

The designer can invoke GAIA's TMKL2 interpreter to execute an agent model and thereby interact with the Freeciv server. The interpreter walks the TMK tree of state machines in an iterative fashion until the agent either succeeds or fails to achieve its top-level Goals. When the interpreter attempts to accomplish a subGoal whose Mechanism is an Operation, it must place into the Operation Request Queue a request to the Freeciv server to execute a game action, encoding parameters as necessary. The Operation must have been previously (and, at present, manually) mapped by the designer to a Freeciv action.

17.4.4 REM

REM [53, 54] is an extensible *meta-reasoner* that reasons over TMK models of software agents. REM supports agent self-adaptation because it is capable of monitoring not only what is happening in the game world, but also the agent's internal state with respect to the accomplishment of its goals and the methods it used to accomplish them. Further, it can redesign the agent to adapt it to better accomplish those goals. When given an agent model and a situation, such as a failed Goal or an altered Environment, REM produces an updated agent model engineered either to successfully accomplish the Goal or to take advantage of the new knowledge in the Environment.

To achieve retrospective adaptation after a failed Goal, REM performs three steps: localization (determining which of an agent's subGoals and associated Mechanisms were inadequate to accomplish the agent's overall Goal), transformation (devising an alternative Goal), and realization (providing/altering a Mechanism to accomplish this Goal). Localization is accomplished in REM using a heuristic to find a low-level State in an Organizer such that the State's Provides condition suffices to accomplish the failing Goal. Further, the detected State must have a failing precondition (Requires condition). The presumption is that the State had not been reached, and, if it had been reached, then the agent would have succeeded. Realization and transformation are accomplished by matching the failing situation against a library of adaptation plans, choosing a candidate transformation from the library and applying the result to the agent's Model to produce a revised Model.

REM sits atop the Powerloom knowledge representation and reasoning system [47] that is available publicly.[3] Powerloom supports classification, deduction and truth maintenance. TMKL2 logical expressions are easily mapped to/from Powerloom, and REM algorithms are easily expressed in Powerloom's variant of first-order logic.

17.4.5 Meta-Reasoning for Agent Self-adaptation

To validate our approach to meta-reasoning for self-adaptation in Freeciv-playing software agents, we have conducted several experiments, each involving variants of the Alice agent depicted in Fig. 17.2. In the experiments, Alice plays a simplified variant of Freeciv against other agents. In particular, the simplified game consists of two agents. Each agent controls a civilization and is responsible for its government, economy, citizen morale, and military forces. Each civilization has one city, citizens in that city, and a number of warriors. All cities, civilians, and warriors are located on one large continent. Each game tile yields a quantity of food, production, and trade points each turn of the game. Food points feed a city's civilians; production points are used to support existing warriors or produce new warriors. Trade points are distributed among luxury, tax, and science resources. Initially both players start out

[3] http://www.isi.edu/isd/LOOM/PowerLoom/.

in year 4000 BC, with fifty gold pieces, zero warriors, and one worker who collects resources from a nearby tile. A city is either producing a warrior or collecting gold pieces on any given turn. Alice can win by collecting 185 pieces of gold, and the other agent can win by capturing Alice's city.

An experiment consists of running Alice against its opponent and noting the results. Then, if Alice loses against the other agent, the human designer can experiment with Alice's design by asking REM to adapt the model of the Alice agent to avoid losing again. REM uses the results of the simulation to adapt Alice's agent mode. The designer now evaluates the new model by running the simulation again. The designer can repeat this Generate and Test cycle and experiment with Alice's design until an acceptable design is produced.

17.4.5.1 Experiment #1

The purpose of the first experiment was to test whether REM could make a simple adaptation to improve Alice's performance versus Freeciv's built-in robot player. In general, a player of this reduced game has to make a decision about allocating resources between collecting gold and creating warriors to defend its city. In the first experiment, Alice's strategy was such that she devoted all of her resources to the former pursuit. An obvious adaptation is to adjust Alice to balance her resource allocation, and the first experiment tested whether REM could make this adaptation.

In the experiment, Alice played against Freeciv's robot player, which we call *Frank*, configured at its highest skill level. Although Alice had knowledge that Frank could win by capturing her city, she was unaware that Frank had more powerful weaponry and more production capacity than she had.

When played against Frank, Alice initially succumbed to his attacking chariots, legions, and horsemen. Before losing, Alice was able to acquire 175 units of gold and lived for 3075 years. However, Alice failed to accomplish her Goal of collecting 185 units of gold, thereby requiring retrospective adaptation.

In this experiment, no transformation was needed. That is, the failure was that an Organizer rather than a Goal was flawed. Realizing a replacement Organizer took place by inserting a new State, whose success would satisfy the preconditions of a problem State. Such States are called *patch states*.

A patch State was created by first, searching a small library of generic Goal patterns, each of which was checked to see if its success would satisfy the preconditions of the problem State. After an instantiated Goal pattern was found it was assigned as the Goal of the patch State. This patch State was then inserted into the localized Organizer just prior to the problem State. This guarantees the problem State's precondition is satisfied upon its visitation.

In Experiment #1, the patch State, was added with a Goal to build additional warriors. This Goal increases the defense of Alice's city if she is visibly outgunned on the game map.

After performing this adaptation, the new agent, Alice', was tested against Frank. While still outgunned, Alice' fared better in longevity and defense. She lasted 3125 years and killed one of Frank's powerful attacking units. Because some of her resources had been allocated to defense, she fared worse in gold acquisition, acquiring only 147 units. The lesson learned was that compensating for a well-understood limitation could be accomplished by making use of a simple heuristic alteration of a TMKL2 Model, a small library of patterns, and knowledge of the Environment at the end of the failing game.

17.4.5.2 Experiment #2

Because of Frank's superior fire power, Experiment #1 was an unfair contest for Alice. To explore how Alice would fare versus a similarly equipped opponent, a second experiment was conducted. This experiment involved two näive agents named *Alice* and *Barbra*. Both play the simplified version of Freeciv described in Experiment #1. Barbra's strategy was to focus on producing warriors to attack Alice's city. By so doing, Barbra wins by overwhelming Alice's defenses. Before succumbing, Alice is able to acquire 93 units of gold and while living through 1450 years.

The same adaptation process in Experiment #1 was used to adapt Alice and resulted in the same Alice' being produced as in Experiment #1. Running Alice' versus Barbra results in Alice' winning. Alice' was able to collect 185 units of gold, while living through 4700 years. The experiment increased our confidence in the approach used in Experiment #1.

17.4.5.3 Experiment #3

The previous two experiments were examples of retroactive adaption in which a failure was mitigated. In Experiment #3, proactive adaption was attempted to take advantage of a slightly altered game rule. In particular, it now takes 189 gold units for Alice to win. Tests were run on Alice to see if Alice's model was still valid after the rule change. REM tested if each Mechanism's Provides condition satisfies its parent Goal's Makes condition; that is, if the Mechanism was capable of accomplishing the new Goal. When this test failed, REM located the responsible Mechanism.

In this experiment, REM localized Alice's GainGold Organizer. Next, a replacement Organizer was created to achieve the new win condition. To do this, REM used an external planning tool called Graphplan [7]. Graphplan is a mature, publicly available planner.[4]

REM translated the initial game Environment into a Graphplan facts file, amounting to over 10,400 facts. Then all Organizers, Operations, and game rules were translated into a Graphplan operators file. After pruning out operators with no effects, the resulting Graphplan file contained 10 operators. Next, REM ran Graphplan on the facts and operator files. Graphplan was able to generate a three-stage plan capable of accomplishing Alice's top-level Goal. This plan was then translated back into an

[4] http://www.cs.cmu.edu/~avrim/graphplan.html.

Organizer to replace GainGold. The lesson learned from this experiment was that for a simple numeric change, an invalid TMKL2 Organizer can be located and adapted using an external planning tool.

17.4.5.4 Experiment #4

The first two experiments described above were *off-line* in the sense that the adaptations were made after a game was completed. The third experiment did not involve running the game at all before adapting Alice. Experiment #4 is an *on-line* adaption in that Alice is changed while she is running. Moreover, her opponent, Barbra is also adapted during the game.

In this experiment, both Alice and Barbra were reconfigured into two parts, one *allopoietic* and the other *autopoietic*. These term are borrowed from the literature of self-organizing systems and denote, respectively, the part of a system that changes and the part that does the changing [48].

Alice's allopoietic part used a parameter, *alpha,* to determine how Alice should divide her resources between obtaining gold or producing warriors. The autopoietic part of Alice adapted the allopoietic part by adjusting alpha to produce gold only if she had sufficient defensive capability to fend off Barbra's visible attackers. Similarly Barbra's allopoietic part used a parameter, *beta*, to determine the number of warriors with which to attack Alice's city. Initially beta is set to 0. The autopoietic part of Barbra adapts the allopoietic part by adjusting the number of warriors Barbra attacks Alice with after every failed attack.

For both agents, the autopoietic part was itself a (meta-) agent. In particular, the meta-agent's Environment consisted of a description of the allopoietic part, including Goals, Mechanisms and (allopoietic) Environment. By monitoring game status, the meta-agent could make appropriate adjustments to the base agent's parameter by executing (meta) Operations.

Running Alice versus Barbra resulted in the agents engaging in an arms race. Eventually Alice was able to defeat Barbra. In winning, Alice collected 186 gold units, Barbra had 6 dead warriors, Alice had 3 live warriors and never lost a battle. Barbra adapted herself 4 times and Alice adapted herself 6 times. The lesson learned was that TMKL2 models allow for simple real-time adaptations by using meta Operations to control the agent's strategy.

17.5 Related Work

In this section we briefly relate this work to similar research on agent modeling, meta-reasoning, game playing, design thinking, and computational creativity.

17.5.1 Agent Modeling

An intelligent agent is an autonomous entity that maps a history of percepts in an external environment into an action on the environment to achieve goals and maximize utility [66]. Thus, a game-playing agent is intelligent if it autonomously selects actions that accomplish its goals and maximize its utility. TMKL2 is a language for modeling agents for the purposes of meta-reasoning, and, in particular, meta-reasoning for agent self-adaptation.

TMKL2 expands and extends the original TMKL language [53, 54]. TMKL is comparable to the Hierarchical Task Network language (HTN) [24, 56] for automated planning. Planning problems in HTN are specified in terms of different kinds of tasks including goal tasks, primitive tasks corresponding to actions in the world, and compound tasks that are composed of simpler tasks. Constraints among the tasks are expressed in the form of networks. Both HTN and TMKL languages emphasize the connection between goals and methods for accomplishing them, with methods in TMKL composing the primitive tasks for achieving higher-level tasks much like compound tasks do in HTN. While the HTN language arose out of AI research on planning, the origins of TMKL2 are rooted in AI research on knowledge-based systems [10], Chandrasekaran, Johnson and Smith [11] and functional modeling of physical systems [73]. While HTN was designed for automated planning, a very common task in AI, TMKL was designed for meta-reasoning. Hoang, Lee-Urban and Munoz-Avila [36] have compared the expressivity of HTN and TMKL. They found that in principle TMKL and HTN [56] have the same expressive power, but that TMKL is more explicit in its some of its constructs than HTN. Molineux and Aha [50] have used a variation of TMKL for modeling game-playing agents in their TIELT game-playing agent benchmarking system.

TMKL2 is also comparable to the Procedural Reasoning System (PRS) language Georgeff and Lansky, Rao and Georgeff [29, 55, 63]. Both languages emphasize the connection between goals and the procedures for accomplishing them, and both support reasoning about the procedures rather than their construction on the fly, as would be done by a planner. While PRS supports asynchronous communication with agents in continuous, real-time settings and reactive control, TMKL2 is discrete and synchronous. Although both languages supports alternative means for accomplishing goals, the operational semantics of TMKL2 does not currently provide support for dynamic choice among them. Although both languages in principle support meta-level reasoning, later versions of PRS drop this feature for reasons of performance. In contrast, as we noted above, TMKL2 is specifically designed to support reflection. Finally, both languages have a database of world knowledge, although there is no mention of PRS supporting inheritance, which TMKL2 does.

While HTN supports planning, it does not support plan adaptation. In contrast, if a planned action taken by an agent results in a failure due to changes in the world, then PRS can modify the action, thus making the planning reactive to changes the world. The goals of GAIA are different. While PRS adapts a plan in the presence of a plan failure, we seek to adapt the *planner* itself. Thus, while the PRS language supports reasoning about how to modify the agent's plan, TMKL2 supports reasoning about how to diagnose and redesign the agent itself.

17.5.2 Meta-Reasoning

Meta-reasoning refers to thinking about thinking [3, 16, 17]. In this article, we have focused mostly on meta-reasoning for agent self-adaptation [32]. Much research on meta-reasoning for self-adaptation has used self-models of agents to help localize and identify modifications to the agent design, e.g., [4, 6, 12, 26, 27, 32, 37, 43, 52–54, 73, 74].

We can trace several themes in model-based self-adaptation in intelligent agents. Firstly, self-adaptations can be retrospective [4, 6, 12, 26, 27, 37, 43, 52, 73, 74], i.e., after the agent has executed an action in the world and received some feedback on the result, or proactive [53, 54], i.e., when the agent is given a new goal similar and related to but different from its original goal. The recent work on goal-driven autonomy [51] appears to be related to proactive self-adaptation. As our four adaptation experiments in Sect. 17.5 indicate, GAIA can handle both proactive and retrospective adaptations.

Secondly, the self-adaptation in an intelligent agent may pertain to the deliberative element in an agent's architecture [4, 6, 12, 26, 27, 32, 37, 43, 52–54, 73], or the reactive element [74]. The experiments described in this chapter pertain to the deliberative element but in principle GAIA should be able to manage both.

Thirdly, self-adaptation in an intelligent agent may pertain to the agent's reasoning [4, 6] Collins et al [12] [43, 52–54], or the agent's knowledge [26, 27, 32, 37, 73]. The experiments described in this chapter pertain to revising the reasoning, but again in principle GAIA should be able to address both.

Fourthly, in general the meta-reasoner can call upon special-purpose reasoners. For example, the original REM system invoked a situated learner when needed [54, 75].

Autonomic computing [40] represents a different but related line of research. Autonomic computing pertains to self-managing software systems, including systems capable of self-configuration, self-optimization, self-healing, and self-protection. The experiments described in Sect. 17.5 exhibit elements of self-configuration and self-healing, and to this extent, they may qualify as autonomic computing systems.

17.5.3 Game Playing

Over the last generation, interactive games have emerged as an important domain for AI research [1, 15, 38, 41, 42, 46, 50, 58, 59, 61, 65, 75, 80]. Laird and Van Lent [42] call interactive games the "killer application" for human-level AI. Love, Hinrichs and Genesereth [46] describe GDL, a general game description language.

Molineax and Aha [50] describe TIELT, an environment for evaluating learning in game-playing agents.

Proposals for developing CAD-like environments for games date at least as far back as Nelson and Mateas [58, 59] and Jones et al. [38]. Nelson [57] describes an interactive game-design environment that supports the game designer by making knowledge of the game mechanics explicit; in particular, it uses first-order logic to capture the mechanics in video games. In contrast, our work focuses on designing game-playing *agents;* thus, it supports the designer by making the knowledge and the reasoning of the game-playing agents explicit. Both lines of work however share the goal of game/agent design at the knowledge-level [60] as opposed to solely the behavioral level.

For complex game worlds, such as that of multi-player turn-based strategy games, it is not possible in general to design a game-playing agent such that it always selects the right action in any state and thus always achieves its goal or maximizes its utility. This is because multi-player, turn-based strategy games are dynamic, only partially observable, non-deterministic, and have huge state spaces. The dynamic behavior of such a game emerges out of the interactions of its numerous, heterogeneous components. Instead, game-playing agents can, and often do, fail to achieve their goals in such game environments. In such cases, retrospective agent adaptation is needed.

Further, if a game designer modifies the mechanics of a game without also revising the software agents against whom the human player plays the game, then the game may lose balance and the human player may lose interest. In such cases, proactive agent adaptation is desirable.

17.5.4 Design Thinking

The term *design thinking* has become quite popular over the last generation in part because of its adoption at IDEO[5] and Stanford University's Design School. Design thinking generally refers to the cognitive processes engaged in design [18, 19, 21–23, 28, 30, 31, 62, 64, 68–71]. The cognitive processes in design thinking include not only the memory, reasoning and learning processes in a designer's mind, but also the cognitive processes situated in and distributed over physical, informational, organizational, and social worlds.

For the purposes of this chapter, we characterize design thinking as follows:

(1) Design thinking pertains to problems that initially are ill defined [64].
(2) Design thinking is iterative, entailing experimentation with multiple designs, with each iteration resulting in a deeper understanding of the problem as well as the space of solutions, and engaging both problem reformulation and redesign [21].
(3) Design thinking is reflective, entailing the construction of cognitive artifacts such as drawings and journals that facilitate sharing, conversation, and, criticism [69].

---

[5] www.ideo.com

(4) Design thinking is human centered.

As we noted in the introduction, CAD environments have been immensely successful in architecture and engineering in large part because they support experimentation, entailing the iterative process of design generation and testing. CAD environments are also human centered: For example, they provide human designers with access to visual representations of designs and allow the designers to visually redesign the designs. Further, they enable humans to make design decisions but they automate low-level design processes such as constraint propagation.

Of course, design thinking pertains not only to production of physical artifacts, but also to the creation of abstract artifacts such as art and music, stories and poetry, and games and movies; hence, our goal of developing GAIA, a CAD-like environment for designing game-playing software agents. As we described in Sects. 17.3 and 17.4, GAIA is human-centered: it provides the designer with a visual representation of the design of the game-playing agent, and supports high-level decision making while automating low-level processes such as translation of the agent model into executable code. GAIA enables the designer to reflect on the agent design and its behavior upon execution in the game world. The REM module within GAIA uses automated reflection for agent self-adaptation.

17.5.5 Computational Creativity

Finally, we directly relate our work to AI research on computational creativity, the topic of this volume. AI research on computational creativity goes back at least as far as the DENDRAL program that identified the structure of chemical molecules responsible for spectroscopic data [45], the BACON program that induced physical and chemical laws from numerical data [44], and the AARON program that drew abstract paintings [49]. The literature typically characterizes creativity in terms of its product or result: creativity results in products that are (i) novel, (ii) useful or valuable, and (iii) non-obvious, unexpected or surprising. Colton, Lopez de Mantaras and Stock [13] provide a review of more recent developments in AI research on computational creativity. These recent developments build in part on psychological studies [19], Sternberg [72] socio-psychological studies [2], socio-cultural studies [67], and philosophical analysis [8] of creativity. For example, Wiggins [78] operationalizes some of Boden's notions about creativity. Colton and Wiggins [14] have called computational creativity the "final frontier" for AI research.

Our work relates to research on computational creativity in two ways. Firstly, we posit experimentation as a core task in the creative process. Thus, GAIA enables a designer to address the creative task of designing a game-playing software agent through experimentation, generating, executing, and revising agent designs until the designer has constructed an acceptable agent design. Likewise, GAIA can use REM to automate this process of experimentation in designing game-playing agents.

Finke, Ward and Smith [25] proposed a model of ideation called *Genexplore* in which "preinventive structures" are first generated and then explored. Importantly, the processes of generation and exploration in Genexplore are subconscious and thus unavailable for introspection. Benami and Jin [5] have operationalized this model to ideation in conceptual design. In contrast, we posit experimentation as a deliberate methodology for creative tasks such that the results of the experiment are available for inspection and reflection. Cook, Colton and Gow [15] view inspection, modification and execution of code of video games as an example of computational creativity. Their work is similar to ours in that both emphasize the importance of evaluation in the design process and in both the program code is generated automatically. However, we view experimentation, not merely evaluation, as a central element of creativity.

Secondly, we postulate meta-reasoning as an important cognitive process of creativity. Meta-reasoning allows the agent to monitor not only its behaviors in the world but also its knowledge and reasoning in producing the observed behaviors. Meta-reasoning enables the agent to reflect and introspect on its behaviors, knowledge and reasoning, and to revise its knowledge and reasoning to modify its behaviors. This process of observation, reflection, and adaptation appears to be a cognitive process of creativity.

17.6 Summary, Conclusions and Future Work

We posited that because most creative tasks, such as design, invention and discovery, tend to be ill defined and open-ended, the creative process is characterized by *experimentation.* The experimentation in creative tasks typically entails the generation of hypotheses and models, and the evaluation of the hypotheses/models. Thus, we posited that a fundamental question for computational creativity is how can we support experimentation.

We investigated this question in the context of the creative task of designing software agents for playing the turn-based strategy game of Freeciv. We described GAIA, a CAD-like interactive environment for designing software agents for playing Freeciv. GAIA provides a visual editor for constructing Task-Method-Knowledge models of an agent in the TMKL2 language. GAIA has the capability of translating the TMK model of the agent's design in TMKL2 into program code and executing the agent in the Freeciv world. The designer may observe the behavior of the agent and monitor its reasoning and knowledge states. If the designer is not satisfied with the agent's behavior, he can revise the TMK model of the agent, and repeat the above process until he finds an acceptable agent design. In this way, GAIA enables the designer to experiment with the design of the game-playing agent.

We explored the role of meta-reasoning in the creative process. Given an agent endowed with a model of its own reasoning and knowledge, the agent may use its self-model to monitor its reasoning and knowledge states of the agent in addition to its behaviors and interactions in the world. Given a failure in the agent's interaction of the world, model-based meta-reasoning also enables generation of hypotheses about the failure and corresponding revisions to the agent's model. Thus, model-based meta-reasoning too enables experimentation, in this case with the agent's experimentation with its own design.

The REM module within GAIA's architecture has the capability of meta-reasoning over the TMK model of any agent designed in the TMKL2 language for adapting the agent design to avoid previously encountered failures as well as achieve new goals similar to the original goals of the agent. This results in GAIA supporting interactive meta-reasoning: once the designer has constructed the TMK model of an initial design of the agent, he may experiment with agent design by executing the agent in the Freeciv world, evaluating its behaviors and revising the agent model. Alternatively, the designer may ask REM to use its meta-reasoning and propose revisions to the agent model. The designer may accept REM's recommendations, reject them, or refine them so that the designer and REM cooperatively revise the agent's model.

The four experiments in self-adaptation described here cover a small range of retrospective and proactive agent adaptations. They demonstrate that (i) it is possible in principle to design game-playing agents so that their teleology can be captured, specified and inspected, (ii) the specification of the teleology of the agent's design enables localization of modifications needed for the four instances of self-adaptation, and (iii) this self-adaption in turn enables the agent to play interactive games, monitor its behavior, adapt itself, play the game again, and so on. The next steps in our work are to (a) empirically investigate many more adaptation scenarios, and (b) generalize from adaptation scenarios to classes of adaptations.

Acknowledgments We thank Lee Martie for his contributions to the construction of GAIA. We are grateful to the US National Science Foundation for its support for this research through a Science of Design Grant (#0613744) entitled "Teleological Reasoning in Adaptive Software Design". An earlier version of this paper appears in Rugaber, Goel and Martie [65].

References

1. Aha, D., Molineaux, M., Ponsen, M.: Learning to win: case-based plan selection in a real-time strategy game. In: Proceedings of ICCBR-2005 (2005)
2. Amabile, T.: Creativity in Context, 2nd edn. Westview Press, Boulder (1996)
3. Anderson, M., Oates, T.: A review of recent research in metareasoning and metalearning. AI Mag. **28**(1), 12 (2007)
4. Anderson, M., Oates, T., Chong, W., Perlis, D.: The metacognitive loop I: enhancing reinforcement learning with metacognitive monitoring and control for improved perturbation tolerance. J. Exp. Theor. Artif. Intell. **18**(3), 387 (2006)
5. Benami, O., Jin, Y.: Creative stimulation in conceptual design. In: Proceedings of ASME IDETC 14th Annual Conference on Design Theory and Methods, Montreal, Canada (2002)
6. Birnbaum, L., Collins, G., Freed, M., Krulwich, B.: Model-based diagnosis of planning failures. In: Proceedings of 8th National Conference on Artificial Intelligence, p. 318 (1990)
7. Blum, A., Furst, M.: Artif. Intell. **90**(1–2), 281–300 (1997)
8. Boden, M.: The Creative Mind: Myths and Mechanisms, 2nd edn. Routledge, London (2004)
9. Bozdoc, M.: The History of CAD. http://mbinfo.mbdesign.net/CAD-History.htm (2003)

10. Chandrasekaran, B., Johnson, T.: Generic tasks and task structures: history, critique and new directions. In: David, J.-M., Krivine, J.-P., Simmons, R. (eds.) Second Generation Expert Systems, pp. 232–272. Springer, Berlin (1993)
11. Chandrasekaran, B., Johnson, T., Smith, J.: Task structure analysis for knowledge modeling. CACM **35**(9), 124–137 (1992)
12. Collins, G., Birnbaum, L., Krulwich, B., Freed, M.: The role of self-models in learning to plan. In: Meyrowitz, A. (ed.) Machine Learning: Induction, Analogy and Discovery. Kluwer Academic Publishers, Boston (1991)
13. Colton, S., Lopez de Mantaras, R.: Computational creativity: coming of age. AI Mag. **30**(3), 11–14 (2009)
14. Colton, S., Wiggins, G.: Computational creativity: the final frontier? In: Proceedings of ECAI (2012)
15. Cook, M., Colton, S., Gow, J.: Nobody's a critic: on the evaluation of creative code generators—a case study in video game design. In: Proceedings of 4th International Conference on Computational Creativity (2013)
16. Cox, M.: Metacognition in computation: a selected research review. Artif. Intell. **169**(2), 104–141 (2005)
17. Cox, M., Raja, A. (eds.): Meta-Reasoning: Thinking About Thinking. MIT Press, Cambridge (2011)
18. Cross, N.: Design Thinking: Understanding How Designers Think and Work. Berg, Oxford (2011)
19. Csikszentmihalyi, M.: Creativity: Flow and the Psychology of Discovery and Invention. HarperCollins, New York (1996)
20. Davies, J., Nersessian, N., Goel, A.: Visual models in analogical problem solving. Found. Sci. **10**(1), 133–152 (2005)
21. Dorst, K., Cross, N.: Creativity in the design process: problem-solution co-evolution. Des. Stud. **22**, 425–437 (2003)
22. Dym, C., Agogino, A., Eris, O., Frey, D., Leifer, L.: Engineering design thinking, teaching and learning. J. Eng. Educ. **94**(1), 103–120 (2005)
23. Dym, C., Brown, D.: Engineering Design: Representation and Reasoning, 2nd edn. Cambridge University Press, New York (2012)
24. Erol, K., Hendler, J., Nau, D.: HTN planning: complexity and expressivity. In: Proceedings of 12th National Conference on AI (AAAI-94), Seattle (1994)
25. Finke, R., Ward, T., Smith, S.: Creative Cognition: Theory, Research and Applications. MIT Press, Cambridge (1992)
26. Fox, S., Leake, D.: Using introspective reasoning to refine indexing. In: Proceedings of 14th International Joint Conference on Artificial Intelligence, pp. 391–397. San Mateo, CA: Morgan Kaufmann (1995)
27. Fox, S., Leake, D.: Introspective reasoning for index refinement in case-based reasoning. J. Exp. Theor. Artif. Intell. **13**, 63–88 (2001)
28. French, M.: Conceptual Design for Engineers. Design Council, London (1985)
29. Georgoff, M., Lansky, A.: Reactive reasoning and planning. In: Proceedings of National Conference on AI (AAAI87), Seattle, Washington, pp. 677–682 (1987)
30. Goel, A.: Design, analogy and creativity. IEEE Expert **12**(3), 62–70 (1997)
31. Goel, A., Helms, M.: Theories, models, programs and tools of design: views from artificial intelligence, cognitive science and human-centered computing chapter 20. In: Chakrabarti, A., Blessing, L. (eds.) An Anthology of Theories and Models of Design, pp. 417–432. Springer, Berlin (2014)
32. Goel, A., Jones, J.: Meta-reasoning for self-adaptation in intelligent agents. In: Cox, M., Raja, A. (eds.) Meta-Reasoning. MIT Press, Cambridge (2011)
33. Goel, A., Joyner, D.: Computational ideation: model construction, evaluation and revision. In: Proceedings of AAAI-2014 Workshop on Discovery Informatics, Quebec City, Canada, June 2014

34. Griffith, T., Nersessian, N., Goel, A.: The role of generic models in conceptual change. In: Proceedings of 18th Conference of the Cognitive Science Society, pp. 312–317. San Diego, July 1996
35. Griffith, T., Nersessian, N., Goel, A.: Function-follows-form: generative modeling in scientific reasoning. In: Proceedings of 20th Second Conference of the Cognitive Science Society, Philadelphia, pp. 196–201. August 2000
36. Hoang, M., Lee-Urban, S., Munoz-Avila, H.: Hierarchical plan representations for encoding strategic game AI. In: Proceedings of AI in Digital Entertainment Conference (AIIDE-05). AAAI Press (2005)
37. Jones, J., Goel, A.: Perceptually grounded self-diagnosis and self-repair of domain knowledge. Knowl.-Based Syst. **227**, 281–301 (2012)
38. Jones, J., Parnin, C., Sinharoy, A., Rugaber, S., Goel, A.: Adapting game-playing agents to game requirements. In: Proceedings of 5th AAAI Conference on Artificial Intelligence and Interactive Digital Entertainment (AIIDE-09), pp. 148–153, Stanford University, California, USA, 14–16 October 2009
39. Joyner, D., Goel, A., Papin, N.: Intelligent generation of agent-based simulations from conceptual models. In: Proceedings of 18th International Conference on Intelligent User Interfaces (IUI2014), pp. 289–298. Haifa, Israel, February 2014
40. Kephart, J., Chess, D.: The vision of autonomic computing. Computer **36**(1), 41–50 (2003)
41. Laird, J.: Research in human-level AI using computer games. CACM **17**(4), 15–16 (2002)
42. Laird, J., Van Lent, M.: Human-level AI's killer application: interactive computer games. AI Mag. **22**(2), 15–25 (2001)
43. Leake, D.: Experience, introspection and expertise: learning to refine the case-based reasoning process. J. Exp. Theor. Artif. Intell. **8**(3–4), 319 (1996)
44. Langley, P., Simon, H., Bradshaw, G., Zytkow, J. (eds.): Scientific Discovery: Computational Explorations of the Creative Process. MIT Press, Cambridge (1987)
45. Lindsay, R., Buchanan, B., Feigenbaum, E., Lederberg, J. (eds.): Applications of Artificial Intelligence for Organic Chemistry: The DENDRAL Project. McGraw-Hill, New York (1980)
46. Love, N., Hinrichs, T., Genesereth, M.: General game playing: game description language specification. Technical Report. Stanford University, 4 April 2006
47. MacGregor, R., Bates, R.: The LOOM knowledge representation language. USC/ISI Technical Report ISI/RS-87-188 (1987)
48. Maturana, H., Varela, F.: The Tree of Knowledge: Biological Roots of Human Understanding. New Science Library/Shambhala Publications, Boston (1987)
49. McCorduck, P.: Aaron's Code. W.H. Freeman, San Francisco (1991)
50. Molineaux, M., Aha, D.: TIELT: a testbed for gaming environments. In: Proceedings of 20th National Conference on AI, pp. 1690–1691 (2005)
51. Muñoz-Avila, H., Jaidee, U., Aha, D., Carter, E.: Goal directed autonomy with case-based reasoning. In: Proceedings of 18th International Conference on Case-Based Reasoning, pp. 228–241, Springer, Alessandria, Italy (2010)
52. Murdock, W., Goel, A.: Learning about constraints by reflection. In: Proceedings of the 14th Biennial Conference of Canadian AI Society, pp. 131–140. Springer, Berlin (2001)
53. Murdock, J., Goel, A.: Localizing planning with functional process models. In: Proceedings of 13th International Conference on Automated Planning and Scheduling (ICAPS-2003) (2003)
54. Murdock, J., Goel, A.: Meta-case-based reasoning: self-improvement through self-understanding. J. Exp. Theor. Artif. Intell. **20**(1), 1 (2008)
55. Myers, K.: A procedural knowledge approach to task-level control. AIPS Proceedings. AAAI Press, Menlo Park (1996)
56. Nau, D., Au, T., Ilghami, O., Kuter, U., Murdock, J., Wu, D., Yaman, F.: SHOP2: an HTN planning system. J. Artif. Intell. Res. **20**, 379–404 (2003)
57. Nelson, M.: Representing and reasoning about videogame mechanics for automatated design support. Ph.D. Dissertation, School of Interactive Computing, Georgia Institute of Technology; forthcoming (2014)

58. Nelson, M., Mateas, M.: An interactive game design assistant. In: Proceedings of 20th Intelligent User Interface Conference, pp. 90–98. Spain (2008)
59. Nelson, M., Mateas, M.: A requirements analysis for videogame design support tools. In: Proceedings of 4th International Conference on the Foundations of Digital Games (FDG) (2009)
60. Newell, A.: The knowledge level. Artif. Intell. **18**, 87–127 (1982)
61. Ontanon, S., Mishra, K., Sugandh, N., Ram. A.: Case-based planning and execution for real-time strategy games. In: Proceedings of ICCBR-2007 (2007)
62. Plattner, H., Meinel, C., Leifer, L. (eds.): Design Thinking: Understand, Improve, Apply. Springer, Berlin (2011)
63. Rao, A.S., Georgeff, M.: BDI agents: from theory to practice. In: Proceedings of the 1st International Conference on Multi-Agent Systems (ICMAS-95), pp. 312–319, San Francisco, CA, June 1995
64. Rittel, H., Webber, M.: Dilemmas in a general theory of planning. Policy Sci. **4**, 155–169 (1973)
65. Rugaber, S., Goel, A., Martie, L.: GAIA: a CAD environment for model-based adaptation of game-playing agents. Procedia Comput. Sci. **16**, 29–38 (2013)
66. Russell, S., Norvig, P.: Artificial Intelligence: A Modern Approach, 3rd edn. Prentice-Hall, Upper Saddle River (2010)
67. Sawyer, K.: The Science of Human Innovation: Explaining Creativity, 2nd edn. Oxford University Press, Oxford (2011)
68. Schon, D.: The Reflective Practitioner: How Professional Think in Action. Basic Books, New York (1983)
69. Schon, D.: Designing as reflective conversation with the materials of a design situation. Knowl.-Based Syst. **5**(1), 3–14 (1992)
70. Simon, H.: Sciences of the Artificial. MIT Press, Cambridge (1969)
71. Simon, H.: Sciences of the Artificial, 3rd edn. MIT Press, Cambridge (1996)
72. Sternberg, R. (ed.): Handbook of Creativity. Cambridge University Press, Cambridge (1999)
73. Stroulia, E., Goel, A.: Functional representation and reasoning in reflective systems. J. Appl. Artif. Intell. **9**(1), 101–124 (1995)
74. Stroulia, E., Goel, A.: Evaluating problem solving methods in evolutionary design: the autognostic experiments. Int. J. Hum.-Comput. Stud. **51**(4), 825 (1999)
75. Ulam, P., Jones, J., Goel, A.: Combining model-based meta-reasoning and reinforcement learning for adapting game-playing agents. In: Proceedings of 5th Conference on Artificial Intelligence and Interactive Digital Entertainment (AIIDE-08), Stanford University, California, October 2008
76. Vattam, S., Goel, A., Rugaber, S.: Behavior patterns: bridging conceptual models and agent-based simulations in interactive learning environments. In: Proceedings of 11th IEEE International Conference on Advanced Learning Technologies, pp. 375–377. Athens, Georgia, USA, July 2011
77. Weisberg, D.: The Engineering Design Revolution. http://www.CadHistory.net (2008)
78. Wiggins, G.: A preliminary framework for description, analysis and comparison of creative systems. Knowl.-Based Syst. **19**, 449–458 (2006)
79. Winston, P.: Artificial Intelligence, 3rd edn. MIT Press, Cambridge (1993)
80. Young, R., Riedl, M., Branly, M., Martin, R., Saretto, C.: An architecture for integrating plan-based behavior generation with interactive game environments. J. Game Dev. **1**(1), 51–70 (2004)

Chapter 18
Collective Discovery Events: Web-Based Mathematical Problem-Solving with Codelets

Petros S. Stefaneas, Ioannis M. Vandoulakis, Maricarmen Martinez and Harry Foundalis

Abstract While collaboration has always played an important role in many cases of discovery and creation, recent developments such as the web facilitate and encourage collaboration at scales never seen before, even in areas such as mathematics, where contributions by single individuals have historically been the norm. This new scenario poses a challenge at the theoretical level, as it brings out the importance of various issues which, as of yet, have not been sufficiently central to the study of problem-solving, discovery, and creativity. We analyze the case of collective and web-based proof-events in mathematics, which share their temporal and social nature with every case of collective problem-solving. We propose that some ideas from cognitive architectures, in particular the notion of codelet—understood as an agent engaged in one of a multitude of available tasks—can illuminate our understanding of collective problem-solving, and act as a natural bridge from some of the theoretical aspects of collective, web-based discovery to the practical concern of designing cognitively-inspired systems to support collective problem-solving. We use the Pythagorean Theorem and its many proofs as a case study to illustrate our approach.

All authors contributed equally to this chapter.

P.S. Stefaneas (✉)
School of Applied Mathematical and Physical Sciences,
National Technical University of Athens, Athens, Greece
e-mail: petros@math.ntua.gr

I.M. Vandoulakis
School of Humanities, Hellenic Open University, Patras, Greece
e-mail: i.vandoulakis@gmail.com

M. Martinez
Mathematics Department, Universidad de Los Andes, Bogotá, Colombia
e-mail: m.martinez@uniandes.edu.co

H. Foundalis
Center for Research on Concepts and Cognition (CRCC),
Indiana University, Bloomington, USA
e-mail: hfoundal@yahoo.com

T.R. Besold et al. (eds.), *Computational Creativity Research: Towards Creative Machines*,
Atlantis Thinking Machines 7, DOI 10.2991/978-94-6239-085-0_18

18.1 Introduction

Theorem proving is only one of possibly thousands of different cognitive activities with which a mind can be engaged. Minds most probably do not invent new architectural principles to treat each cognitive domain in a special way, because the architecture of the underlying hardware (the brain) is fixed, honed by millions of years of evolution. It has been hypothesized that just as brains are architecturally fixed, so are minds that arise as emergent properties of brains [10, 12]. That is, there is an "architecture of mind" which is as fixed and unchanging as the architecture of brain. When a mind confronts a cognitive problem it uses "tools" from a fixed repertoire, which however are flexible enough to adapt themselves to and be useful in the solution of any problem. One such set of architectural tools of minds are the *codelets* [3, 10].

The purpose of this paper is to examine the feasibility of using the idea of codelets as agents who actively participate in seeking and discovering proofs of theorems. To this end, after clarifying the notions of codelet and proof-event, we look at some software-assisted projects for collaborative Web-based mathematical problem solving. Then we discuss why, in our view, Goguen's [6] understanding of proofs as events, enriched with the notion of codelets, provides an adequate framework for analyzing this sort of Web-based collaborative activity. Finally, we illustrate the workings of a hypothetical system for collaborative activity by showing how such activity could be applied to the problem of proving the Pythagorean Theorem, which we selected for reasons of simplicity and universal knowledge of its statement. To the best of our knowledge, the idea of applying notions from cognitive architectures to Web-based collaboration has not yet been explored.

18.2 Problem Solving with Codelets

Codelets can be conceived of as short pieces of programmed code, but in an abstract sense. In brains, codelets can be implemented at the lowest level by means of neurons; in computers, they can be short pieces of programming instructions.

The purpose of codelets is to build conceptual structures in working memory, given some input. Sometimes they can demolish structural pieces, or even whole structures. However, the bulk of their work is constructive rather than destructive. Codelets work in parallel, ignoring each other's existence. Each one has a specific and simple task to complete, and is allocated a given amount of time. If a codelet fails to finish its work within a reasonable time, it "dies" and another codelet of very similar nature makes a "fresh start", working on the same task anew.

Sometimes a codelet may spawn a number of other, different codelets that are deemed useful by it in working on various aspects of the task. Thus, the generator codelet becomes a higher-level "supervisor" of the sub-codelets that it generated, waiting for them to finish their sub-tasks in order to continue with its "main" task.

This generates a hierarchy of codelets, in which those at a certain level have "knowledge" of the codelets they generated, but ignore both their "superior" codelets and their "peers".

Some differences between the way that codelets work and more traditional programming are the following:

1. the structures built by codelets can be both dispensable and redundant, whereas programs usually have non-redundant code, and whatever they build is rarely destroyed;
2. codelets that are "peers" (i.e., at the same level in the hierarchy) work in parallel, whereas programs are usually written to run sequentially; and
3. there is no "supervisor" with a total knowledge of which codelets run at any moment and what they will eventually achieve, except at a very local level, when a higher-level codelet becomes a supervisor, but only of the codelets of the immediately lower level that it spawns; instead, in traditional programs the programmer has an overall view and knowledge of which pieces of code exist and can run at any time. Thus, the system of codelets is dynamic and distributed, bound only by the constraints of its hierarchical structure.[1]

An example might clarify the question of how codelets build structures, solving problems at the same time. Let us consider the problem of visually perceiving a written piece of a sentence, and attempting to understand its meaning. Suppose the phrase (i.e., fragment of a sentence) is:

"meaning that he failed to discover it".

We can imagine a highest-level codelet the task of which is: "to understand the given phrase". This codelet spawns a number of other codelets, some of which have tasks as: "to understand one given word"; others: "to put words together in a syntactically correct structure"; and so on. Codelets of the former kind spawn other codelets assigned the task "to read a word". If the perceiving agent is a human being, then the task "to read a word" entails signaling the muscles that move the eyes to perform eye saccades and sample a few spots within the word; whereas if the perceiving agent is a program it could do something analogous but by processing letters within a string, or by processing the pixels of an image, if the above phrase was part of one. Occasionally, some codelets might produce a wrong result. For instance, the "e" of "failed" might be seen as a "c", but this will not make sense in the context of the output of other codelets because there is no word like "failcd"; thus, another codelet can re-perceive the letter in a way that makes sense. The destruction of already-built structures can be seen more explicitly at the level of syntax: the reader might perceive the word "meaning" as a noun, interpreting the phrase in this sense: "it was that kind of meaning which he failed to ..."; but, alas, after the word "discover" comes the pronoun "it", which either is redundant or—if this is a correctly written fragment of

[1] Hofstadter [10] does not assign a hierarchical structure to codelets; the idea of a hierarchy of codelets is introduced in the present article.

a sentence—destroys the perceived syntactic structure and necessitates a re-reading of the phrase. Indeed, with a fresh set of codelets working at the syntactic level from scratch, the word "meaning" can be seen as a participle, in which case the pronoun "it" cannot refer to "meaning" but to something else prior to the given phrase. (For example, the phrase could be part of this sentence: "He claimed that the islet did not exist, actually meaning that he failed to discover it.")

Now take the case of collaborative theorem proving. A person engaged in solving a particular task toward the completion of a proof can be thought of as a codelet. The task could be proposed by a supervising codelet (some person, but with a somewhat wider view of the project), and could be taken by a pool of people who have volunteered their services and availability to the proving project, as long as they feel that the task is suitable for their abilities. Similarly, a person working on a codelet could assign sub-tasks as other codelets, of simpler nature and of an ever-narrower view and scope, which can be taken by less qualified or less specialized proving agents. At the highest level could stand a person of qualified knowledge who gave the initial broad strokes, i.e., decided the highest-level tasks and placed them in the "codelet pool" to be undertaken by qualified agents. The tacit assumption is that perhaps in this way proofs of greater complexity can be achieved than is possible by the faculties of a single person-prover. In the rest of this chapter we shall examine this idea more thoroughly.

18.3 Mathematical Problem-Solving as Proof-Event

Goguen introduced the concept of proof-event in an attempt to formulate a wider viewpoint on proof, designed to incorporate traditional mathematical proofs (both constructive and non-constructive ones), but also non-mathematical proofs (apodictic, dialectical, ontological, etc.) as well as new kinds of proving practice, such as computer proofs and proof steps.

> Mathematicians talk of "proofs" as real things. But the only things that can actually happen in the real world are proof-events, or provings, which are actual experiences, each occurring at a particular time and place, and involving particular people, who have particular skills as members of an appropriate mathematical community....
>
> A proof-event minimally involves a person having the relevant background and interest, and some mediating physical objects, such as spoken words, gestures, hand written formulae, 3D models, printed words, diagrams, or formulae (we exclude private, purely mental proof-events...). None of these mediating signs can be a 'proof' in itself, because it must be interpreted in order to come alive as a proof-event? we will call them *proof objects*. Proof interpretation often requires constructing intermediate proof objects and/or clarifying or correcting existing proof objects. The minimal case of a single prover is perhaps the most common, but it is difficult to study, and moreover, groups of two or more provers discussing proofs are surprisingly common (Goguen [6]).

From this point of view, a proof-event is a social event: it takes place at a certain location and lasts for some time interval; it also involves a public presentation or communication of a narrative that is addressed to particular social groups, who are assumed to have particular skills and expertise, enabling them to perceive the narrative and formulate a conclusion on it. Proof-events are not confined necessarily to communication of mathematical truths, since a proof-event may concern the presentation of an incomplete proof, a false proof, or a proofless, intuitive exposition of ideas on a particular problem.

Following Goguen's idea, proof-events presuppose the involvement of at least two types of agent: a *prover*, which can be a human or a machine or a combination of them (in the case of hybrid proving), and an *interpreter*, who generally can be a human (or group of humans) or a machine (or group of machines) or a combination of them. Both types of agent are situated in an environment that might be a real *physical* one, like a material world inhabited by various kinds of things (objects), or a *virtual*, or a *simulated* one, when for instance human agents interact with physical tools and software entities or the Internet. An agent is equipped with past background knowledge and past learning experience.

A human prover may experience an insight (intention) that something in mathematics is true and produce an item (choosing some semiotic code) to encode their experience. This item may be not a complete proof, but an outline of a proof or even a conjecture. This initiates a proof-event. The prover may feel so confident in the truth of their experience, that they may decide to present the produced item publicly, expecting that a potential reader (interpreter) will be persuaded easily in the truth of their experience and will succeed in understanding (decoding) it, as the prover understands it. Thus, the time of public communication of the item can be considered as an ending point of a proof-event. In the case of a machine, this is the time that the machine halts and produces an output.

Thus, agents enact different *roles*, according to their type. The roles enacted by agents determine their tasks with respect to a specified *problem* (defined by certain *conditions*). The task of a prover is to set a problem and develop an argument that aims at providing support or even a proof to a given problem; the task of an interpreter is to understand or interpret or verify a suggested argument or proof. The implementation of this task may involve the use of a computer program to verify a suggested proof, or to check certain special cases of a suggested proof, or to find a gap in a suggested proof or cross-check the proof. Moreover, the interpreter may undertake the task to fill the gap in the proof by advancing a new proof, or to invent a counterexample that would ultimately refute the proof, or reformulate the gap as a new problem. In these cases, the interpreter enacts as prover, rather than interpreter. The tasks of the prover and the interpreter have different goals and can be realized in different proof-events that are successive in time forming a *sequence of proof-events*. A sequence of proof-events evolves in time with reference to a fixed underlying problem and represents the *history of a proof*. An agent can freely change roles in the course of history and enact as prover at a certain time and as interpreter at another time of a sequence of proof-events. Moreover, the definition of concepts and standards of rigor are not invariant during the unfolding of a sequence of proof-events, since the task of an interpreter includes possible gap-filling operations that may be fulfilled by modification of definitions or explication of implicit concepts in order to achieve higher standards of rigor in the purported proof.

The prover and the interpreter, involved in a sequence of proof-events, may act in different geographical locations, be surrounded by different environments at different times and belong to different mathematical cultures. However, they are in *communication* by sharing a common *interpersonal space*, so that communication becomes possible. The environments in which the agents are located provide the *means of communication* between the agents, i.e. written texts, conditions for verbal or non-verbal communication, as well as for communication through practices. The "text" communicated by a prover can be encoded in any (already available, or ad hoc invented) semiotic code (usually in a blend of the natural language and some formal language) and is articulated in a specific *style* that serves communication functions. Styles can be personal for provers or for the schools they belong to or for a whole mathematical tradition; it may be also mimicry of the style of a renowned authority. The style is a *meta-code* that determines the *selection* of a particular code and the *combination* of blending principles to produce the "text" by the prover [18].

Different agents can ascribe different meanings to the same "text", communicated by a prover. The meaning that the prover may ascribe to the "text" (the *intended meaning*) is generally different from the *perceived meaning* that the interpreter may ascribe to the same "text". Moreover, the prover and the interpreter may follow *different kinds of logic* in their reasoning. Understanding is achieved when the meaning perceived by the interpreter corresponds to the intended meaning of the prover, that is, whenever a mapping called *semiotic morphism* (or "translation") can be established between the semiotic space of the prover into the semiotic space of the interpreter, as defined in [4, 5].

The interpreter's understanding of a "text" is a prerequisite to its interpretation; yet this does not guarantee the validity of the proof. *Interpretation* means the determination of the meaning of the signs in which a transmitted proof or what is thought of to be a proof is encoded. Interpretation is an active process, during which the interpreter may amend the initial proof by adding new concepts (definitions) or fill possible gaps in the proof by elaborating new parts of the proof, etc. In some sense, interpretation is a reconstruction of meaning or conscious reproduction of the information content conveyed by the "text". During this process, the interpreter may also choose a new, different code to express more adequately the meaning of the prover's "text". However, during this transformation the original "text" loses its stylistic peculiarities. The style of the interpreter may be completely different from that of the prover (although it may incorporate some elements of the style of the prover). In general, the item produced by a prover may lead to different communication outcomes:

> For a mathematician engaged in proving, the most satisfactory outcome is that all participants agree that 'a proof has been given.' Other possible outcomes are that most are more or less convinced, but want to see some further details; or they may agree that the result is probably true, but that there are significant gaps in the proof event; or they may agree the result is false; and of course, some participants may be lost or confused (Goguen [6]).

A sequence of proof-events is considered as terminated (finite) when the agents (interpreters) involved in it conclude that they have understood the proof and agree that a proof has actually been given, i.e. when a consensus on the validity of the purported proof is reached among the relevant community of agents. This marks the termination of the sequence of proof-events (the history of the proof). Otherwise, the proof is not considered valid and the relevant community remains in a state of indecision. Therefore, the community of relevant agents is the ultimate truth-maker of the (sequences of) proof-events.

Thus the truth of the proof is ultimately declared by the relevant mathematical community, which is competent enough to check the proof. However, in the case of computer-generated proofs, their validation by humans can be practically impossible, since no human or group of humans is capable of checking a huge number of proof steps. What can be considered as validation in such a case is the validation of the correctness of the theorem-prover itself (as a program). Therefore, in this case, validating the program can guarantee that all possible proofs generated by this software tool have been "validated". One way or the other, either by human or by automated means, the final output of the validation needs the recognition or consensus of the mathematical community to be regarded as true. This marks the termination of the sequence of proof-events at this particular time.

18.4 Web-Based Mathematical Problem-Solving as Proof-Event

The conceptual framework outlined above is adequate to describe Web-based mathematical problem solving as proof-events as is practiced in the Kumo [7] and Polymath projects [8, 17].

Web-based proving appears to be a novel kind of proving practice, having new important features. First, it is characterized by a *change of the communication medium*: the Web serves as both an information source (a repository of information, ideas and methods available) and a communication medium (creating global interest-based communities). Web-based mathematical problem-solving is *open* to all, and communication is transformed from one-to-one or one-to-many into *many-to-many*. Second, the use of the Web as a medium of communication between agents opens new possibilities of interactivity between the agents involved in a proof-event. The agents are located in a *virtual* environment, when for instance human agents interact with software entities through the Internet. For example, agents could interact through a *blog*, in the case of Polymath project, or in a more sophisticated way, through *proofweb* sites (displayed as *proof pages* or *Tatami pages*) in the case of Kumo. This facilitates the formation of new metaphors for information visualization and proof representation. Interactivity also enables the formed group of agents to collaborate on the problem by behaving like a goal-directed agent system, in which agents enact interchangeably both roles of prover and interpreter. This results in astonishing intensification and speed of the mathematical problem-solving activity.

Interactivity in Web-based mathematical problem-solving, as practiced, for instance, in the Polymath project, enables the use of a group problem-solving technique known as brainstorming [13]; in particular, (asynchronous) computer-mediated or Web-based (group) brainstorming [2], by which a group tries to find a proof for a posed mathematical problem by culling a list of spontaneously generated ideas contributed by its members. This feature makes it possible to take advantage of the complementarity of mathematical background and capabilities of the different agents involved in a Web-based proof-event. In a recent study of the third Mini-Polymath project [19], Alison Pease and Ursula Martin [15] have explored the kinds of mathematical reasoning behind the online collaboration and classified them into four kinds of comment that focus on mathematical concepts, examples, conjectures and proof strategies and relate them, wherever relevant, to corresponding theories of mathematical practices elaborated by Pólya [16] and Lakatos [11].

Subsequent to the brainstorming phase or the collection of ideas, parts of proof or incomplete proofs (as practiced in the Polymath project) is the phase of their categorization and integration. This task is allocated to a member of the group of agents or is undertaken by a supervising agent. Thus the roles of the supervising agent and the integrator are crucial. The supervising agent is the master of the code of ethics, and monitors the whole process of the mathematical problem-solving activity. The integrator acts as proof checker or (global, above individual provers) validator and integrates the different parts of the proof into a unified whole.

In Web-based proof-events the agents involved suggest possible contributions to the proof, which can potentially be integrated into the final proof, or disregarded. Thus, it is a *modular* contribution to proving that resembles the *open-source* cooperative activity or *crowdsourcing* (in case the proof-event is initiated by a single agent). These features of Web-based proof-events expand the capacity of the agents involved (i.e. the effectiveness of the agent system), ensuring deeper insight into the problem. The final outcome of a Web-based proof-event cannot be credited to its initiator or the supervising agent; it is a kind of collective intellectual property, in which the individual contribution of every agent is clearly identifiable in the history of the proof, represented as a sequence of proof-events.

18.5 A Codelets-Based Model for Web-Based Proof-Events

Web-based mathematical problem-solving is a process based primarily on the prover-interpreter interaction over the Web. During this process, an initial interpreter inserts into a *pool of unresolved issues* a list of issues that, if resolved, amount to the solution of the initial problem. For example, if the problem to be solved is the proof of a theorem, then the list of unresolved issues that are inserted into the pool can be the highest-level pieces of the proof, as envisioned by the initial interpreter. The pool is immediately available to the Web-based community of participants, who are informed of its existence by mediating system software and may start acting as *codelets* when they select pieces that they deem solvable. (Henceforth, for simplicity, the participants in the Web-based problem-solving event will be referred to as "codelets".) When codelets form by selecting an unresolved issue they do so either because they feel capable of solving it by their own means, or because they see how to decompose it further into constituent sub-issues, which are then also entered into the pool. The "system software" (henceforth: "system") keeps track of which issues are parts of which larger ones; i.e., the system knows the *hierarchy* of the problem decomposition.

As soon as a prover-codelet feels that the solution of the issue that the codelet was working on is available, it informs the system of this fact. The system informs the supervisor codelet (who had inserted that issue into the pool), and the latter acts as an interpreter of the solution. If the solution is validated by the interpreter-codelet, the system is informed, so that other codelets do not attempt to assign to themselves the solution of the same issue, which appears as "solved" in the pool. The system is also informed when the interpreter-codelet finds that all the sub-issues of the undertaken issue have been solved. Thus, each codelet acts as both a prover (for the single issue that the codelet selected from the pool), and an interpreter (of all the sub-issues that the codelet entered into the pool, after decomposing the selected issue).

In addition to parts of a proof (sub-proofs), codelets may make various other contributions, such as incomplete or even false proofs, ideas, comments, suggestions, opinions and methodology transfer rooted in past experience and expertise. These contributions are also entered into the pool, each distinguished by its type, and are conceived as directed toward the solution of a stated problem. Hence, the contributions are independent, goal-directed processes that evolve over the Web space and time and accumulate as building blocks or modules of a generated Web proof-event.

Particular contributions may turn out to be blind, i.e. to lead in due time to a recognizable deadlock situation. This may entail a change of approach towards the problem, change of methodology applied, etc.; that is, it may give rise to a new contribution and the abandonment of the unfruitful undertaking. After all, as explained in the introductory sections, some codelets might act destructively, invalidating particular sub-proofs and contributions (e.g., when they find an error in a sub-proof, or that an idea is unfruitful, etc.). However, such destructions are local, whereas overall the system proceeds to a coherent solution, if one can be found.

In Web-based proof events codelets have certain specific features:

1. Each codelet, acting as a prover, knows neither who its supervising interpreter-codelet is, nor its "peer" codelets who might be working on other sub-issues of the same problem. However, when that prover-codelet becomes a supervising interpreter-codelet (due to having decomposed the issue and entered its parts into the pool), then it can keep track of which codelets work on which sub-issues. This information is made available to the supervising interpreter-codelet by the system.
2. When agents see that an issue is marked as "being worked on" (or "taken" by a codelet), they are not prevented from taking it as well, forming a new codelet. This is because some codelets may feel they can give a neat solution that might be missed by other codelets. If, eventually, two or more solutions arise for a given issue, it is up to the supervising interpreter-codelet to choose one and inform the system, which incorporates it into the overall solution.

3. As already mentioned, the work of some codelets may turn out to be superfluous or even useless. The outputs of such codelets are not ultimately integrated into the final structure of the formal mathematical proof. Nevertheless, they cannot be considered totally irrelevant, because they might have revealed unexpected relationships with other mathematical concepts or statements or elucidate the independence of some assumption of the mathematical statement to be proved or uncover the need of a weaker or refined assumption.
4. Particular codelets, or their derivative contributions, may vary in location and weight in a process of generation of a Web-based proof-event. A codelet may turn out to be prerequisite, refinement, simple correction or even counterexample for the contribution of another codelet. Therefore, they are arranged neither in parallel, nor in sequential order. They have a complex, graph-like structure that follows the eventual formal structure of the provisional mathematical proof.
5. Administrators do not know in advance the final outcomes of Web-based proof events, so they can't provide deterministic guidance. They are trusted by the community of the contributors in view of their reputation in the academic world. At the final stage of Web-based proof events administrators can potentially intervene, evaluate, correct, filter and integrate all kinds of contributions.

18.6 Collective Codelet Generation: An Example Using the Pythagorean Theorem

Consider Pythagoras' Theorem: $a^2 + b^2 = c^2$ whenever a and b are the lengths of the legs of a right triangle and c is the length of the hypotenuse. While not an example of a research problem in contemporary mathematics, we will use this widely known property of right triangles to illustrate our approach. Here, for the purposes of illustration, we imagine an unrealistic scenario in which there exists a mathematical community that has all the skills and training that mathematicians usually have, but which lacks the knowledge of the truth of the Pythagorean Theorem. This is necessary for our discussion because we want to avoid discussing some other, sophisticated problem, which is understandable only by the real mathematical community of our times. Thus, suppose that the problem of finding proofs of the Pythagorean Theorem is publicly posted, and the members of this hypothetical mathematical community may participate in the quest for a proof (or proofs!). In what follows, we will draw from material from the webpage of Bogomolny [1], a collection of over one hundred different proofs of the theorem from various sources.

Initially there should be a single task: PT (proving the Pythagorean Theorem), initiated by a single agent: A. This marks the beginning of a proof-event that is going to evolve in time, as long as other codelets (acting as provers) would attempt to present a proof of PT. We shall use the notation $\langle agent, task\rangle$ for codelets that enter the system pool. Thus, $\langle A, PT\rangle$ will be the initially single codelet in the pool. This codelet is the overall supervisor of the task PT, with A being the "supervisor agent" who will decide at some future point whether the task was handled satisfactorily.

It would be a normal occurrence in this setting that agents at this high level in the hierarchy of codelets have ideas of "strategies" that might work to obtain a proof of the theorem. For example, agent A might decide to ask for the solution of four different and independent approaches to PT, which we denote as $PT_i, (i = 1, \dots, 4)$ and will explain soon. Alternatively, it could be that A introduces only one or two of the PT_i, whereas other agents, after being informed of the availability of PT in the system, add the rest of the PT_i (up to $i = 4$ for our purposes of illustration). Whenever an agent sees one of the PT_i and deems it interesting—reserving it for solution—a corresponding codelet $\langle A_i, PT_i\rangle$, who undertakes the role of a prover, is introduced in the system pool. Thus, assuming all four PT_i are reserved, we now have four more codelets in the pool: $\langle A_1, PT_1\rangle$, $\langle A_2, PT_2\rangle$, $\langle A_3, PT_3\rangle$, and $\langle A_4, PT_4\rangle$, the status of all of which is monitored by the system software. We assume that each of the four agents A_i has different background and skills and is thereby motivated by different ideas about how to tackle the PT_i task at hand. Each PT_i might lead to the creation of new subtasks, and every time a member of the community accepts working on a subtask a new codelet is created. But let's see what each of the tasks PT_i might look like:

1. $\langle A_1, PT_1\rangle$ might be dealing with PT as a purely visual problem, i.e., as one of dividing the two smaller squares into pieces that then add up to cover exactly the largest square. This decision means that a choice was made to use some "diagrammatic" language to express the argumentation in favor of the validity of the problem. Thus, A_1's proof would be formulated in a "diagrammatic" style. This choice might have been encouraged by A_1's background and expertise, or by A_1's belief that such a proof would be intuitive and convincing. Such a "proof" would consist of not only showing that such a division is visually possible, but also that it can be carried out in the "classic" manner, by compass and ruler alone—a condition imposed upon the solution. An example of such a division is shown in Fig. 18.1 (proof #28 in [1]):
   Here, the diagram might be termed "self-explanatory", and agent A_1 might stop at this point thinking that this is sufficient as a proof. In other words, the agent, acting as a prover might assume that the item (i.e., the diagram) that A_1 communicates to other codelets—which may undertake the role of interpreters—contains sufficient information, and thereby might succeed in convincing a potential interpreter. However, the overall supervisor agent A (acting as interpreter) might ask for a further elaboration. Specifically, A might ask to be convinced that the above division is possible in all cases of shapes of the triangle ABC. i.e., A might evaluate the item of the prover as insufficient and unconvincing in establishing the general validity of PT_1. The overall supervisor A may be suspicious of the language chosen by the prover to present the argument, believing that the prover was misled by the language. This would create a new codelet, $\langle A_{1.1}, PT_1\rangle$, where $A_{1.1}$ is either the same agent as A_1, or someone else (possibly one having more advanced expertise in logical reasoning), who accepted the challenge. To show

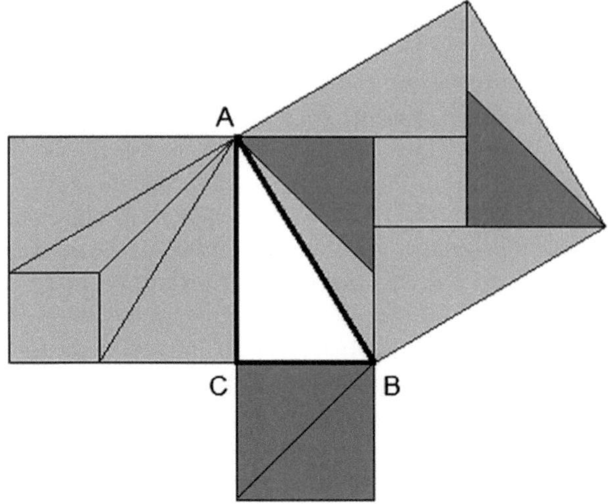

Fig. 18.1 Proof #28 in [1]

that such "visual proofs" are not always valid, $A_{1.1}$ may propose, for instance, the item shown in Fig. 18.2 (false proof #3 in [1]), on the left.
On the left side of Fig. 18.2 we see a division as could be proposed by an agent who thought that the problem is thus solved. However, this division depends on the shape of the initial triangle. As the right side of Fig. 18.2 shows, the number of the constituent areas increases as the angle BAC decreases.

In general, once a "proof" is suspected, there follows a process of evaluation by which either the agent working on the problem, or the supervising agent, may

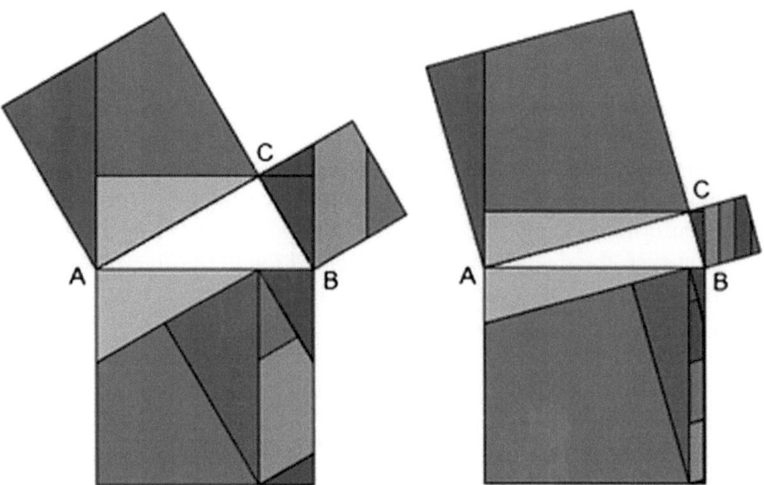

Fig. 18.2 False proof #3 in [1]

(1) accept the proposed answer as correct, (2) reject it as incorrect, or (3) decide that it is insufficient as a proof, i.e. that it contains gaps that need to be filled. In the case of the proof shown in Fig. 18.2, the agent (e.g., the overall supervisor A) may decide that the diagram alone is not a proof and create a new subtask based on it, asking for a proof that the division of the small squares can be made in all cases of triangles, and also with ruler and compass (the condition imposed upon the solution). Then, for each such request, a codelet would be added in the system pool, and get involved in the proof-event as soon as an agent (prover) undertook the requested task of formally proving that the pieces of the division do add up to the large square.

Notice that agent A_1 may be simultaneously supervising (i.e. acting as interpreter) several possible divisions that have been proposed and are being checked by various agents (provers) in various codelets (children of $\langle A_1, PT_1 \rangle$). We have conceived of the supervising role of A_1 as including the duty of filtering some purported solutions to subtasks as correct or not. However, in reality the final evaluation of a proof and its acceptance as such is granted not by any single member of the mathematical community, but by the community as a whole. Thus, in a system like the one we envision here a final subtask should always be created by A_1 whenever a full proof of PT_1 seems to have been successfully completed; namely, the subtask that posts the proof and asks the interpretation agents to "verify the correctness of this proof".

Finally, consider the case that the same agent A_1 finds two or more partitions of the squares. Thus, a single codelet like $\langle A_1, PT_1 \rangle$ may report two or more solutions to its supervising agent A (in a communication facilitated by the system), who then posts an equal number of codelets in the pool, asking for the corresponding verifications of the correctness of proofs.

2. $\langle A_2, PT_2 \rangle$ might instead be asked to come up with ideas of using other theorems of geometry, already accepted as true by the community, so as to give a geometrical proof of PT. Thus, A_2 might post two (or more) tasks of the kind "use Theorem $[X]$ in order to prove PT". For a concrete example, A_2 might create the following two tasks: "Use the Intersecting Chords Theorem in order to prove PT" and "Use the Socrates' Theorem in order to prove PT". The former theorem was first proved by Euclid in his *Elements* (Proposition 35 of Book III). It represents the outcome of a proof-event that took place during the Hellenistic period, in which Euclid, as an agent (prover), communicated it to interpreters. The name of the latter theorem is due to the right isosceles triangle used by Socrates in Plato's Meno (84c) in an argument communicated to a young slave, during a proof-event of solving the problem of Duplication of the Square.[2]

As in the previous case, each agent that is engaged with one of those subtasks (i.e., engaged in the proof-event) will lead to the creation of a codelet. Proofs

[2] The reason Socrates discusses a geometrical result in a philosophical text is in order to argue that knowledge pre-exists in minds, and so all that is needed is a suitable "midwife" who will deliver the knowledge to the consciousness of the mind.

Fig. 18.3 Proof #59 in [1]

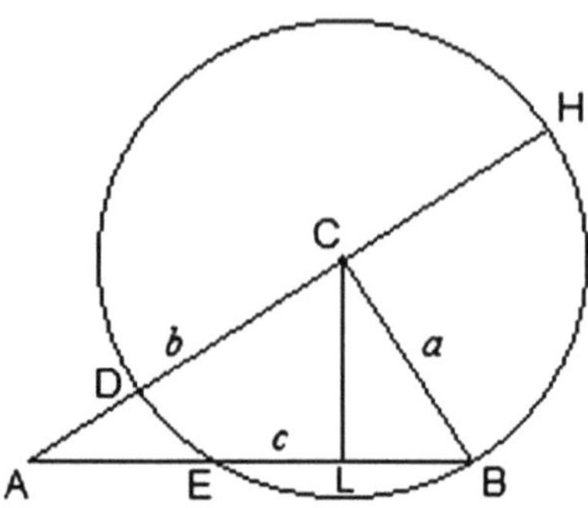

#59 and #60 in [1] use the "Intersecting Chords Theorem" in two different ways, and proof #64 uses Socrates' Theorem. Let us follow proof #59 as given in [1] (see Fig. 18.3):

Let $BC = a$ be the shorter of the two legs of the right angle of the triangle, and $AC = b$ be the longer one. Also, let the hypotenuse $AB = c$. With C as center and radius a describe a circle. Let D be the intersection of AC with the circle, H the other one obtained by extending AC beyond C, and E the intersection of AB with the circle. Draw CL perpendicular to AB. L is the midpoint of BE.

By the Intersecting Chords Theorem, we have: $(AH) \cdot (AD) = (AB) \cdot (AE)$. In other words: $(b + a)(b - a) = c(c - 2BL)$. Now, the right triangles ABC and BCL share an angle at B and, therefore, are similar. Hence, $BL/BC = BC/AB$, so that $BL = a^2/c$. Combining all together we have: $b^2 - a^2 = c(c - 2a^2/c)$. Hence, $a^2 + b^2 = c^2$, i.e., the Pythagorean relation obtains.

Proof #60 proceeds in a similar fashion, based on the idea that the circle has radius b, the length of the longer leg. However, both proofs #59 and #60 fail if triangle ABC is not only right but also isosceles, with $CA = CB$ (or: $b = a$). In that case, the problem reduces directly to the above-mentioned "Socrates' Theorem", referring to the simple statement that the area of a right isosceles triangle with hypotenuse k equals $k^2/4$. Socrates' proof is independent of the Pythagorean Theorem, so there is no circular reasoning in it.

At this point, however, it could be that agent A_1 misses the fact that the previous proof fails in the above special case, i.e., if triangle ABC is not only right but also isosceles. Once A_1 posts the proof on the system, codelets (interpreters) that verify its correctness will hopefully spot the problem and spawn new ones, with the task of solving it. The solution could be given immediately by a codelet (prover) that observes that the problematic special case is merely an instantiation of Socrates' Theorem.

Let us now consider Proof #64 in [1], explicated by Fig. 18.4. Here, the sides of the given triangle ABC have been extended in the manner suggested in Fig. 18.4, forming the square $CDFH$. In addition, given the fixed lengths a and b on the sides of the square, lines BE, EG, and GB have been drawn. The proof proceeds as follows:

The equal triangles ABC and ADE have areas equal to $ab/2$. The isosceles right triangle ABE has area $c^2/2$ (which is given by Socrates' Theorem, not by

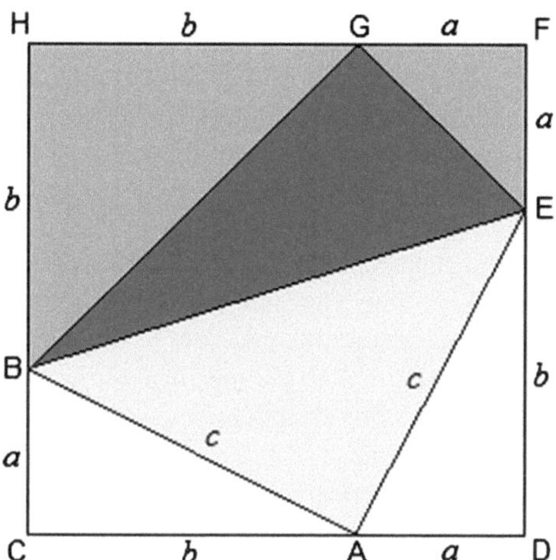

Fig. 18.4 Proof #64 in [1]

the Pythagorean one!). For the same reason, triangle EFG has area $a^2/2$, and triangle GHB has area $b^2/2$. Finally, triangle GEB has area $\sqrt{2}a\sqrt{2}b/2 = ab$. Now, the quadrilaterals $BCDE$ and $EFHB$ are equal (as trapezoids with equal bases and altitude). So, adding the triangles they are made of we have: $ab/2 + ab/2 + c^2/2 = a^2/2 + b^2/2 + ab$, which yields the sought for relation: $c^2 = a^2 + b^2$.

We observe now that all three proofs (#59, #60, and #64) use Socrates' Theorem at some point in their arguments. Thus, the latter is a recurring theme. It is desirable that such recurring themes be identified as potentially important, as they can prove useful in any nontrivial task or sub-task. The question is how to make the system be of any help in this identification.

Recall that codelets are transient, but tasks are more permanent. Thus, if the system keeps tasks in a repository, special codelets could be generated with the sole purpose of identifying recurring tasks. This identification could be mediated semi-automatically, if the tasks are codified somehow and then allowed to form groups (categories) automatically based on their properties; alternatively (or: additionally), human agents might help in this identification. Thus, identified recurring tasks can receive an "activation" by the system, so that repeated tasks will stand out, and this will help in their reuse when they appear again as components of codelets in the system.

3. $\langle A_3, PT_3 \rangle$ might be asked to propose what A_3 thinks is a simple "trigonometric proof", i.e., a proof expressed in trigonometric language, and create only the public subtask of verifying the correctness of the proof.

   The existence of trigonometric proofs sounds suspicious at first thought, because the Pythagorean Theorem asserts that in a triangle ABC the equality $\sin^2(A) + \sin^2(B) = 1$ is equivalent to the angle C being right. In reality, any trigonometric "proof" of the theorem that relies on the basic identity $\sin^2(x) + \cos^2(x) = 1$ must be circular, and thus invalid. Indeed, remark 8 in [1] points to a page with an incorrect proof ("false proof #4"), which proceeds as follows: as usual, let

a and b be the lengths of the perpendicular sides of the right triangle ABC, c the length of its hypotenuse, and B the angle opposite to side b. Then, by the definition of the trigonometric functions, $\sin(B) = b/c$ and $\cos(B) = a/c$. From the fundamental trigonometric identity we have: $\sin^2(B) + \cos^2(B) = 1$. Consequently, $(b/c)^2 + (a/c)^2 = 1$, from which we get: $a^2 + b^2 = c^2$. The problem with this "proof", as mentioned above, is that the derivation of the fundamental trigonometric identity is based on the geometrical fact that the Pythagorean Theorem expresses; thus this is an example of a circular proof. The "proof validation" codelets of the system should "perceive" that the same fact is expressed in both geometric and trigonometric languages and spot this circularity; but rather than simply rejecting the proof, they could spawn further codelets asking for a derivation of the trigonometric identity without the use of the Pythagorean Theorem. Indeed, as pointed out in [1], it is possible to derive the trigonometric identity using solely the subtraction formulas for sine and cosine:

$$\sin(a - b) = \sin(a)\cos(b) - \cos(a)\sin(b)$$
$$\cos(a - b) = \cos(a)\cos(b) - \sin(a)\sin(b)$$

This derivation establishes also that the mathematical fact proven is independent of the language (style) used in the proof, being it geometric or trigonometric. This terminates the proof-event under consideration.

4. Finally, $\langle A_4, PT_4\rangle$ could be asked to take an algebraic approach, which is exhaustive and thus suitable for a computerized system. The choice of algebraic language makes possible the use of algebraic identities standing for geometric relations. Specifically, consider Fig. 18.5 (proof #56 in [1]):
Assuming the usual assignments $AB = c$, $AC = b$, $BC = a$, and further: $CD = x$, $CE = y$, $AF = z$, $EF = v$, and $ED = w$, the following relations hold:

1. $b(b + x) = cz$	10. $zw = y(b + x)$
2. $b(v + w) = az$	11. $xz = y(v + w)$
3. $c(v + w) = a(b + x)$	12. $x(b + x) = w(v + w)$
4. $bw = cy$	13. $z(a - y) = v(b + x)$
5. $bx = ay$	14. $z(c - z) = v(v + w)$
6. $cx = aw$	15. $(c - z)(b + x) = (a - y)(v + w)$
7. $b(a - y) = cv$	16. $y(a - y) = vw$
8. $b(c - z) = av$	17. $y(c - z) = vx$
9. $c(c - z) = a(a - y)$	18. $w(c - z) = x(a - y)$

The purpose is to find combinations of the above equations from which x, y, z, v, and w can be eliminated, thus leaving a relation that holds between a, b, and c. Clearly, no single equation suffices for this task. A brief examination shows that no pair of equations suffices, either. Thus, equations must be examined in sets of three, four, or more. In addition, such sets must contain only independent equations; for example, Eqs. 1, 2, and 3 comprise a dependent set; so do Eqs. 1, 4, and 10; and so on.

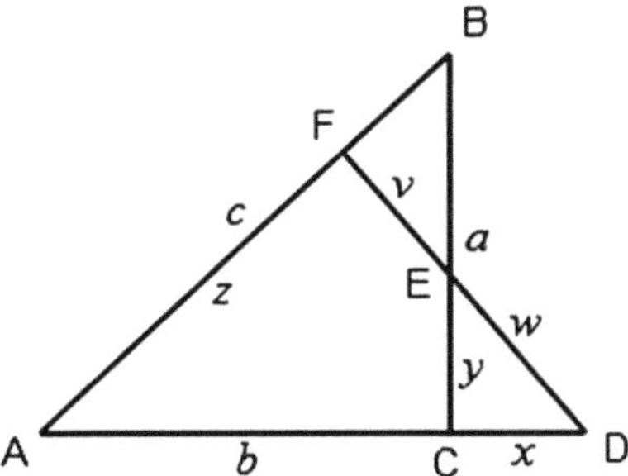

Fig. 18.5 Proof #56 in [1]

> The above problem can be approached thoroughly and most efficiently with the help of computers, as the combinatorial number of systems of equations that must be examined exceeds the abilities of the average person. Actually, Bogomolny [1] reports that B.F. Yanney and J.A. Calderhead found 4864 proofs from the above set as early as 1896—at a time when computers were unavailable. There is no claim made, however, that this was an exhaustive search, nor that it was all correct. Only computer searches can be reliable enough to draw trustworthy conclusions in such cases, which brings us to this important observation: in a codelet-aided system such as the proposed one, some of the codelets can be computer processes; and in some cases, they must be computer processes. In general, whatever can be automated, it better be automated and examined mechanically. Thus, the model we envision should be flexible enough to allow for a mixture of human and computerized agents.

In summary, this entire example regarding the Pythagorean Theorem suggests that, for complex problems such as those in mathematics, the agents associated with codelets at the highest level of the hierarchy will generally possess a higher level of expertise or sophistication, as the tasks are likely to consist of things such as reformulating a problem or task in a way that is more suitable for being treated by a wider audience.

Also, a system like this would actually keep track of the whole collective, dynamic, non-linear process of making of a proof, and of its acceptance or rejection by individuals and the community.

18.7 Collective Intelligence as an Emerging Property of Codelets in Web-Based Proof-Events

Collective creative thinking and collective memory are essential components of a Web-based mathematical problem-solving approach. The Kumo assistant and the Polymath project [7] are Web tools facilitating the collaboration of codelets, whereas the Tatami and the Tricki projects [9, 20] serve as repositories of the acquired collective memory. The image of an individual mathematical mind, which is intelligent enough to cope with hard mathematical problems, is replaced in Web-based problem-solving by the image of a "collective" mathematical mind, which is more efficient to handle difficult problems in shorter time. The new picture is vividly outlined by Nielsen [13], as an epoch-making type of "networked science".

Collective intelligence in Web-based problem-solving is characterized by openness, i.e., unrestricted sharing of ideas and intellectual property among codelets, peering of codelets and joint goal-directed action. Thus, collective intelligence can be understood as an emergent distributive property over numerous codelets of a "collective mind" that uses a set of flexible and adaptable tools from a Web-based repository in facing mathematical problems.

Such tools have a double nature: on the one hand they are objects readily available to be used for any specific purpose, i.e., they are objects "ready-to-hand" (to use Heidegger's terminology), just lying there; on the other hand, when these tools are activated (for instance, when *Mathematica* is used) they may initiate processes and produce contributions that even a prover might fail to reach, although they lack the intelligence of a prover. From the latter standpoint, they act as (intelligence-less) codelets, insofar as they actively work on data and follow the architecture of the Web.

18.8 Conclusion

A system for Web-based cooperation among people for the handling of proof events and mathematical problem-solving was proposed in this paper. The main advantage of this approach over the more traditional proving methods is the interesting possibility that mathematical problems that are far too complex to be solved by a single person might become solvable by a community of mathematicians who cooperate following the system outlined in the present text. It is our firm belief that the limits of group thinking and cooperation among members of a community lie far beyond those of individuals, and that such limits need to be further explored.

References

1. Bogomolny, A.: Pythagorean theorem and its many proofs from interactive mathematics miscellany and puzzles http://www.cut-the-knot.org/pythagoras/index.shtml, Accessed 25 Feb (2014)
2. Dennis, A.R., Valacich, J.S.: Computer brainstorms: more heads are better than one. J. Appl. Psychol. **78**(4), 531–537 (1993)
3. Foundalis, H.E.: Phaeaco: a cognitive architecture inspired by bongard's problems, Ph.D. Dissertation, Computer Science and Cognitive Science Departments, Indiana University, Bloomington, Indiana, (2006)
4. Goguen, J.A.: Social and semiotic analyses for theorem prover user interface design. Form. Asp. Comput. (Special Issue on User Interfaces for Theorem Provers) **11**, 272–301 (1999)

5. Goguen, J.: Semiotic morphisms, representations, and blending for interface design. In: Proceedings AMAST Workshop on Algebraic Methods in Language Processing, Conference Held in Verona, Italy, 25–27 August 2003, pp. 1-15. AMAST press
6. Goguen, J.A.: What is a proof? Informal Essay. University of California at San Diego. (2014). from: http://cseweb.ucsd.edu/goguen/papers/proof.html. Accessed 25 Feb 2014
7. Goguen, J., Lin,K., Rosu,G.,Mori,A., Warinschi, B.: An overview of the Tatami project. In: K. Futatsugi, K., Tamai, T., Nakagawa, A. (eds.) Cafe: An Industrial-Strength Algebraic Formal Method, pp.61–78. Elsevier (2000)
8. Gowers, T., Nielsen, M.: Massively collaborative mathematics. Nature **461**, 879–881 (2009)
9. Gowers, T.: Tricki now fully live. (2014). http://gowers.wordpress.com/2009/04/16/tricki-now-fully-live. Accessed 25 Feb 2014
10. Hofstadter, D.R.: Fluid Concepts and Creative Analogies: Computer Models of the Fundamental Mechanisms of Thought. Basic Books, New York (1995)
11. Lakatos, I.: Proofs and Refutations. Cambridge University Press, Cambridge (1976)
12. Minsky, M.L.: The society of mind. Simon and Schuster, New York (1988)
13. Reinventing discovery. The New Era of Networked Science, Princeton, Princeton University Press (2011). (e-Book)
14. Osborn, A.: Applied Imagination: Principles and Procedures of Creative Problem Solving, 3rd edn. Charles Scribner's Sons, New York (1963)
15. Pease, A., and Martin, U., Seventy four minutes of mathematics: An analysis of the third Mini-Polymath project. In: Proceedings of AISB/IACAP World Congress 2012 - Alan Turing (2012), Symposium on Mathematical Practice and Cognition II(2–4), July (2012), pp. 19–29. http://www.mrtc.mdh.se/gdc/work/AISB-IACAP-2012/NaturalComputingProceedings-2012-06-22.pdf. Accessed 21/07/2014
16. Pólya, G.: How to solve it. Princeton University Press, (1945)
17. Stefaneas, P., and Vandoulakis, I.M.: The web as a tool for proving, metaphilosophy (Special issue: Philoweb: Toward a Philosophy of the Web: Guest Editors: Harry Halpin and Alexandre Monnin, vol. 43(4), pp 480–498, july (2012)). Reprinted in the collection: Harry Halpin and Alexandre Monnin (Eds.) Toward a Philosophy of the Web, pp. 149–167, Wiley-Blackwell (2014)
18. Stefaneas, P.: and Vandoulakis, I.M.: Proofs as spatio-temporal processes. In: Pierre Edouard Bour, Gerhard Heinzmann, Wilfrid Hodges and Peter Schroeder-Heister (eds.), Proceedings of the 14th Congress of Logic, Methodology and Philosophy of Science, Philosophia Scientiae, vol. 18(3), pp. 111–125 (2014)
19. Tao, T.: Mini-polymath3: Solving a problem from the 2011 International Mathematical Olympiad. Proposed Jun 9, (2011); launched and solved, Jul 19, (2011). http://terrytao.wordpress.com/2011/06/09/mini-polymath-3-2011-imo-question/; http://terrytao.wordpress.com/2011/07/19/mini-polymath3-discussion-thread/; http://polymathprojects.org/2011/07/19/minipolymath3-project-2011-imo/. Accessed 21 Jul 2014
20. Tao, T.: Tricki now live. What's new. (2014). http://terrytao.wordpress.com/2009/04/16/tricki-now-live. Accessed 25 Feb 2014
21. Vandoulakis, I.M., and Stefaneas, P.: Conceptions of proof inmathematics. In: Bazhanov, V.A., Krichevech, A.N., Shaposhnikov, V.A. (eds.) Proof. Proceedings of the Moscow Seminar on Philosophy of Mathematics, [Dokazatel'stvo. Trudy Moskovskogo seminara po filosofii matematiki], pp. 254–278. Moscow, URSS (2014)

Part III
Postface

Chapter 19
A Personal Perspective into the Future for Computational Creativity

Pablo Gervás

Abstract Looking into the future is like standing where the road has led you in a fog and trying to imagine where that same road might lead on from there. One may guess based on several things: where one wanted to go in the first place, where roads that traversed a similar landscape have led, and where other people heading down the same road think they are headed. I explore these possibilities for computational creativity, based on 15 years of experience travelling down that particular road.

19.1 Introduction

When I first started writing programs that could write poetry, back in the 90s, my only motivation was a personal intuition that it should be possible to model computationally the processes by which humans carry out the task. I was gratefully surprised to discover that there were other people interested in the general problem of modelling processes with a similar creative ingredient. Seduced by the possibilities suggested by this vision, I joined in the effort of creating a research community to nurture this topic, and I have been involved, on and off, in many of the initiatives over the years to make computational creativity into a rigorous and well-founded scientific discipline. Over this period of time, the field has expanded considerably, acquiring very valuable inputs from neighbouring fields, developing a substantial body of knowledge, and putting together a set of best practices. As a result, my perception of the problem has shifted, becoming more abstract, broader in range, and more ambitious in its aims. Section 19.2 outlines a part of this vision, which is then used in Sect. 19.3 to articulate a description of how I see the future of Computational Creativity (CC).

The various discussions below have been illustrated with examples drawn mostly from poetry and narrative generation, due to the author's extensive familiarity with these particular topics. Similar examples might be found for existing efforts at automated generation of music, graphic art, or stories, but an exhaustive review along these lines has been considered beyond the scope of the present chapter.

P. Gervás (✉)
Universidad Complutense de Madrid, Ciudad Universitaria, Madrid, Spain
e-mail: pgervas@ucm.es

T.R. Besold et al. (eds.), *Computational Creativity Research: Towards Creative Machines*, Atlantis Thinking Machines 7, DOI 10.2991/978-94-6239-085-0_19

19.2 Computational Creativity as Computational Generation of Novel Artefacts

Over the years working in this field, as a result of many hours digesting models, understanding practical applications and discussing theories, and being exposed to many bright minds engaged in this field, I have come to develop an abstract mental schema that acts as underpinning structure on which to hang ideas about computational creativity. When a new idea appeared that did not match the existing structure, I adjusted the conceptual schema a little. In order to accommodate the wealth of material that the field has produced over time, this structure became more and more abstract, to encompass new ideas, new views, and new relevant aspects that we had not considered at the start. As one does when abstracting away from specifics, this process has also created a number of gaps in the structure, where it has become clear that certain aspects, equivalent or parallel to existing ones, need to be developed. I will describe this conceptual structure in order to use it both for exploring the potential for evolution of existing initiatives in CC, and to outline some additional avenues of research that may be of interest in the future.

19.2.1 A Conceptual Schema for Computational Artefact Generation

There is no consensus of what creativity is or how it may be defined[1]. But we can build some kind of consensual understanding of what computational creativity might involve by exploring our shared intuitions by means of some simple thought experiments.

Let us focus for the time being on the computational aspects that would be involved in autonomous automated artefact generation.

Assume we want to create apples. From a computational point of view, we would have to identify a set of necessary ingredients, and establish a constructive procedure for putting them together into "apple candidates". The set of all possible "apple candidates" that can be produced by a given set of ingredients and a constructive procedure is what may be called the conceptual space being explored [1]. Our initial attempts might produce artefacts that do not look like apples at all. As we progressively refine our selection of ingredients and our constructive procedure, we may arrive at a point where some of the artefacts we are producing start to resemble apples. These can now be picked out as successful results. There are actually two ways of doing this.

[1] Throughout the present chapter I have deliberately tried to make restricted use of the word "creativity" in the description of the schema employed as reference. Yet it is my sincere belief that the schema as presented captures many of the fundamental aspects that people have in mind when they mention the word "creativity". Either way, I have superimposed this schema as a grid on the current landscape of computational creativity, to provide a reference that can be used to explain what I think has been achieved already, what is being addressed by ongoing work, and what is work that remains for the future. My hope is that, having avoided the C-word—that triggers so many spurious discussions—in favour of more fundamental aspects of the process, it might be easier for people to accept the basic view and concentrate on the arguments beyond it.

One involves establishing some means for spotting apple-like candidates among system output. This could be achieved by defining a set of desirable properties of apple-candidates, and using those to filter system output. Another involves modifying further the set of ingredients and/or the constructive procedure, constraining our conceptual space to include only those artefacts that resemble apples.

Think of it as five separate processes. An initial process of selecting a constructive procedure, followed by a process of selecting ingredients to feed the constructive procedure. A third process would combine those—for instance, successively applying the constructive procedure to the set of ingredients—to produce a set of apple-candidates. A fourth process would establish the selection criteria for picking desirable candidates. A final process would look at this set of candidates and pick out the one most likely to be accepted as an apple.

It is important to note that the last two processes are conceptually similar—in as much as they both involve selection among a set of candidates—but at the same time different in that one rules out unacceptable candidates and the other picks out top performers. Think of them as lower and upper bounds on the fitness of the candidates: candidates below the lower threshold get rejected outright, candidates above the upper threshold effectively determine when the generation process can be stopped if only one result is desired.

The procedure described in our thought experiment is essentially cyclic in nature: note the phrase "as we progressively refine" in the paragraph above. This progressive refinement may be applied to any or all of the five progresses described. In truth, the procedure implies additional process of reflection upon the results of each of the five processes already described, with a view to refining them as described.

The schema resulting from this thought experiment involves 7 basic processes:

1. **Procedure Selection**: selection of constructive procedure
2. **Ingredient selection**: selection of ingredients to use
3. **Lower bound definition**: establishment of candidate rejection criteria
4. **Upper bound definition**: establishment of candidate success criteria
5. **Construction**: application of the procedure that resulted from 1 to the ingredients that resulted from 2
6. **Selection**: application of results of 3 and 4 to results of 5
7. **Reflection**: reflection upon the accumulated results with a view to refining steps 1 to 4.

From a procedural point of view, this conceptual schema could be applied to most processes of artefact generation, including those carried out by humans. It can be applied to an engineer trying to put together a robot to carry out a particular task. Or to an artist considering what his next creation might be. Or a programmer building a poetry generator. Or a poetry generator at work.

I want to make two important observations at this stage.

First, there are significant differences between the case of the engineer and the artist. The engineer is very unlikely to make any drastic changes to his constructive procedure (process 1 of procedure selection)—which would correspond to the design of the robot—or his set of ingredients (process 2 of ingredient selection) beyond selecting those that fit the task at hand. He will probably reuse an existing design, and he is almost sure to rely on existing robotic components. He is more likely to concentrate on establishing the appropriate criteria (processes 3 and 4 of lower and upper bound definition) to ensure that the resulting robot is fit for the task, possibly reflecting (process 7 of reflection) on the results of tests (processes 5 and 6 of construction and selection) with a view to refine the overall process. The task that the robot is to achieve comes very clearly specified with the commission, and the engineer basically translates it into the selection criteria that will be applied to test the robot, which becomes the specification of the robot. In contrast, the artist is likely to explore much more widely in terms of constructive procedures and ingredients, trying to innovate more radically. The task or goal that the artistic piece intends to fulfill is very much open, and a very significant amount of reflection on the part of the artist may go into establishing what it may be. The selection criteria developed as a result of this reflection become not so much a specification but more of an aesthetic.

Second, there are also significant differences between the poetry generator and the other three. Both the engineer and the artists can be considered to carry out all of the processes. If they decide to delegate any, these are more likely to be construction and selection of construction and selection, which involve systematic application of the decisions arising from the others. This is also the simplest option open to the programmer: carry out processes 1 to 4 to build a program, run the program to cover processes 5 and 6, then iterate including a process 7 of reflection. But of all of those, only processes 5 and 6 are actually carried out by the machine. The big question is: how many of these processes is a poetry generator applying? Most existing systems include computational procedures that could be considered equivalent to process 5 of construction. Since the concern about creative systems being able to evaluate their own output became prevalent, some systems now include procedures akin to process 6 of selection. In recent developments, systems have started considering selection among a set of possible constructive procedures, or even dynamic construction of criteria for selecting among candidates. But even these improvements are still hard-wired solutions thought up by the programmer, rather than the results of reflective processes implemented in the system.

19.2.2 Novelty as a Virtue

The considerations above apply to artefact generation even if the said artefacts are to be mass produced to a set standard. If they are, the process of creation would take place as described, to be followed by repeated application of process 5 and 6 to produce many copies of the desired artefacts. In such a setting, processes 5 and 6 of construction and selection could be seen to correspond to manufacture and quality control. Under these circumstances, the creation process is carried out to identify the particular solution that meets the requirements, and once it has been found, the remaining processes of manufacture and quality control proceed systematically. The overall process described above for the engineer corresponds to this type of situation.

The situation changes significantly if one starts attributing relative merit to the construction process itself, beyond and above the intrinsic value of the artefact in terms of practical use. This is very much what happens in the realm of intellectual and artistic endeavours, such as for instance literature, music, painting, or art in general. In fields of this nature, the value of artefacts becomes related to their ability to provide stimuli to the mind of the perceiver that enrich their experience of the real world. This is very much what happens in the case of the artist, and it plays a fundamental role in setting the differences described between the process for the engineer and that of the artist. From the conceptual point of view that we want to consider, this peculiarity introduces a very dramatic perturbation into the schema that needs to be considered for computational implementations of solutions to artefact generation in these cases.

For a start, the creation process that we were considering can no longer be considered in isolation. The specification of the goal or the task of the artefacts to be generated under this new approach is relative to a concept of mind that reacts to a stimulus, intended as an abstraction of the audience that receives the artefacts produced and provides some feedback on whether they prove satisfactory. Although this characterization is but one of many possible abstractions to describe the situation, it captures the basic idea that the specification of the desired result is no longer phrased in terms of specific properties of the artefact—as, in for instance, requiring that an apple be red and shiny—but instead it is phrased in terms of possible reactions from a very complex entity—a mind. Now a model capable of reproducing the reactions of a mind is in itself a challenge even greater than that of artefact generation. This premise underlies the main argument presented by Bringsjord and Ferruci [2] against the possibility of ever achieving computational models of literary creativity. According to these authors, as machines will always be unable to model the extremely rich range of reactions of a mind—in terms of imagery and evocation of sensual or emotional experiences—to the simplest stimulus provided in textual form, creativity as carried out by humans will always be beyond their reach. This is certainly an important obstacle that cannot be dismissed outright. It is clear that modelling the type of reaction being considered here would introduce significant complication into the set of criteria as considered in processes 3 and 4 above for establishing lower and upper bounds to the quality of artefacts desired. This complication may be challenging enough for these processes to start meriting some kind of breakdown into simpler processes. This is clearly an open problem at present, and one that has received little attention from the computational creativity community. Some recent efforts point at the interactions between computational creativity and cognitive computing [19]. The need for more detailed research along these lines has recently been argued for the case of story generators in [11].

A second issue is introduced by the fact that the perceived stimulus often arises as a contrast between a newly produced artefact and previous instances of similar artefacts. When repeatedly exposed to similar stimuli, the mind gets bored and reacts less and less. This implies that a model is required to account for the moment in which the audience is exposed to a particular result, as this determines which other results have been seen before. This may involve a further selection process (process 8 of publication), so that only one of the top performers produced in a given run is chosen to be "published" or made public. Some kind of record of published artefacts—or their relevant characteristics—should be kept, so that it can be taken into account when deciding what are valid candidates for publication. This involves additional processes to be considered during artefact generation. As a result of this additional constraint, perfectly acceptable candidates in terms of the criteria resulting from selection processes 3 and 4 may be rejected, not on the grounds of their intrinsic quality but on the grounds of not being different enough from previously produced artefacts. In addition, the degree of similarity with previous results may also be applied during reflection (process 7) to trigger modifications of the constructive procedure (process 1) or the set of ingredients (process 2). Acceptable candidates produced but not "published" may be mined for valuable features not already present in previously published results, with a view to guiding the modification process. The consequences of this change are significant. Even when a single producer of artefacts is considered it implies a transition from a stateless process to one that needs to keep track of what has been produced before. But if the system is to operate in a wider context, artefacts produced by other creators would need to be considered as well. This corresponds to the difference identified by Boden between *p-creativity* and *h-creativity* [1]. Whereas artefact generation as considered previously could be optimised for a particular target, and then satisfactory instances could be mass produced, the new situation requires that the set of previously produced results be taken into consideration. The production of one instance of an optimal solution in principle rules out all other solutions built along similar lines. This shifts the focus of the overall procedure much more onto the reflection process.

From a computational point of view, the need to consider the state of a given generation procedure introduces the need for an additional process that filters results based on their similarity with previously seen results. The similarity used here may be established at various levels, from the physical to the conceptual, including similarity in terms of procedures or ingredients employed, emotions induced, or effects produced. The already argued complexity of modelling the reactions of a human mind is therefore implicit within this process, making it a significant challenge for the field. By analogy with the other ingredients of the conceptual schema, we should consider two separate process that a creator may need to contemplate: one for identifying the criteria for this filtering operation—we can identify this as process 8 of selection novelty-based filtering criteria specifically—and one for applying these criteria to the set of results—which may in fact be considered an extension of process 6 of selection, where, therefore, the various selection criteria resulting from processes 3, 4 and 8 are applied together to the results of the production process 5.

It is important to note that reflection (process 7) may also need to be extended to refine process 8, in view of the accumulated results of all the other processes.

19.2.3 Current Efforts in Computational Creativity with Respect to the Conceptual Schema

The conceptual schema presented above is intended as a compilation of the set of decisions required for the instantion of a process of artefact generation, including a refinement to account for cases where the suitability of the artefacts is based on their ability to create an impact on a hypothetical human audience. This schema can now serve as a yardstick for how far Computational Creativity has progressed along the road to developing models of the type of process considered. It has already been argued in Sect. 19.2.1 how existing poetry generators tend to focus on a very small subset of the processes involved in the simple generative act. Process 8 introduced in Sect. 19.2.2 is not better covered. Some effort at addressing this point was made at the theoretical level in [8].

The process most commonly addressed in systems designed to generate artefacts traditionally considered creative tend to focus on processes 5—application of a constructive procedure to a selected set of ingredients to obtain a set of candidates—and 6—application of selection criteria to the resulting candidates to establish a preferred result. The motto of the ICCC 2012, "Scoffing at mere generation for more than a decade" highlighted the generally accepted need to go beyond construction to consider evaluation as part of the remit of creative systems.

Very few systems can be said to include instantiations of process 1 of selection of a constructive procedure. The poetry generator described in [4] includes a process for defining what is called an aesthetic, which involves selecting between a number of different ways of constructing the final poem, based on a set of pre-established criteria. This could be considered as a basic approximation to the process of selecting a computational procedure to apply. Systems based on evolutionary [9] or genetic programming [13] do in some way include some means for the system to determine dynamically how the artefact is constructed, in terms of combinations of very basic evolutionary or genetic operators.

Process 2 of selection of the ingredients to be employed is also poorly represented in existing systems. The MEXICA story generator [17] received as input a set of previous stories, from which it built the internal resources it would use to generate further stories. Some recent poetry generators [4, 10] fetch texts from the Internet that they use to extract the ingredients they will later use to construct new poems. The PoeTryMe system [12, 15] is designed as a framework that can consider different linguistic resources as sources for its ingredients. These are instances of systems designed to be able to consider various inputs, but the decision of which input set to consider in each case is usually taken by the designer or programmer.

Processes 3 and 4 involve establishing selection criteria for identifying valid candidates. A fundamental work on this subject is the paper by Graeme Ritchie [18]. This work combined criteria for validity of candidates—referred is the paper as typicallity—and for novelty. The criteria proposed by Ritchie cover together processes 3 and 4 for establishing lower and upper bounds on quality and process 8 for determining the degree of novelty of the artefact. In his paper, Ritchie presented his criteria as a toolkit out of which developers could choose to suit particular purposes. Implementations of such a selection process, if included in a particular system, would correspond to instantiations of the processes 3, 4 and 8 we have discussed. Although Ritchie's criteria have been applied in practice, the instances known to me [5, 6] involve their application as a suite of criteria for the evaluation of given systems, rather than as criteria for selecting candidates among the results. To date, I am aware of no effort to select particular criteria to achieve specific goals. A system capable of doing this on its own would be a significant advance.

Process 8 of establishing selection criteria based on the novelty of the results have also been addressed for poems [8] and for narratives [16]. The need for a component to model human cognition during the assessment phase of computational creativity has been defended in [19].

Process 7 of reflecting on the accumulated results of the various other processes described with a view to revising the processes involving design decisions (1, 2, 3, 4, 8) has to my knowledge not been addressed in any existing system, though the need for iteration involving reflection over accumulated results is an integral part of the ICTIVS model of narrative generation [11]. In many ways, the need to evaluate and reconsider not just the artefacts that result from a given system but also the processes and the evaluation functions employed within the system was originally suggested by Wiggins [21] as part of his framework for understanding creative systems. This procedure of refining the processes and functions he referred to as "creativity at the meta-level", and he described it as a possible way of considering what Boden [1] termed *transformational creativity* within the same framework as *exploratory creativity*. The concept of *framing* [3] included in the FACE descriptive model addresses a static view of this reflective procedure as perceived by an external viewer if a computational creative system is provided with the means for explaining its actions. However, the concept of framing as described is perfectly compatible with such explanations being hard-wired in by the programmer, either as justifications of the procedures actually implemented—as done for instance in [4]—or even as flights of fantasy unrelated to them. The reflective process as described here would require the system to actually take decisions with respect to the procedure or criteria to be employed. The framing could then be produced by providing articulate justifications of these decisions. Proper framing would then evolve beyond a template-based explanation of hard-wired constructive procedures, to become a transcription of complex decision processes actually carried out dynamically by the machine.

19.3 Looking into the Future

When asked to provide a personal perspective on the future of Computational Creativity I had to think carefully about what I could do, but most of all, about what I should not do. Nobody can foretell the future. So from the start I was determined not to include wild guesses about what might be the best-performing technologies in CC in the future, or what the killer applications will be 10 years from now. All efforts in the past to foretell this type of information have generally failed. What I thought I might do is to see how the road travelled so far, and the experience accumulated in the process, might give me some hints as to what the path ahead may bring. I have done this in three different ways One is to try and explain which of the processes of the conceptual schema presented above are still outstanding, and how likely I think they are to be addressed in the near future. Another is to consider what were the factors that played a significant role in the success (or lack thereof at given moments of time) of disciplines that have faced similar challenges. Finally, a different way is to consider the visions expressed by others, and try to value where they are likely to lead in terms of the initially stated aims.

19.3.1 Processes of the Conceptual Schema to be Addressed in the Future

Of the basic proceseses for computational generation of novel artefacts described in Sects. 19.2.1 and 19.2.2, I have outlined in Sect. 19.2.3 which ones have been addressed so far in existing efforts on Computational Creativity. With respect to the future, the important consideration is which of those processes are likely to undergo further evolution, which ones may be good candidates for the development of practical applications, and which of them have more potential in terms of long term impact on society.

It is clear that those processes that have received less attention in the past—1, 2, 3, 4, 7 and 8, covering selection of constructive procedure or ingredients, validity and novelty criteria, and, most importantly reflection over results leading to improvement—are likely to experience significant growth in the future. For a start, these processes constitute virgin territory, where low hanging fruit may be waiting for any pioneers brave enough to consider them. But remember that, as argued in Sect. 19.2, these are the subtasks that human creators would least consider delegating. They correspond to the role of the engineer designing a robot, rather than the factory manager supervising its mass production. Or to the role of the sculptor conceiving a piece, as opposed to the foundry manager that supervises the production of the final cast in bronze. In a way, they embody the core of the creative task, and it is very possible that a large percentage of the mistique of creativity arises from decisions taken in these processes. The remaining processes 5 and 6 correspond to mass production and quality control, and are much less likely to involve essential ingredients of creativity. There are some aspects of process 8 for establishing novelty of particular artefacts that have particular potential. In contrast to other processes that rely on human abilities that are hard to model in a computer [2, 19], the simple establishment of novelty may be significantly helped by massive storage capabilities and high rates of comparison, abilities that computers can achieve with significantly more ease than humans. This has already been demonstrated by the success of computer solutions for detecting plagiarism, which can be seen as a related problem. In contrast, those aspects of process 8 requiring more complex assessment methods for novel artefacts, based on models of human cognitive abilities [19], present a greater challenge. They are nevertheless very likely to see a significant surge in the future, as our knowledge of human cognition improves. I suspect that the implementation of all these design processes in working computational systems will go a long way towards improving the perception of creativity of the systems and/or their resulting outputs. This is much in line with some of the existing descriptive models for creative systems [3, 21], that capture in their theoretical formulation concepts such as framing [3] or meta-level creativity [21] which correspond, as argued above, to instantiations of these design processes.

This should not be taken to mean that the productive process—5 and 6, for applying the constructive procedure to the ingredients and the selection criteria to the resulting candidates—are likely to be abandoned. By their very nature, these productive process have a significant potential for applicability. A combination of processes 5 and 6 is the most likely candidate for finding commercial applications, where instances of the produced artefacts can be sold to the general public, fulfilling the need for personalised artefacts at low cost. It is important to note that the development of new constructive procedures and new evaluation functions would correspond to research on these processes. As such research is specific to each type of artefact, and, within those, to genres or particular aims, it is unlikely that the potential for innovative and valuable research on these processes be ever exhausted.

19.3.2 Factors that Will Influence the Future of Computational Creativity

A very close discipline to Computational Creativity is Artificial Intelligence. Both are based on computational principles, both attempt to emulate higher abilities of humans, both were met at the start with mixed feelings of significant scepticism and high hopes. There are some important lessons to be drawn from the evolution of AI over time. The evolution of a discipline seems to be constrained by the expectations it raises, the dreams it can inspire, the fears it invokes, and the financial profit it can generate. All these come together to determine the amount of funding and effort that can be invested in it. Another important point is that different sets of people react to these various aspects differently. Researchers tend to be fuelled by dreams. Funding agencies look more closely at expectations and financial profit. The general public tends to consider expectations, and can be very much affected by any possible fears.

In the case of Computational Creativity, researchers are motivated by dreams of machines capable of autonomous creativity, and of understanding human creativity. These dreams should not be allowed to cloud the significant potential for impact arising from other possibilities. Machine-supported human creativity, with machines playing the role of colaborators or co-creators, may be very profitable in the medium term [14]. Machine-creativity, different and beyond human creativity may prove to be a very profitable outcome of Computational Creativity in the future [7].

The general public is worried about the possibilities of machines taking away the jobs of many humans now involved in the creative industries. Funding agencies have over the past few years been trying to project the criteria traditionally applied to research in technology to Computational Creativity. In doing so, they are handicapped by the fact that research investment in technology has in the past been focused on industrial processes and/or endeavors known to be profitable. This would correspond to computational systems that instantiate processes 5 and 6 as described above. A significant portion of the innovative aspect of CC lies in the fact that technology is being shifted from a tool for simplifying (even undertaking) repetitive tasks that machines can perform better than humans, to a tool for suggesting new procedures or even new types of artefacts that humans might not have considered before. To achieve this, more systems should be developed that address the design processes. This constitutes a significant conceptual leap along different axes. First, because while a technology for undertaking a repetitive task may replace a human worker, any new procedures or new types of artefact produced by CC technologies are very likely to need human validation before being deemed acceptable. Whereas industrial technology replaces humans (thereby including a potential for generating significant increases in financial profit), CC technologies are more likely to serve as tools for extending the creative range of human operators. It is unlikely that CC technologies lead to increased productivity to the extent that they result in layoffs. By definition, CC techniques should focus on decisions to produce artefacts of type A rather than artefacts of type B, rather than on producing artefacts of any given type faster or in a larger scale.

Some members of the general public are also uncomfortable at the possibility of certain artefacts that they associate with creativity (such as paintings, music, or poems) being generated by machines without a soul. This particular argument says more about how people justify the impact of these artefacts on themselves than about the properties of the artefacts. This brings up again the already mentioned challenged of modelling a mind's reaction to given artefacts. As more becomes known about how these processes operate, such fears should progressively fade away.

19.3.3 Open Avenues for Computational Creativity

Over the years researchers in the field of Computational Creativity have expressed their thoughts about what might be important directions for the future. In this section I consider some of these statements and analyse them with respect to the arguments presented in the rest of this chapter.

Wiggins [21] presented a framework for description, analysis and comparison of creative systems which described the operation of such systems as a search over a conceptual space, defined in terms of a number of functions for traversal of such a space and evaluation of candidate points. This was intended as a conceptualization of the system rather than a description of their operative procedures. Wiggins mentioned that transformational creativity, as described by Boden [1], could be achieved computationally if one extended the framework so that search could range over the possible traversal and evaluation functions, as well as the conceptual spaces defined by each such choice. With respect to the processes described in Sect. 19.2, such an extension of the search would correspond to instantiation of the design processes. This has therefore been known to be a theoretical possibility but has only been considered in implemented systems in a very limited fashion, as described above (see Sect. 19.2.3). If Wiggins was right, future generations of computationally creative systems will include more and more of the design processes, resulting in an improved perception of their creativity.

Colton et al. [3] postulated Computational Creativity Theory as a term to encompass efforts to describe generic characteristics of what it implies for a computer to be creative. To my knowledge, the term has not been adopted by a wider community and remains applicable only to their FACE and IDEA descriptive models. The motivation described for these models is "to suggest future directions for computational creativity research". The models presented by Colton et al. had the merit of bringing attention to a number of issues that have indeed given rise to valuable research efforts. The conceptual schema described in this chapter could be considered a contribution along similar lines. The relation with the framing component of the FACE model, and how the conceptual schema presented here goes beyond it to describe the reflection process in relation to all the others has been described in Sect. 19.2.3.

Veale [20] campaigns in favour of a future for Computational Creativity as a connected network of creative web services. He argues that such a service-oriented architecture would allow researchers to pool creative technologies in a robust interoperable framework, and industry developers to exploit novel results of CC research in a robust, low-risk form, without having to re-implement algorithms from a quickly moving field. With respect to the set of processes described above, it is clear that functions representing the selection criteria subject of processes 3 and 4, and the constructive procedures subject of process 1, could be very well implemented as web services. Web services could also be good vehicles for implementing providers of ingredients subject of process 2. In each case, if a number of alternative versions where available for each service, software components that implemented decisions of which particular versions to use would constitute instantiations of the processes themselves. In such a setting, processes 5 and 6—applying a constructive procedure to ingredients and selecting valuable results out of the constructed candidates—could be developed by composition of simpler services so selected. It is less clear how process 7 of reflection on the results could be implemented in such an architecture, but it might take the form of a service broker, explicitly handling the interconnections between the corresponding services. One notable exception is the likelihood of instantiating process 8 for establishing the novelty of particular results. By its nature, such a process requires the consideration of state—in terms of the set of artefacts of the same kind already contemplated. This is contrary to the definitions of a service most often considered, which includes statelessness as one of its requirements.

Varshney et al. [19] argue along the lines that a creative system without a human cognitive component for the purpose of evaluation is severely handicapped. This matches the prediction by Bringsjord and Ferrucci [2]. If one accepts this prediction, the future should see an increase in the number of creative systems that include components based on cognitive informatics and computing.

19.4 Conclusions

The arguments presented in this chapter indicate that there is a significant number of the elementary processes involved in computational generation of novel artefacts that have yet to be addressed by computational creativity systems, or that have only been touched upon very lightly. These processes are also known to be particularly relevant to the perception of creativity in processes carried out by humans. The future is likely to see an increase of research efforts into these less explored processes.

Additionally, research into the best explored processes concerned with artefact generation and selection is likely to keep growing, due to their relative importance within the overall procedure and their significant potential for commercial application.

The future of Computational Creativity may also be significantly affected by the subjective perceptions of the various stakeholders. Researchers should make an effort not to create unreasonable expectations. The fear of creative machines sometimes expressed by journalists covering the field is not justified and every effort should be made to dispell it.

Let us hope that the joint efforts of so many valuable researchers continue to lead the field into ever expanding horizons, that they progressively clarify our understanding of the field and of human creativity in general, and that they allow us to explore without prejudice new grounds like human-machine symbiosis for creativity or pure machine-creativity.

Acknowledgments The author of this chapter acknowledges the support of the project PROSECCO 600653 funded by the European Commission, Framework Program 7, the ICT theme, and the Future and Emerging Technologies FET program.

References

1. Boden, M.: Creative Mind: Myths and Mechanisms. Weidenfeld & Nicholson, London (1990)
2. Bringsjord, S., Ferrucci, D.A.: Artificial Intelligence and Literary Creativity: Inside the Mind of BRUTUS, A Storytelling Machine. Lawrence Erlbaum Associates (1999)
3. Colton, S., Charnley, J., Pease, A.: Computational creativity theory: the FACE and IDEA descriptive models. In: Ventura, D., Gervás, P., Harrell, D.F., Maher, M.L., Pease, A., Wiggins, G. (eds.) Proceedings of the Second International Conference on Computational Creativity, pp. 90–95. México City, México (2011)
4. Colton, S., Goodwin, J., Veale, T.: Full-FACE poetry generation. In: Maher, M.L., Hammond, K., Pease, A., Pérez, R., Ventura, D., Wiggins, G. (eds.) Proceedings of the Third International Conference on Computational Creativity, pp. 95–102. Dublin, Ireland (2012)
5. Gervás, P.: WASP: Evaluation of different strategies for the automatic generation of spanish verse. In: Proceedings of the AISB-00 Symposium on Creative & Cultural Aspects of AI, pp. 93–100 (2000)

6. Gervás, P.: Exploring quantitative evaluations of the creativity of automatic poets. In: Workshop on Creative Systems, Approaches to Creativity in Artificial Intelligence and Cognitive Science, 15th European Conference on Artificial Intelligence (2002)
7. Gervás, P.: Engineering linguistic creativity: Bird flight and jet planes. In: NAACL HLT 2010 Second Workshop on Computational Approaches to Linguistic Creativity, Los Angeles, June (2010). Association for Computational Linguistics, Association for Computational Linguistic
8. Gervás, P.: Dynamic inspiring sets for sustained novelty in poetry generation. In: Ventura, D., Gervás, P., Harrell, D.F., Maher, M.L., Pease, A., Wiggins, G. (eds.) Proceedings of the Second International Conference on Computational Creativity, pp. 111–116. México City, México (2011)
9. Gervás, P., Computational modelling of poetry generation. In: Artificial Intelligence and Poetry Symposium, AISB Convention, University of Exeter. United Kingdom 04(2013), (2013)
10. Gervás, P., Evolutionary elaboration of daily news as a poetic stanza. In: IX Congreso Español de Metaheurísticas, Algoritmos Evolutivos y Bioinspirados—MAEB, Universidad Complutense de MAdrid. Madrid, Spain 09/2013 (2013)
11. Gervás, P., León, C.: Reading and writing as a creative cycle: the need for a computational model. In 5th International Conference on Computational Creativity, ICCC 2014, Ljubljana, Slovenia, 06/2014 (2014)
12. Gonçalo Oliveira. H., PoeTryMe: a versatile platform for poetry generation. In: Proceedings of the ECAI 2012 Workshop on Computational Creativity, Concept Invention, and General Intelligence, C3GI 2012, Montpellier, France, August (2012)
13. Machado, P., Correia, J.: Semantic aware methods for evolutionary art. In: Genetic and Evolutionary Computation Conference, GECCO'14, Vancouver, Canada, 12–16 July 2014
14. McQuaid, S.: Report on the Proceedings of the Workshop: The Future of ICT for Creativity and Creative Industries. Technical report, Creativity Unit in DG CONNECT, European Commission, Luxembourg, March 2014
15. Oliveira, H.G., Hervás, R., Díaz, A., Gervás, P.: Adapting a generic platform for poetry generation to produce spanish poems. In: 5th International Conference on Computational Creativity, ICCC 2014, Ljubljana, Slovenia, 06/2014 (2014)
16. Peinado, F., Francisco, V., Hervás, R., Gervás, P.: Assessing the novelty of computer-generated narratives using empirical metrics. **20**(4), 588, 10/2010 (2010)
17. Pérez y Pérez, R.: MEXICA: a computer model of creativity in writing. PhD thesis, The University of Sussex (1999)
18. Ritchie, G.: Some empirical criteria for attributing creativity to a computer program. Mind. Mach. **17**, 67–99 (2007)
19. Varshney, L.R., Pinel, F., Varshney, K.R., Schrgendorfer, A., Chee, Y.M.: Cognition as a part of computational creativity. In: ICCI*CC, pp. 36–43. IEEE (2013)
20. Veale, T.: A service-oriented architecture for computational creativity. J. Comput. Sci. Eng. **7**(3), 159–167 (2013)
21. Wiggins, G.A.: A preliminary framework for description, analysis and comparison of creative systems. Knowl.-Based Syst. **19**(7), 449–458 (2006)

MIX
Papier aus verantwortungsvollen Quellen
Paper from responsible sources
FSC® C105338

If you have any concerns about our products,
you can contact us on
ProductSafety@springernature.com

In case Publisher is established outside the EU,
the EU authorized representative is:
Springer Nature Customer Service Center GmbH
Europaplatz 3, 69115 Heidelberg, Germany

Printed by Libri Plureos GmbH
in Hamburg, Germany